Product Design

Product Design

A practical guide to systematic methods of new product development

Mike Baxter

Design Research Centre
Brunel University, U.K.

CRC PRESS

Boca Raton London New York Washington, D.C.

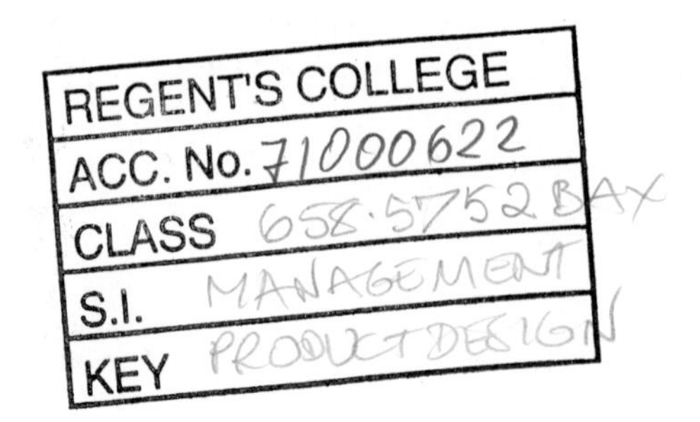

Library of Congress Cataloging-in-Publication Data

Catalog record is available from the Library of Congress

Direct all inquiries to CRC Press LLC, 2000 N.W. Corporate Blvd., Boca Raton, Florida 33431.

Visit the CRC Press Web site at www.crcpress.com

Originally published by Chapman & Hall

International Standard Book Number 0-7487-4197-6

Dedication

To the
Design Research Centre team;
this book couldn't have
happened without you

Contents

Preface		xiii
Acknowledgements		xv
Chapter 1	INTRODUCTION	1
	Innovation - risky and complex	2
	Ground rules for systematic design: the 3 very unwise monkeys	4
	Notes on Chapter 1	8
Chapter 2	PRINCIPLES OF NEW PRODUCT DEVELOPMENT	9
	Success and failure in new products.	9
	Risk management funnel.	11
	Stages of the risk management funnel	14
	Risk management - theory into practice	16
	Managing design activities within the risk management funnel	18
	Quality control of product development	20
	Quality targets	23
	Meeting targets in product development	24
	Key concepts in Chapter 2	26
	Notes on Chapter 2	29
Chapter 3	THE PRINCIPLES OF PRODUCT STYLING	31
	Visual perception of products	32
	Perception of product style	35
	The rules of visual perception	36
	General rules	36
	What do the rules mean for product styling?	39
	Specific rules	44
	The fabulous Fibonaccis	46
	Bisociative attraction	50
	Social cultural and business effects	51
	The determinants of style	54
	Attractiveness and product style	55
	The four faces of attractiveness	55
	The styling process	57
	Key concepts in Chapter 3	58
	Notes on Chapter 3	59
Chapter 4	THE PRINCIPLES OF CREATIVITY	61
	The importance of creativity	61
	The psychology of creativity	62
	First insight	62

Chapter 4 (Cont'd)

Archimedes' Eureka 63
Preparation 64
Faraday's electricity 64
Incubation and illumination 65
Buckyballs! 65
From hymn books to Post-it Notes 67
The nature of incubation and illumination 67
Bisociation and lateral thinking 68
Creative thinking in practice 71
Preparation 71
Preparation toolkit 72
Idea generation 73
Idea generating procedures 75
Idea generation toolkit 76
Synectics - a critique 77
Idea selection 78
Idea selection toolkit 79
Reviewing and improving creative thinking procedures 79
Key concepts in Chapter 4 80
Design Toolkits
Brainwriting 81
Parametric analysis 82
Problem abstraction 84
Collective note-book 87
Orthographic analysis 88
SCAMPER 90
Analogies 91
Dot sticking 93
Cliches and proverbs 94
Evaluation-PIPS 96
Notes on Chapter 4 100

Chapter 5 THE INNOVATIVE COMPANY **101**

Key measures of product development effectiveness 103
How well do companies do? 105
Strategy for product development 106
The ball-point pen - blurred distinctions between pioneers and responders 109
Elements of strategy 110
Steps towards strategy development 111
Corporate planning 114
Corporate strategy 116
Implementation 119
Strategic planning of product development 119

Chapter 5 (Cont'd)

Steps towards a product development strategy 120
Implementing product development strategy 122
Individuals and teams for product development 122
The Psion Series 3: case study 124
Key concepts in Chapter 5 127
Design Toolkits
SWOT analysis. 128
PEST analysis 130
Tracking study 131
Product maturity analysis 132
Competitor analysis 134
Product development risk audit 136
The design team 139
Notes on Chapter 5 140

Chapter 6 PRODUCT PLANNING - OPPORTUNITY SPECIFICATION **141**

The product planning process 142
Commitment - the aim of product planning 144
What is an opportunity specification? 146
Opportunity justification 148
Researching and analysing the opportunity 150
Product triggers 151
Competing product analysis 152
Market needs research 154
Market research: the Marito factor 157
Technological opportunities 158
Plasteck Ltd: from strategy to opportunity research 161
Selecting a product opportunity 166
Systematic opportunity selection 169
Price positioning the new product 171
Style planning 174
Contextual styling factors 175
Intrinsic styling factors 176
Styling specification 177
Plastek Ltd: opportunity specification 177
Competing potato peeler analysis 179
Potato peeler market research 180
Potato peeler opportunity specification 184
Key concepts in Chapter 6 188
Design toolkits
Delphi technique 189
Market needs research 191
Opportunity specification 197
Notes on Chapter 6 200

Chapter 7 CONCEPT DESIGN 201
The concept design process 201
Establishing the aims and scope of concept design 202
Force-generation of concepts 205
Task analysis 206
Plasteck-peeler task analysis 208
Product function analysis 210
Life cycle analysis 213
Value analysis 214
Concept styling 217
Product semantics 218
Product symbolism 218
Lifestyle, mood and theme boards 221
Concept selection 229
Psion case study 231
Key concepts from Chapter 7 235
Design Toolkit
Product function analysis 236
Life cycle analysis 239
Notes on Chapter 7 241

Chapter 8 PRODUCT PLANNING - CREATING QUALITY, ADDING VALUE 243
Product quality 243
Specifying product quality 248
Translating customer needs into technical objectives 249
Quality function deployment 250
Stage 1: The heart of quality function deployment 251
Stage 2: Competing product analysis 253
Stage 3: Setting quantitative targets 254
Stage 4: Prioritising the design targets 255
Quality function deployment - beyond product planning 256
The design specification 259
Product development - project planning 261
Key concepts in Chapter 8 267
Design Toolkit
Design specification 268
Notes on Chapter 8 270

Chapter 9 EMBODIMENT AND DETAIL DESIGN 271
Product architecture 273
Feature abstraction 274
Product feature permutation 275
Design integration - the heart of embodiment design 278
Embodiment prototyping and testing 285

Chapter 9
(Cont'd)

Principles for prototype development 285
Testing for product failure 288
Failure modes and effects analysis 290
From embodiment to manufactured product 292
Key concepts in Chapter 9 293
Design Toolkit
Failure modes and effects analysis 294
Notes on Chapter 9 298

References 299
Index 305

Preface

The **purpose** of this book is to give an in-depth understanding of the entire process of new product development, as it should operate within a modern manufacturing company. This means design and development, not only of the visual appearance of products but also design for manufacture, design to meet market needs, design for cost reduction, design for reliability and design for environmental friendliness.

The **scope** of the book is intended to cover all aspects of product development for mass-produced products. It is concerned with industrial and engineering design rather than architectural design or design for craft production. The design problems range in complexity from the up-dating and minor improvement of an existing product to the introduction of a completely new innovative product. Commonly, books on new product development are written either from a marketing or an engineering viewpoint; how to find out and subsequently deliver what consumers want, on the one hand, and how to design and create the product, on the other hand. **A key belief** underlying this book is that these two aspects of product development cannot be divided and dealt with separately. The discovery of market needs and the embodiment and manufacture of a product to meet those needs are vital and integral parts of the same process. This book, therefore, sets out with the important but ambitious aim of covering the entire new product development process, from market research, through concept design, embodiment design, design for manufacture to product launch.

A great deal of information has been derived from the findings of academic research into innovation and product development yet the book remains predominantly industry-oriented. It is **written for** managers, designers, engineers and technologists working in manufacturing companies and students studying engineering, design and technology prior to starting work in industry. The ideas contained within the book have been developed during the development of commercial products, whilst working with commercial companies.

The **central theme** running through this book is that managing and controlling product development is fundamentally difficult. By its very nature, innovation deals with uncertainty and requires decisions to be made on the basis of factors which are difficult, if not impossible, to predict. This means that product development is often done badly, or even worse, not done at all. Since new technologies are both encouraging and facilitating a faster rate of innovation in most manufacturing sectors, 'Innovate or Perish' is going to be a familiar notion to increasing numbers of companies in coming years. The difficulties of innovation management must, therefore, be overcome.

The **spirit** of the book is, however, optimistic. Innovation **can** be managed well. The innovation process **can** be described rationally and systematically. Reasons **can** be found for why some innovations succeed whilst other fail. The development of new products **can** be budgeted, strictly timetabled, and rigorously quality-controlled. Risks will always remain but with good management they can be minimised. Most importantly, the costs of failure can be contained by identifying the unsuccessful products before too much money is committed to them.

The **objective** of the book is to make a convincing case that the thorough use of systematic methods is worth the effort. Specifically, it aims to give an understanding (a workable and directly applicable understanding) of both a structured framework for the management of innovation and the systematic design and development methods which fit into that framework.

The **style** of the book attempts the usually-unwilling combination of simplicity and easy reading on the one hand and technical accuracy and academic rigour on the other. In pursuit of simplicity, references are not cited in support of every technicality; to maintain some academic rigour, and for the benefit of the more enthusiastic reader, the main propositions within the book are referenced and further reading suggestions are given in the chapter notes. Industrial design is becoming increasingly jargon-laden, a tendency I have tried to avoid. In particular, designers have become obsessed with the use of acronyms as labels for design methods. I cannot do much to cure them of this disease but I have tried to keep clear of the symptoms.

The **best way to read the book** is to treat it as your companion and helper during a live product development project. This is not a book that demands cover-to-cover reading. Browse or dip into chapters 1 to 5 at your leisure and make sure you are, at least, familiar with the keynotes at the end of these chapters. Then, when you start a new product development project, begin at chapter 6 and let the time-course of your project lead you through to chapter 9.

Acknowledgements

Without the support and encouragement of Brunel University, the Design Research Centre would never have existed and without the DRC this book would never have been written. Prof. Eric Billett deserves most of the credit but my thanks go also to Prof. Mike Sterling, Linda Cording and Dr. John Kirkland. The DRC has been the intellectual boiler-house of this book. Most of the ideas that are new or exciting in the pages that follow have been created, developed or refined by DRC staff. My particular thanks go to Chris McCleave for his help with strategic planning and Tom Inns for many insights into design management. Richard Bibb helped with much of the early research, and Paul Veness helped with later research. My thanks also to Paul for the design of the cover and for his detailed revisions of the manuscript. To Dan Brady and Mike White, my thanks for their highly productive time in creating artwork. Thanks also to Sally Trussler for many of the photographs she took specially for the book. Lesley Jenkinson tolerated my endless page shuffling with unfailing good humour. And, last but not least, my thanks to Aileen for putting up with me for the last, hectic and anti-social, six months and also for bringing me down to earth when my ideas for the book began to head skywards.

Most of the research upon which this book is based has been conducted under the Government's Teaching Company Scheme. This scheme, which supports knowledge- and skills-transfer from Universities to commercial companies has allowed the DRC to work closely with several small manufacturing companies over two-year design and development projects. The experience gained, and the research required to support these projects, provides the main source of knowledge for this book. Many of the examples in the book, and in particular the Plasteck example which runs through several chapters, are based on real product development case studies within the DRC Teaching Company Schemes. Often, for the sake of commercial confidentiality, the products and the design data have had to be disguised.

1 Introduction

Innovation is a vital ingredient of business success. Free market economics depend upon companies competing with each other in the marketplace and trying to close any lead established by another company. In terms of selling products, companies must continually introduce new products to prevent their more innovative competitors eating away at their market share.

In more recent years, the pressures to innovate have grown. As manufactured products have become more global, the competitive pressures from overseas companies have increased substantially. This is not only a threat from the multi-national industrial giants. International licensing agreements can spread products around the world through networks of small or medium sized companies. Looking over your shoulder requires much longer vision that it used to! To make things worse, the average life span of products is shrinking fast. New technologies, such as CAD, rapid prototyping and tooling are reducing product development times. Customers can therefore be more choosy and change their demands more frequently, stimulated and fuelled by whatever product or new idea has been introduced most recently. Pity any poor manufacturer who does not move fast enough, or worse still, tries to stand still in such a rapidly changing marketplace. The statistics on business success show clear

> 'The managerial tactic of consciously shortening the market life of products, through rapid new introduction is a strategic weapon against slower-moving competitors. Pioneered by the Japanese, this viciously effective approach is now being emulated by a growing number of Western countries. As a result, every competitor is having to scramble to produce a greater and faster flow of new products than in the past.'
> Christopher Lorenz [1]

Year	% Total sales	% Profits
1976–1980	33	22
1981–1986	40	33
1985–1990	42	–
Projected 1995	52	46

Figure 1.1
The percentage of total company sales and profits generated by new products [2]

trends in product innovation (Figure 1.1).

For designers this is exciting news. Design is the vehicle for product change and the more products change, the more design will be needed. But it is not all good news. The failure rate for new products is another often-quoted business statistic. The figures vary a great deal because they use different definitions for what constitutes a new product and what constitutes success. But the consensus suggests that for every 10 ideas for new products, 3 will be developed, 1.3 will be launched and only 1 will make any profit [3]. Considering that the cost of new product development can run well into six-figure sums, these are terrifying odds. No one in their right mind would start – except, as we have seen, it happens to be essential. This makes design a very different vehicle to drive. Reaching the finishing line, with a fully developed new product is no use at all if the product is unsuccessful. Choosing the right destination, picking a good course, changing course when necessary, negotiating hazards, avoiding accidents – and all whilst driving fast enough to be sure not to lose sight of your competitors. This is what 'driving' the design process is all about in the world of modern manufacturing.

'....for every 10 ideas for new products, 3 will be developed, 1.3 will be launched and only 1 will make any profit...'

Innovation – risky and complex

The secret to successful innovation is, therefore, the management of risk; the subject at the heart of this book. Risk management can be looked at in two ways:

'There are only two important functions in business: marketing and innovation; everything else is cost'
Peter Drucker

- First, it is about targeting; getting the new product on to the market, ensuring that it appeals to, and satisfies the customer, is fit for its intended purpose, is of a quality to last its design lifetime and can be made at an acceptable cost. Innovation methods must take account of all these factors and minimise the risks of the new product failing.
- Secondly, it is about killing off products as soon as they fall short of the set targets; innovation must be examined critically at every stage in order to stop the development of unsuccessful products as early as possible. If only a small proportion of new products are going to succeed it is vital to the financial health of any company that the wastage of resources on failure is minimised.

Already, a few paragraphs into this book, the complexity of new product development is becoming clear. Just **how** can new products be developed in order to meet the multitude of different, and often conflicting requirements?

- Customers want improved product performance and better value
- Marketers want a competitive edge and product differentiation
- Production engineers want simple production and easy assembly
- Technologists want to try new materials, designs and processes and
- Accountants demand costs and margins for minimum investment.

New product development is inevitably a compromise process. At the very least, compromise will be required between factors which add extra value to the product (e.g. improved product quality, the addition of secondary features) and those which constrain its development costs (e.g. product price, development lead time, design and development man-hours). This compromise must end up with a product which will compete effectively against other products in an ever-changing market. An ill-informed, ill-considered or just plain unlucky compromise may cause that product to fail and all the resources invested in its development to be wasted.

New product development is never simple and straightforward. It requires thorough research, careful planning, meticulous control and, most importantly, the use of systematic methods. Systematic design methods are not, at present, used to anything like their full potential in new product development. To do so requires a multidisciplinary approach, embracing marketing methods, engineering methods and the aesthetic and styling methods of industrial design. This marriage of social science, technology and applied art is never going to be easy but the imperative to innovate demands that it must be attempted. The best designers of the future will be multi-skilled and will feel as comfortable discussing market research as they do colouring up a rendering of a new product or selecting which plastic it should be made from. This is not to suggest that the design profession will be limited only to naturally gifted polymaths. In-depth expertise in all areas is not required; that can always sought from others within the company or from external consultants. But what is required is a basic understanding of a wide range of approaches and methods for tackling the development of new products. The ability to use basic methods in each of the three main disciplines – marketing, engineering and industrial design – will remove the fear factor and will give the designer both an empathy and a working understanding of new product development in all its aspects.

'A good workman is known by his tools!'

Proverb

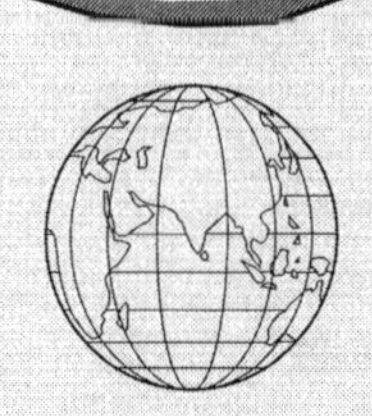

Global Manufacturing PLC requires an

Industrial Designer of the Future

- **Must be:**
- Multi-skilled
- Fanatically 'customer-oriented'
- Deeply committed to systematic design methods
- Knowledgable about a wide range of manufacturing business
- Comfortable in marketing, design and engineering disciplines
- Accomplished at creative problem solving

Ground-rules for systematic design: The 3 very unwise monkeys

The 3 Wise Monkeys: See no evil, Hear no evil, Speak no evil

See no evil! Seeing evil, which in the new product development dictionary means spotting the new products which will fail in the marketplace, is vital in new product development. In fact it is probably the second most important task a designer can perform, second only to being able to create products which will not fail! This is far too important a task to be left to clairvoyance and must, therefore, be tackled systematically. Setting clear and realistic targets for a new product provides the vision of what that product must achieve to be successful. The most important targets are those demanded or wished for by customers. Other important targets include compatibility with the skills and facilities of the manufacturer, suitability for the intended marketing, sales and distribution channels and conformance with relevant statutory or industry standards. Designers who fail to set targets will indeed 'see no evil' but they also will fail to see what they must achieve in order for the new product to succeed. And as the old adage says 'if you do not know where you are going then every road is the right one'.

Hear no evil! Setting targets for the product is of little value unless its progress throughout the development process is monitored. Sounding out the product at periodic intervals and checking these soundings against the set targets is the only way to spot when the product has started to go off course. If a new course can be set, the product development may continue, otherwise it must be abandoned before it has cost too much. Designers who are deaf to the tell-tale signals of a new product going wrong can only hope they are deaf enough to miss the crash when the product is launched!

'Setting clear and realistic targets provides the vision of what a product must achieve to be successful.'

Speak no evil! Freedom of creative expression is at the heart of design. And it is one of the most widely held myths that the use of systematic design methods strikes that creativity dumb. Nothing could be further from the truth. Creativity, as Thomas Edison once said is 1% inspiration and 99% perspiration. The perspiration arises in preparing your mind – building the foundations upon which the building blocks of creativity are set (Creative thinking is described in greater depth in Chapter 4). The historical accounts of truly great discoveries tend to focus on the final leap of intuition which made the breakthrough. Little mention is made of the months or years of researching the problem, wrestling with existing solutions and exploring all the ideas which eventually proved useless. In many of these cases it was the articulation of incorrect solutions which led, step by painful step, to the breakthrough. So it would seem to be with new product development. If, as it was said earlier, it takes ten ideas to come up with one successful product, creativity must be given the freedom to come up with unsuccessful ideas in order to discover the successful ones. Indeed, the number and quality of the ideas which are rejected is probably the best measure of a person's idea-generating capabilities. When someone has a single idea and proceeds on the basis of that idea, it might be a good idea but, equally, it might be mediocre or even completely useless. When someone has selected the best of 10 ideas, the chances are much greater that this idea will be good. Taken to its logical extreme, the closer you get to thinking up every possible solution to a problem, the closer you will be to finding the best possible solution. Freedom to express creative ideas, including those which ultimately prove worthless, is, therefore, a virtue in new product development. Since it is difficult enough to come up with a single solution to most problems, the prospect of having to generate many solutions is formidable. The use of systematic idea-generating techniques is, however, ideally suited to the task. Failure to use systematic methods is, in Edison's terms, equivalent to a large dose of anti-perspirant!

'...the closer you get to thinking up every possible solution to a problem, the closer you will be to finding the best possible solution.'

Complying with these rules throughout the innovation process is easier

Ground-rules for systematic design

- Establish targets for new product development. These should be clear, concise, specific and verifiable
- Compare the emerging new product against these targets at several stages during development. Kill the product off as soon as it becomes clear that it will fail to meet essential targets
- Be creative. Generate many ideas and select the best. Don't be afraid of producing ideas which may later prove non-viable.

said than done. If, however, you have faith that new product development can be managed systematically and have the will and commitment to try to do so, the remainder of this book will act as your guide and tutor. Unlike many books on product design, this one does not try to present any single correct way of developing new products. Rather, it offers a structure for the management of product design, based to a large extent, upon the design management procedures first brought to widespread attention by Pahl and Beitz [4]. In the UK these procedures have been developed into the pioneering and, to my mind, excellent guidelines for the management of design, BS 7000 [5].

Within this management structure, 'toolkit files' describe methods and techniques for the design and development of new products. Used together, these design methods provide:

- a systematic approach to new product development
- a strong market-focus throughout the design process
- techniques for stimulating creativity and coming up with more innovative solutions to design problems

Design Toolkit

What is a design toolkit?

A concise description of a systematic method for use in the design and development of new products. It should be thought of as 'a collection of tools' to stimulate ideas, analyse problems and structure design thinking. In total, there are 38 methods presented in toolkits, ranging from general ways of analysing the innovation strategy of a company to detailed analytical techniques for predicting product failure or reducing product cost. Taken together, these methods make up a 'toolkit', from which the most appropriate tools can be selected for the job at hand.

Rarely will you need to use **all** the methods, hence the idea of a 'toolkit', from which you select the appropriate 'tools' for the job at hand. The 'toolkit' approach has been found to work well within many design offices. Design projects never present exactly the same problems and hence do not require the same methods for their solution. The descriptions of the different design methods are located throughout the book according to which stage in the design process they are most commonly used. Wherever possible, the principles described in the book, and the toolkit files in particular, are backed up with practical examples.

'....unlike many books on product design, this one does not try to present any single correct way of developing new products...'

- Chapter 2 provides an overview of the product development process. It begins by reviewing the research into the determinants of success and failure in new product development. It portrays product development as a risk management process and explores how risk can be reduced effectively. Most importantly, it introduces the framework within which product development is managed and also introduces the concept of convergent and divergent thinking within each stage of that framework.

- Chapter 3 explores product styling. It begins with an overview of the psychological factors by which we judge objects as attractive or otherwise. It then moves on to explore how these factors lead to general styling 'rules' and then reviews the implementation of these within the design process.

- Chapter 4 examines creativity, in general at first and then within the design process. Several general techniques for stimulating creativity are described in Design Toolkits.

- Chapter 5 is on the innovative company. What factors make one company consistently innovative whilst the next cannot make the simplest of improvements to an existing product? Company strategy and the management of people within the product development process are examined.

- Chapter 6 covers the product planning stage of product development. It reviews the aims of product planning and then describes methods for market research, competing product analysis, identifying new product opportunities and preparing design specifications.

- Chapter 7 gives an overview on the design process from concept design to detail design and examines how concurrent engineering manages the operation of these stages concurrently.

- Chapter 8 is on concept design and examines methods for concept generation and concept selection.

- Chapter 9 covers both embodiment and detail design. It focuses on designing the product to be fit for purpose and examines, in depth, the role of failure analysis in relation to prototyping and testing.

Notes on Chapter 1

1. Lorenz C. 1986, *The Design Dimension.* Basil Blackwell, Oxford.
2. Page A.L. 1991, *New Product Development Practices Survey: Performance and Best practices*. Paper presented at PDMA's 15th Annual Conference, October 1991.
3. Some data product attrition rate is given in Page 1991 (see 2 above). See also Hollins B. and Pugh S. 1987, *Successful Product Design.* Butterworth & Co. London and Cooper R.G. 1993, *Winning at New Products*. Addison-Wesley Publishing, Boston.
4. Pahl G. and Beitz W. 1987, *Engineering Design: A Systematic Approach.* Design Council, London.
5. BSI 1989, *Guide to Managing Product Design, BS 7000*. British Standards Institution, London.

2 Principles of new product development

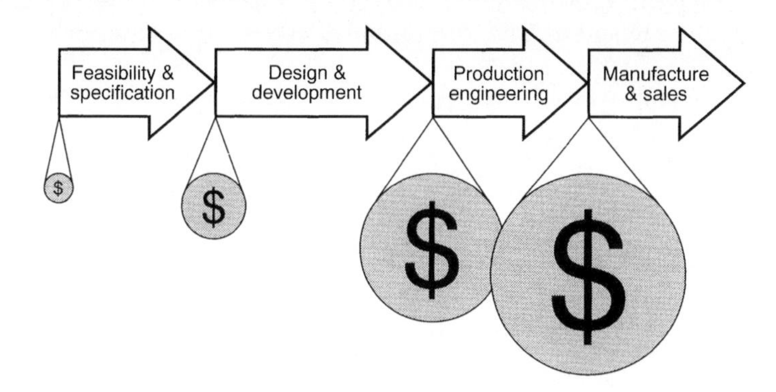

Developing new products is both important and risky, this much is clear from the last chapter. But why is it so risky? What makes so many products fail? Perhaps if we knew that we could see more clearly how to succeed. Major studies in the UK, USA and Canada [1] have undertaken the laborious task of back-tracking through completed product development projects to find out how they were done and how this related to their commercial performance. In total, over 14 000 new products in 1000 firms were studied, some of which went on to be commercially successful and others which failed. By studying what was done differently during the development of successful products, on the one hand, and failed products, on the other, key factors for new product development have been identified.

Success and failure in new products

A number of factors in three broad areas (Figure 2.1) made the most important differences between success and failure [2].

Market orientation The biggest, and probably most obvious, single factor determining commercial success was market differentiation and customer value. Products which were seen by customers as i) being substantially better than competing products and ii) being better in ways which were highly valued had 5.3 times the success rate of those that were only marginally different. This may seem obvious but it raises two important points. Firstly this difference is huge and so, if you are going to focus attention on any particular aspect of a new product it must be its market orientation. Secondly, if you can only achieve marginal differentiation with a new product, it may be wiser to kill it off during development – the chances are that it may be a commercial failure when it is launched anyway. Other factors contributing to market orientation were found to be ability to market the product early and to having a highly effective product launch.

Early feasibility assessment and specification Two factors are important here. Products which underwent a thorough and stringent **assessment** prior to development were 2.4 times as likely to succeed as those that had not. Products which are sharply and well **defined** in a design specification prior to development were 3.3 times as likely to be successful as those that were not. The message – put lots of effort into getting the product right at the start before beginning the design work.

Quality of the new product development process (and the team that does the work). Where technical activities are consistently carried out to a high quality, products have, in general, a 2.5 times greater success rate. Specifically, where the company's technical skills were well matched to the activities needed to develop the new product, the chances of success were 2.8 times greater. Where the company's sales and marketing skills were well matched to the new product the chances of success were 2.3 times greater. Where there was a high degree of 'working harmony' between the technical and marketing staff within a company, the chances of product success were 2.7 times greater than where there was 'severe disharmony'.

A great many conclusions have been drawn from these types of studies and the nature of new product development has changed considerably as a result. One of the most generic conclusions is Robert Cooper's 'gambling rule' [2]. This says that when the uncertainties are high, keep the stakes low; as the uncertainties reduce, increase the stakes. In new product development terms the uncertainties are highest at

Figure 2.1
Success factors in new product development [2]

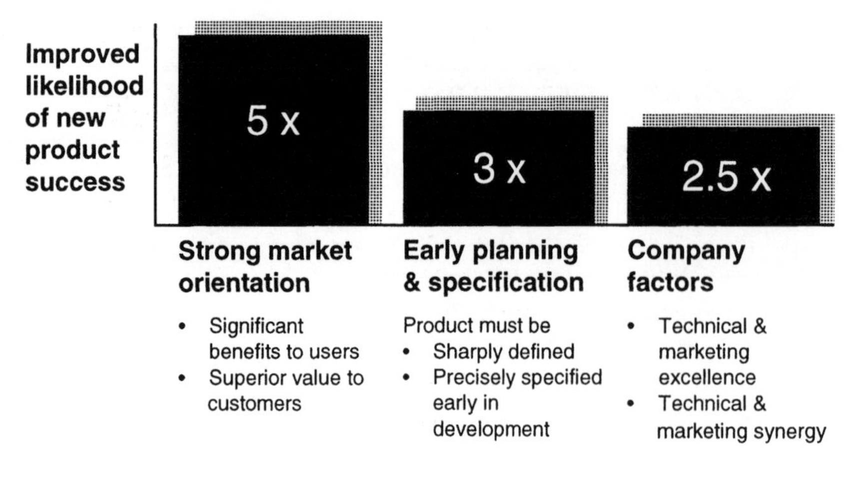

the start. You do not know what the product is going to look like, how it will be made, what it will cost or what customers will think of it. As a result, you must keep the stakes low. So, avoid investing heavily (e.g. in prototyping or commissioning production tools) until the preliminary stages of product development have reduced some of the uncertainty. This can be done relatively cheaply. Doing the early design work, producing sketches or models, estimating costs and talking to customers requires only time and a minimum of materials. If this proves successful, the uncertainties have reduced and the stakes can increase. Expressing all of this in a more useful and systematic way we can present new product development as a risk management funnel (Figure 2.2) .

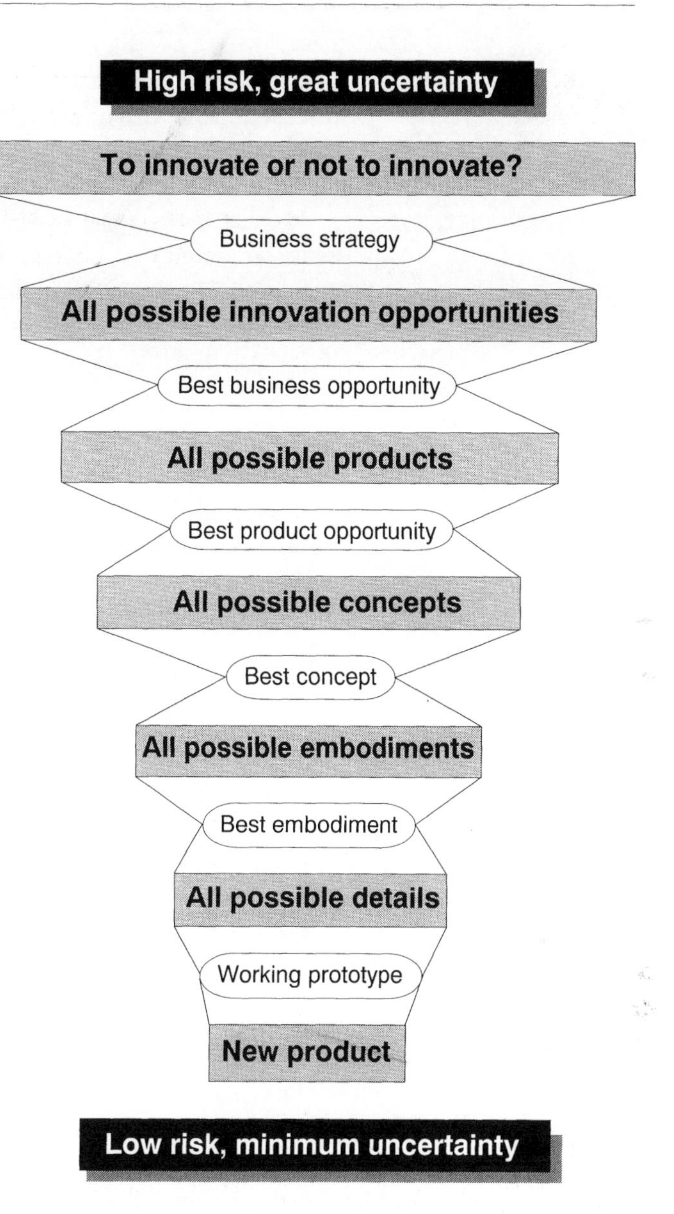

Figure 2.2 Risk management funnel [3]

Risk management funnel

A risk management funnel [3] is a way of thinking about new product development which shows how risk and uncertainty change as new products develop. It is, essentially a decision-making process in which the shaded boxes represent the options and the open rounded boxes represent the decisions made from these options.

Business strategy As a first step, companies must decide whether or not to innovate. From what has been said about innovation so far, this might seem a rhetorical question. Surely, companies need to innovate in order to cope with the increasing pressures of competition and the faster turn around of new products. Most probably do, but not all. Many companies have a traditional range of products selling well in a traditionally-minded marketplace. For them, innovation may not only be unnecessary, it might jeopardise their business. The 'innovate or not' decision is very risky and carries great uncertainty. Deciding to innovate runs the risk of substantial expenditure with no return. Deciding not to innovate could see the company forced out of business by more innovative competitors.

Business opportunity Once innovation is decided upon and incorporated into the company strategy, the next step is to examine all the possible innovation opportunities. The aim here is to select the best business opportunity presented by product innovation. This has nothing to do with individual products; it is not a decision between making a new kettle or a new coffee maker. Rather, what type of innovation, in general, is best suited to the company? Is it, for example, i) introducing reduced-price products by cutting costs of production, ii) introducing enhanced-value products by improved styling or use of new materials or iii) introducing extensions to an existing product line in order to boost turnover and spread overhead costs.

This innovation opportunity should ideally be pursued over the development of several new products. This establishes medium- to long-term goals for the company, allowing skills and expertise to develop in specific aspects of innovation. On-going innovation according to an agreed strategy has been shown to make new product development much more efficient than intermittent projects lacking such continuity [4]. Pursuing this one type of opportunity, however, still carries a high risk. What if you have decided to reduce costs and cut prices at a time when the market is looking for enhanced value and new features? It may take several product failures before the mistake is realised.

Product design and development Next come the stages of development of a particular new product. This involves less risk and uncertainty than either of the preceding stages (deciding strategy and the innovation opportunity) [10]. It also further reduces the risk and uncertainty, step-wise and progressively as decisions are made on i) the specific **product opportunity** to be exploited ii) the principles of operation of the product (**concept** design) iii) how the product is going to be made (**embodiment** design) and finally iv) the **detail** design and production engineering.

Clearly, risk and uncertainty will still remain, even when the new product is stockpiled in the warehouse ready for sale. But minimising that risk and uncertainty is the essence of effective product development. And developing new products through the risk management funnel is a lot safer than the alternative approach, which is to 'make-one-and-see-if-it-sells'!

Risk management funnel in action: decision-making by a small electronics company

Decision/Action	Risk	Risk Management
Innovate or not?		
Yes, the company is going to develop new products.	Wrong strategy? Risks wasting money discovering that the company is unsuited (technically, commercially or managerially) to developing new products.	Analyse the company's strengths and weaknesses
Possible innovation opportunities		
The company will develop leading edge products based on state-of-the-art technology	Wrong opportunity? Return on investment too long term for a small company.	Analyse the market and the company's present products in that market.
Possible products		
A new remote control device for use with emerging interactive TV technology	Wrong product? Dependent upon the success of interactive TV.	Form strategic partnership with key technology provider.
Possible concept		
The product will work like a cordless computer mouse, driving pull-down on-screen menus and having a single select button	Wrong concept? Dependent upon software not under the company's control.	Prove value of concept through market research. Establish joint venture with software developer.
Possible embodiment		
IR communication, sealed micro-switch button, injection moulded clam-shell casing in ABS, 9V battery.	Wrong embodiment? Inadequate operating range, battery life too short , plastic casing breaks when dropped.	Failure modes and effects analysis (FMEA) followed by prototyping and testing.
Possible details		
Full working prototype produced.	Manufacturing problems? Incorrect assembly, faults in supplied components, manufactured parts outside specified tolerances.	Repeat FMEA and introduce quality control procedures.

Stages of the risk management funnel

'....the risk management funnel progressively and systematically reduces the risk of product failure.'

The stages which make up the risk management funnel provide a useful and sensible way to divide up the process of new product development. They are not, however, cast in tablets of stone. Other authors writing about new product development consider that the process is better described with either more stages or fewer stages. The precise definition of each stage, what it comprises and where it starts and stops relative to adjacent stages is not particularly important. What is important is that the risk management funnel, as a whole, is seen as process which progressively and systematically reduces the risk of product failure. This extends from identifying a strategic need for innovation within a company's business plan through to the commercial launch of the product. Since this is a lengthy and complex process, subdividing it makes quality control more planned and structured. The need for flexibility, however, becomes obvious when we consider different aspects of quality control. Financial control over new product development, for example, will usually be exercised in 4 stages (Figure 2.3).

- Initial financial approval, for a small commitment of resources. This allows the commercial feasibility of the proposed new product to be explored and, if it looks promising, a product design specification will be prepared.
- Upon acceptance of this specification, a further financial commitment will allow the design and development of the product to proceed. This requires significantly more resources than the previous stage, but is still only a small proportion of the total product development costs.
- Once designed fully on paper, and usually tested in prototype form, the next commitment is to tool-up for manufacture. This can be a substantial expenditure if new tools and assembly lines are required. Hidden costs can also be significant at this stage: equipment, factory space and labour may have to be taken off the production of an existing (and presumably money- making) product in order to prepare for the new product.
- The final commitment is to commence production. This will involve the initial production of stock, its distribution to suppliers and the marketing effort to launch the product. Huge amounts of money can be involved here. If, for any reason, the product should fail at this stage, damage to the company's reputation can be an immense indirect cost.

'The secret is to fail with small dollars, learn from your mistakes and succeed with big dollars. Innovation and risk-taking are the critical skills.' H.B. Atwater, Chairman, General Mills Inc. [5]

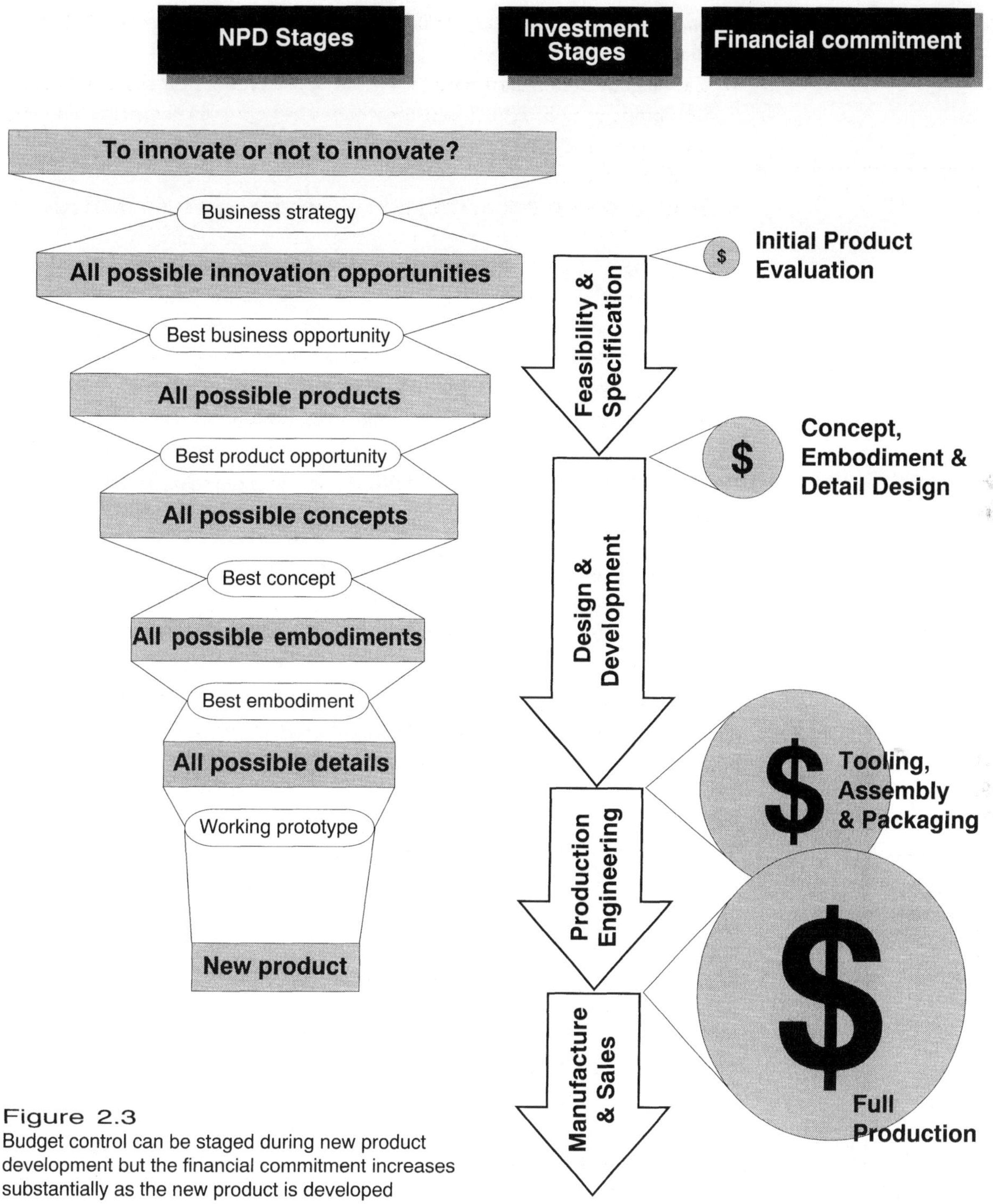

NPD Stages

Investment Stages

Financial commitment

To innovate or not to innovate?

Business strategy

All possible innovation opportunities

Best business opportunity

All possible products

Best product opportunity

All possible concepts

Best concept

All possible embodiments

Best embodiment

All possible details

Working prototype

New product

Feasibility & Specification

Design & Development

Production Engineering

Manufacture & Sales

$ Initial Product Evaluation

$ Concept, Embodiment & Detail Design

$ Tooling, Assembly & Packaging

$ Full Production

Figure 2.3
Budget control can be staged during new product development but the financial commitment increases substantially as the new product is developed

So, subdivisions of the new product development process are important for planned and structured quality control. The stages into which the process is divided are not immutable. They can be selected to suit the product and the company environment within which it is being developed.

Risk management - theory into practice

Some designers feel uneasy when they see the design process divided into stages at all! In the real world, they argue, design is chaotic. Design thinking explores some ideas at a conceptual level, whilst, at the same time, thinking about others in detail. Ideas spring from all sorts of sources; they are not just funnelled down from the previous stage of the design process. And, in the course of producing a finished design, the design process will have been looped through several times as the design is iteratively improved. So, where does this leave the risk management funnel?

It is important, at this point, to be absolutely clear about what the risk management funnel **is** and what it **is not**. The risk management funnel represents the decision-making about a new product during its development. It shows how the options available and the decisions made change as the product develops. Its value is that it reduces risk progressively over the entire process of product development. It does not wait until the product is launched to see if it is going to be a commercial success - the 'make-one-and-see-if-it-sells' approach. The risk management funnel is not, and never can pretend to be a representation

Risk management funnel - an analogy

The map below is of a small section of one of Britain's motorways. It is clearly schematic and simply shows the junctions and the intersecting roads. It does not suggest that the motorway is entirely straight any more than it suggests that every junction is at a perfect right angle and that the services area is in the middle of the motorway, blocking all traffic. It simply shows the options offered on this stretch of motorway upon which the driver must make decisions. The activities necessary to drive along this motorway are however much more complex. There are corners to be negotiated and other traffic to be avoided. When you miss a junction you must proceed to the next junction and loop back to the one you missed.

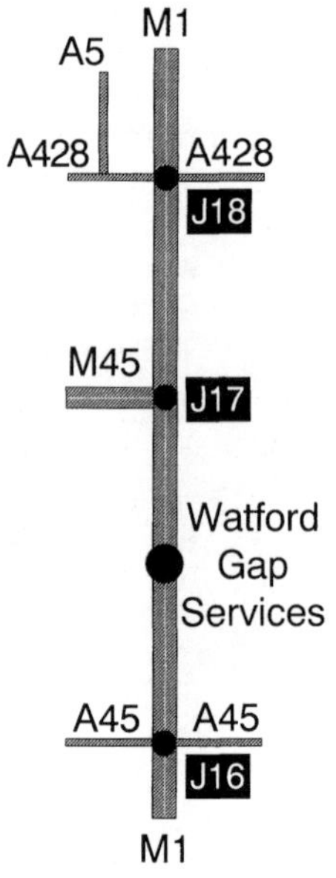

None of this is shown on the schematic map. But the map is still useful and it is all that an experienced driver needs to navigate a motorway journey. My argument is identical for new product development. The risk management funnel does not show all the design activities needed to develop a new product. But it does show the options and decisions available to the designer. And it is all that an experienced designer needs in order to navigate the process of new product development.

of design activities. How can the inspiration of a new idea be represented by a series of boxes in a row? What about the products that only emerged after having explored several different concepts and repeated the design process many times. The diagram of this type of design activity would be a whirl of loops within loops. But the risk management funnel is not trying to say that these activities are, or should be, neatly ordered and structured. It is the decisions which result from these activities which the risk management funnel represents (See analogy opposite). This does not constrain the way designers think or even how designers design. It simply provides a framework for the organisation of decision-making about new products.

Another way of emphasising this distinction is to look at what product development activities are occurring at different stages of the risk management funnel. Figure 2.4 shows a wide range of business, marketing, design and engineering activities running alongside the risk management funnel. It is worth drawing attention to two aspects of these activities. Firstly, many design activities continue across more than one stage of the risk management funnel. This means that they are quality controlled as they progress and move towards progressively more detailed aspects of the product's design or evaluation. Secondly, the design activities do not follow a 'hand-over' or 'over-the-wall' development procedure. Back in the 1960's it was common to think of new product development as a three-stage process during which: i) marketing established the product requirements and handed these over (or threw them

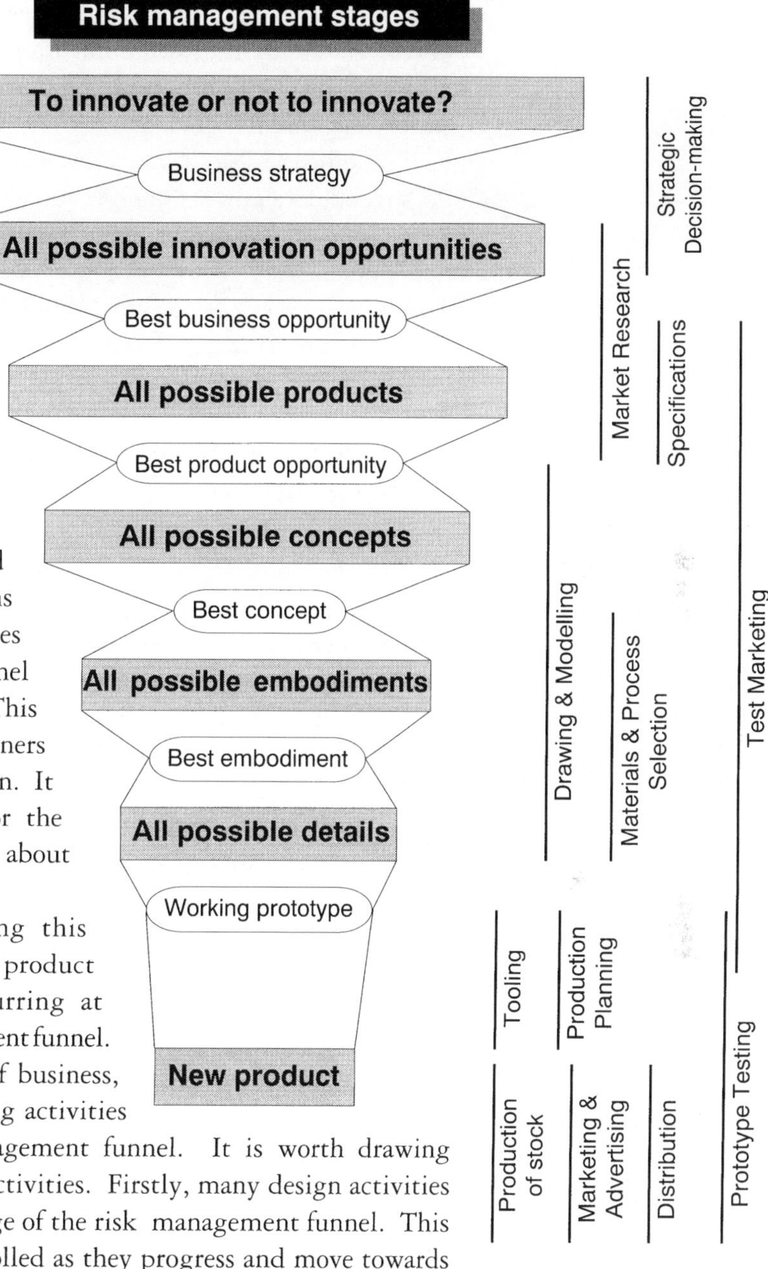

Figure 2.4
Marketing, design and engineering occur concurrently

over the wall) to design and development; ii) design and development then developed the product to the point of a working prototype and handed it over to production engineering; iii) production engineers then sorted out how the product could be manufactured [6]. Since then, common sense seems to have prevailed and the obvious conclusion has been reached that marketing, design and development and production engineering must work together. There are two reasons for this. Firstly, by operating concurrently, the development lead times for new products are shortened. Secondly, and probably more importantly, working together jointly ensures that the product does everything it must to be a commercial success.

Managing design activities within the risk management funnel

The organisation of design activities within product development is always complex. In the figure opposite (Figure 2.5), the development of a relatively simple product is shown schematically. The product is developed through four iterations; the first moves straight from the development trigger to concept design in order to explore product ideas for initial market testing. It is likely that, at this stage, the product would be presented as a simple rendering and shown to a small number of potential customers or salesmen. If this proves successful the second iteration develops an opportunity specification, a design specification and then moves back into concept design. The selected concept will then be market tested, starting the third iteration. If market testing is again successful, the product will move directly to embodiment design. At this stage it would not be uncommon to discover design options not previously considered or to make a technical breakthrough of some sort. This immediately throws the design process back one or more stages in order to review the implications for the developing product. If the breakthrough is of relevance to key aspects of the product's business prospects it will be necessary to go all the way back to the opportunity specification as shown in Figure 2.5. Having been through the process before it should not take long to revise. Of course, the more thorough the work was the first time around, the easier it will be to go through a second time. Upon returning to embodiment design, a preferred embodiment will be selected against the design specification and this would normally be market tested. The development process then moves of to detail design and design for manufacture, ending up with a production prototype. The official 'signing off' of this prototype as satisfactory and complying with the design

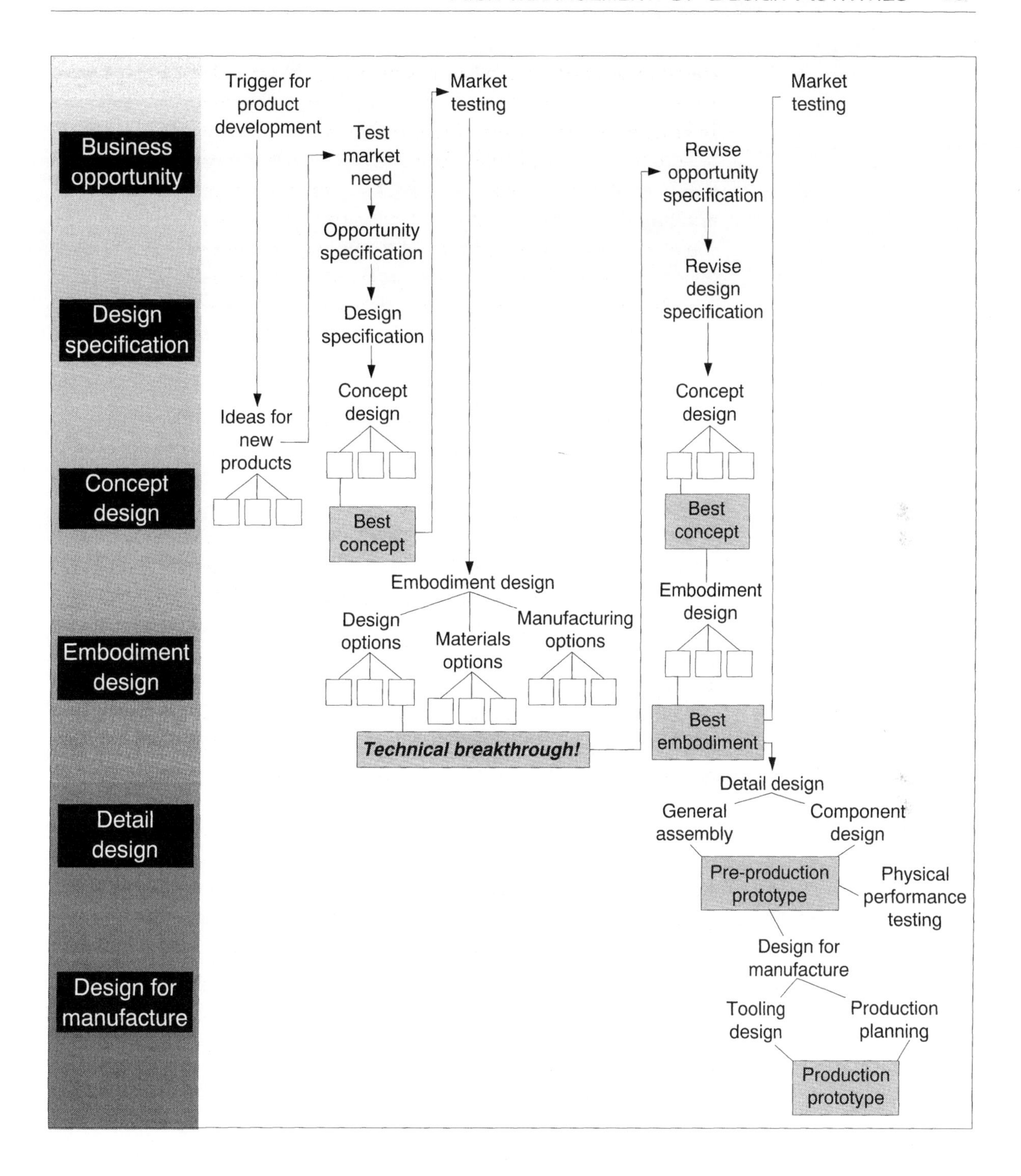

Figure 2.5
Design activities at different stages of product development

specification ends the product development process. This gives the green light for the start of production and product launch.

The pattern of design activities is, therefore, iterative, looping through the design process several times before producing the finished product. The iterations serve two purposes. Firstly they ensure that the product is optimised by successive approximations. By sketching some initial product concepts, market testing them and then going on to repeat the process in greater detail, the product concepts are better and more market oriented than they would otherwise have been. Secondly, iterations allow previously unforeseen opportunities and problems to be fully integrated into the development process. It is all too tempting, when something new emerges during product development, to simply incorporate it into your design thinking without stopping to consider its full implications. By repeating the previously completed design stages, the full impact of this new idea can be explored, hopefully leaving no nasty surprises for discovery closer to the product launch.

There is also an important projective element to design activities. Often during product development it is necessary to project forward through the development stages, for either technical or business reasons. If, for example, approval from company management is needed to commence a new development project you may be asked to present one or more estimates of its eventual production costs. This would involve projecting forward through the design process and giving your best estimate of its principles of operation, how it would be made and what it would cost for production and assembly. If approval was given, you would then, of course, commence the design process afresh.

So, product development **can** be thought of as a structured and ordered process. Each stage of this process comprises a cycle of creative idea generation followed by systematic idea selection. At times you will have to repeat certain stages to be sure you have the best design. At other times you will have to jump forward and project a finished product from your current ideas. But this is all entirely consistent with a structured and ordered decision-making process. Just because the decisions are structured and ordered does not mean that the activities leading to these decisions need to be similarly structured and ordered!

Quality control of product development

Quality control has been mentioned several times in the discussion so far and it is time to consider the issue in more depth to find out how it can be applied to new product development. Most people feel comfortable with the idea of quality controlling manufacturing or administrative tasks. This is because targets can be clearly and objectively specified and the outcome of the task measured against these targets. When it comes to developing a new product, the notion of quality control seems incongruous. At the start of product development, when the target needs to be set, the product does not yet exist.

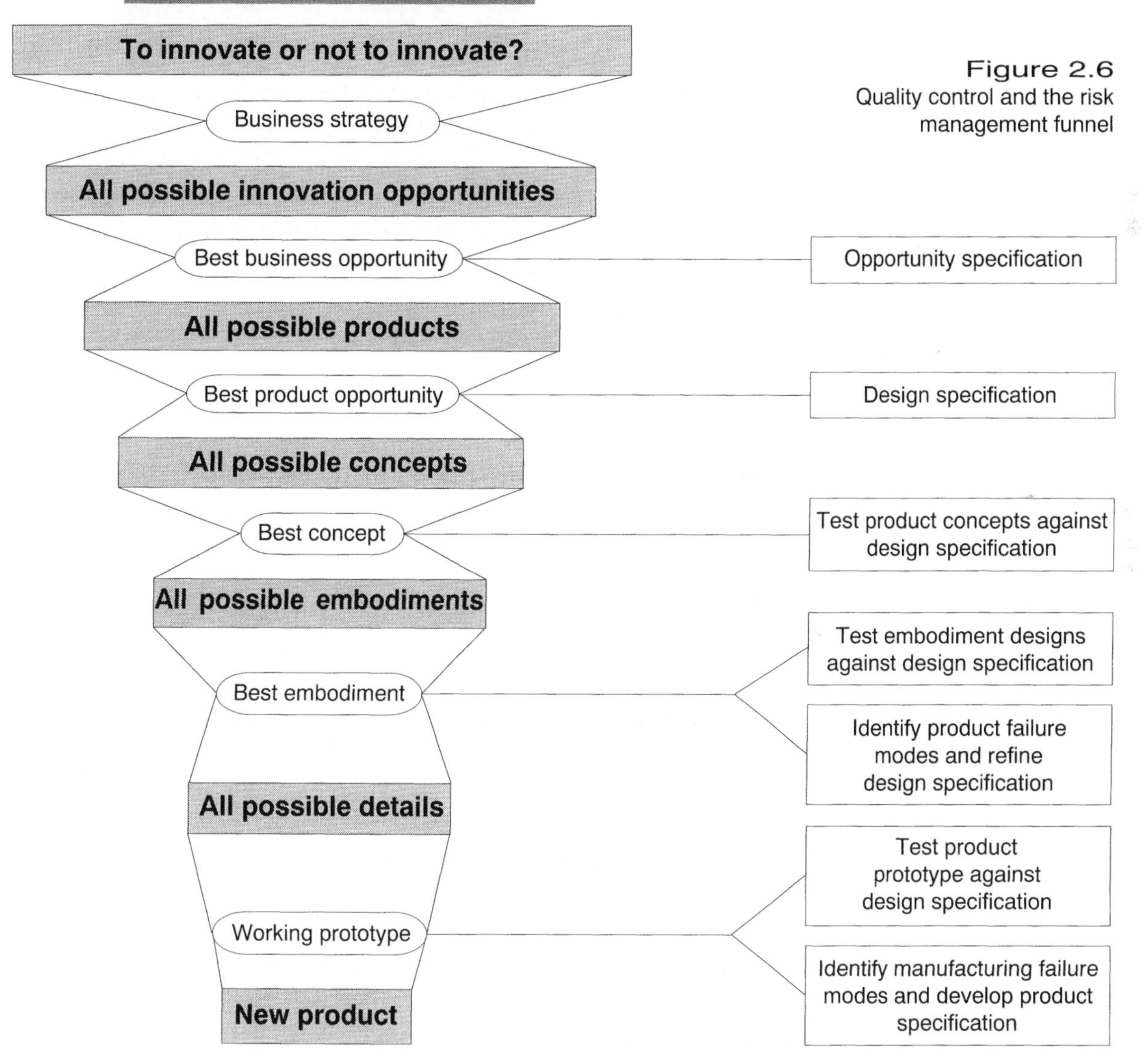

Figure 2.6
Quality control and the risk management funnel

How, then can you decide upon quality control targets before you even know what the product is that you are trying to quality control? This is a fundamental difficulty which appears to place product development beyond the reach of conventional quality control procedures. To overcome this difficulty we need to go back to the first principles of quality control.

These first principles, which are at the heart of quality management standards [7] are simply:

- Say what you are going to do,
- Do it,
- Check that you have done it.

There is no doubt that you cannot specify exactly what a new product is going to be before you have developed it. But that does not mean that you cannot specify anything about that new product. Whenever an opportunity is spotted for developing a new product, certain targets are inevitably set. A new product which is intended to be cheaper than competing products has cost and price targets. A new product which is intended to work better than competing products has functional targets. Even a product intended to catch up with a more innovative competing product must work at least as well as that competing product at no extra cost. So targets for new products can be set at the start of development, albeit less precisely, less quantitatively and less comprehensively than the targets for more routine quality control tasks.

This gets us back to the risk management funnel. Quality control targets can be made to evolve both in terms of how specific they are and in how directly they relate to the characteristics of the finished product, as product development progresses through the risk management funnel (Figure 2.6). In this way, the quality control targets can become more specific and progressively more like manufacturing quality control targets as development proceeds. An opportunity specification is the earliest quality control document and it contains basic commercial targets for the proposed new product. What core benefit will customers perceive to persuade them to buy the new product rather than a competing product. What are the first estimates of manufacturing costs and sales margin over costs for the new product? What are the expected development costs for the product? What is the projected sales volume and hence what is the predicted return on investment over the sales life of the product?

Next comes the design specification - the most important quality control document in product development. This sets technical targets for the new product, covering everything from its appearance and its main functions to how it is to be packaged and shipped to sales outlets. The design specification needs to be a consensus document, reflecting the interests of marketing and sales, design and development and production engineering staff within the company. It also needs to be a document reflecting the criteria for commercial success for the product. Any product which ends up meeting the design specification should be expected to succeed commercially in the market. Any product which fails to meet the design specification should be expected to fail commercially and hence should have been killed off during its development.

This design specification, therefore, becomes the quality control standard against which the acceptability of the product is judged during its development. Thus, concepts, embodiments and working prototypes can all be evaluated against the design specification in order to select the best. Following both embodiment design and detail design it is necessary to refine the design specification to make sure it keeps up with the increasing degree of detail with which the product is now designed. This is usually done by means of a Failure Modes and Effects Analysis, a technique which sets out to anticipate and control for ways in which the product might fail in the future. Eventually, as product development pushes closer to manu-facture, so the design specification changes into a product specification to quality control the manufacturing process.

Figure 2.7
Effective product development kills off non-viable ideas as early as possible

Killing-off product development

The prompt recognition and killing-off of non-viable products is a vital part of successful new product development. The data below [8] shows the new product survival curve, from first idea to profitable sales. The success rate of only 5 per cent shows clearly that making a profit out of new product cannot be left to chance. It is, therefore, vitally important to screen new ideas thoroughly and kill off those which are not going to be profitable. This must be done at the earliest possible stage in product development. The further through the development process a product is allowed to go, the greater the loss if it fails. Paradoxically, a key indicator of the effectiveness of a new product development procedure is how many new product ideas it kills off and how quickly it does so.

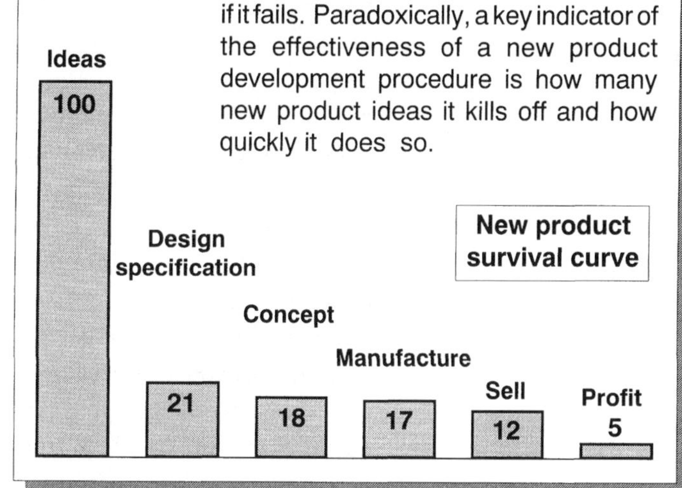

Quality targets

Quality targets begin as simple statements of the business objectives, become refined into technical targets in the design specification and ultimately turn into a full product specification for manufacturing at the end of the product development process. A quality target is something aimed for in the appearance or function of the new product. It can be specified in one of two ways. Firstly, there are 'demands'; these are the product features which must be included if that product is going to be commercially successful. These could include product requirements laid down in law or in industry standards. They also include everything that a customer demands of the product before they would consider buying it. In buying a pair of scissors, for example, the sharpness of the blades and comfort of the hand grips are likely to be demands for most customers. The colour, style and the duration of the 'stay sharp' guarantee are less important and are not absolute pre-requisites for buying the scissors. They are not, therefore, demands within the design specification but, as we will see below, wishes.

Demands within a design specification can be seen as quality control thresholds . If at any time during the development of the product, it is found that one or more of the specified demands cannot be met then that product should be killed off. The rationale behind this is that the demands within a design specification represent the minimum criteria for commercial success once the product is launched on to the market. If the product fails to meet these demands it falls below these minimum criteria for commercial success. And if it is going to fail commercially anyway, it is best to acknowledge this as soon as possible and kill the product off during development to minimise the waste of development resources (Figure 2.7).

The second way that quality targets can be specified is in the form of 'wishes'; these are the product features which are desired in order to maximise the product's advantages over other competing products on the market. The wish list for a product may have marketing, design and engineering inputs. The marketing 'wishes' may be for secondary features, of benefit to the customer, the design 'wishes' may be for improved ergonomics or the use of better materials and the engineering 'wishes' may be to reduce component numbers and hence reduce assembly times. The wishes within a design specification can be seen as quality control aims, and the number of these wishes satisfied during product development can be a measure of the value added to the product, over and above its essential 'demanded' features.

Meeting targets in product development

Setting targets for product development is only useful if there are also procedures in place for reaching these targets. Reaching a target can be seen as a two stage process; firstly think of all the possible ways in which the target might be reached and then, secondly, select the best of these possibilities. In practice, this means that creative idea generation is followed by systematic idea selection, using the design specification as the basis for that selection. This, of course, happens repeatedly, at each stage of the risk management funnel. So, selecting the best concept design for the product involves, firstly, thinking of all the possible principles of operation for the product and, secondly, choosing the best of these on the basis of the design specification. Later in the design process, selecting the best embodiment design for the product involves, firstly, thinking of all the ways in which the product could be made and, secondly, choosing the best of these on the basis of the design specification. So, the same cycle repeats itself throughout product development, operating within the progressively narrower boundaries established during the previous stage of the procedure.

The key role is played by the design specification. Its contents steer the development of the new product by determining which of all the alternative design options is selected for further development in subsequent stages. In addition, the design specification plays the equally important role of quality controlling the developing product. This kills off product development as soon as the product can no longer meet all of the 'demands' within the design specification.

This chapter began by showing how the development of new products is risky. It ends on a similar note. New product development will always be risky. But there are ways of stacking the odds in favour of success. Being systematic and thorough in following the principles covered so far will give you a huge advantage over any competitors who adopt a more haphazard approach to product development. Ignore them at your peril! From Chapter 6 onwards the entire product development process is described, showing how these principles are put into practice.

Key Concepts

1. New products are developed for customers to buy

Product development must be fundamentally and comprehensively customer-oriented. The successful product designer must learn to 'live' in the mind of the customer: every decision and new direction should take the product closer to realising the dreams, needs, wishes, whims and hopes of that customer. Introducing a new product is always difficult. Customers' existing buying habits have to be changed and often they need a good reason to do so. Offering a new product with clear differentiation from existing products and with obvious added customer value provides that good reason. As a result, such products are, on average, more than five times as successful (see Figure 2.1) as poorly differentiated products with only minor added value. So, market-orientation is vital in new product development.

2. Developing new products is a fundamentally difficult problem solving activity

Developing new products is a multi-factorial problem; success or failure is determined by many factors (e.g. attraction to customers, acceptability to retailers, engineering feasibility, product durability and reliability). The problem is also fuzzy; the definition of the problem is incomplete at the start of the development process. Because of this, the way product development is tackled has a significant effect on the success of its outcome.

- Tackle the problem in stages. Product development is a process of translating an idea for a product into a set of instructions for its manufacture. This can only be done effectively in stages, with each stage focusing on progressively more detailed aspects of the product's design. Make sure that the product works in principle before designing it in detail.
- Design to specification. Write down what the product must do (demands) and what it should do (wishes) in order to be commercially successful. Make sure that this specification is a consensus document throughout the company (agreed by marketing, design and production engineering). At each stage in the design process think of all possible solutions and select the best one according to how well it fits the specification. Kill off products as soon as they cannot be made to meet the demands in the specification.

Key Concepts

3. Invest in the front-end of the design process

Of all the stages in new product development, the most important are the initial stages. By the time concept design is completed, the business opportunity will have been established, the technical targets will have been specified and the product's principles of operation will have been finalised. A great many decisions about the product have already been made and a considerable amount of the eventual cost of the new product has been committed. Yet the cost of actually doing the development work so far has been small - the research is mostly desk research and the design work is relatively inexpensive drawing and modelling. So, as the bar chart (Figure 2.8) shows, the cost benefit ratio for the early stages of development is much better than for later stages. A key to success in product development is, therefore, to invest time and effort during these early stages to save having to make costly changes later. Figure 2.9 takes this point further and shows how the costs incurred during the development process are minimal to begin with, start to become significant during embodiment and detail design and rise steeply during production engineering. The costs committed, however, follow a very different pattern. Committing to costs means making a decision which, at some stage of development, will have to be met. The most important of these decisions are made early in the development process. Deciding, during product planning, for example, to make a luxury car rather than an economy car makes a huge commitment to cost. Deciding during concept design to develop a new electric power system for the car rather than use a standard internal combustion engine makes another substantial cost commitment. By the time you reach detail design, most of the cost is committed, even although it is not yet spent. As a result, of this pattern of cost commitment, the potential cost reductions are greatest early in the design process. Another result is that the cost of making alterations increases sharply as development continues. If you change the design at the concept stage you only need to revise your sketches and models. Change during production engineering can involve

Figure 2.8 The investment: return ratios for different stages of product development

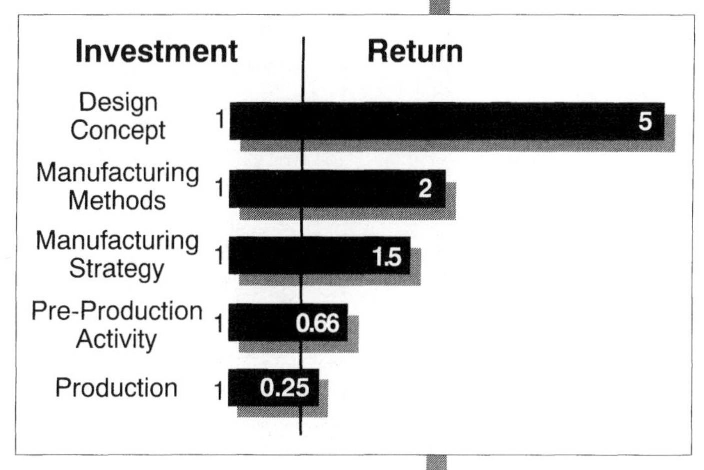

Key Concepts

re-commissioning tooling, at enormous cost.

Proof of the importance of the early stages in the development process is that products which were thoroughly assessed and stringently specified in early development were three times more successful than those less effectively assessed and specified (see Figure 2.1). So, it is vital to get the product right in the early stages of new product development.

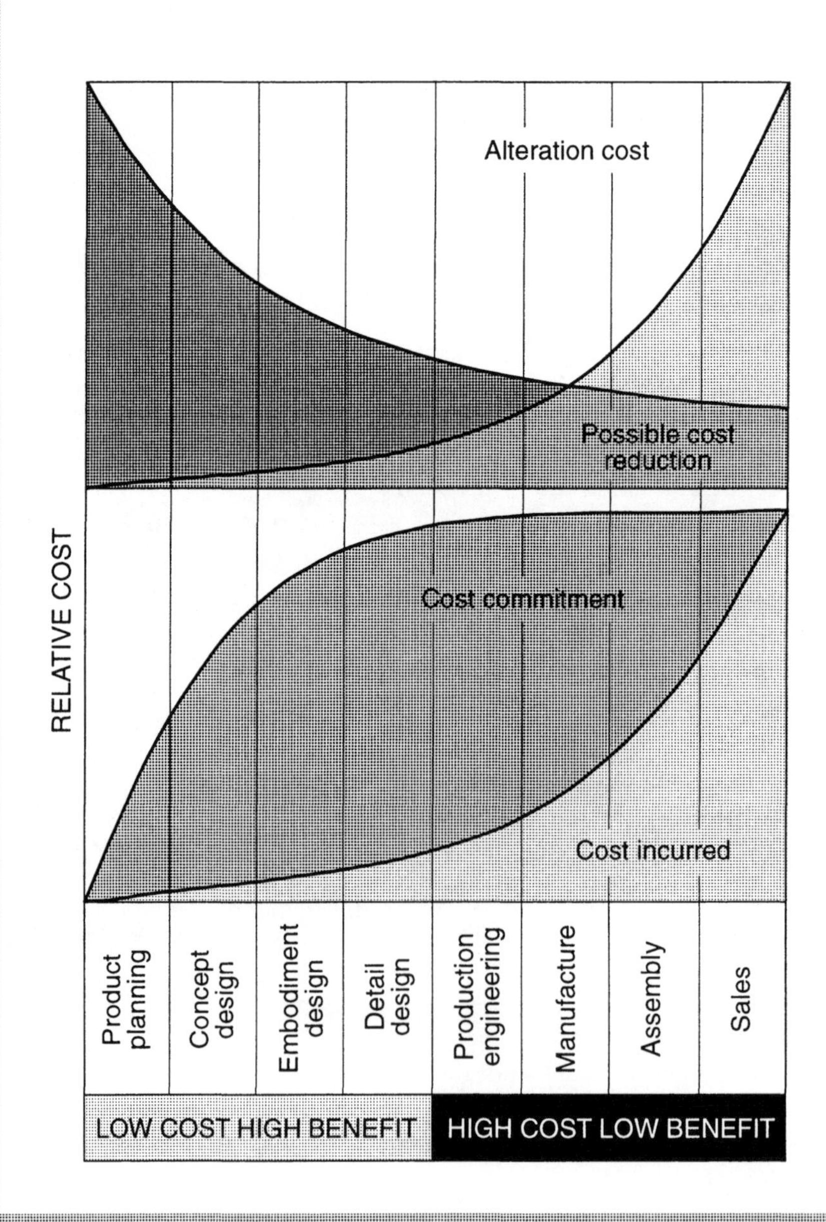

Figure 2.9
Costs and benefits of different design stages [9]

Notes on Chapter 2

1. Studies on the success and failure of new products:

UK	-	Freeman C. 1988, *The Economics of Industrial Innovation* (2nd Edition). Francis Pinter (Publishers), London.
USA	-	Booz-Allen and Hamilton Inc. 1982, *New Product Management for the 1980's*. Booz-Allen & Hamilton Inc, New York.
Canada	-	Cooper R.G. 1993, *Winning at New Products*, Addison Wesley Publishing Co. Boston.

2. This categorisation is the author's own interpretation of how the success and failure data (cited in 1 above) fall into broad categories. The data is mostly derived from Cooper 1993 (see 1 above).
3. The risk management funnel is a concept developed by the author from ideas of the 'stage gate' management of product development in Cooper 1993 (see 1 above) and Wheelwright S.C. and Clark K.B. 1992, *Revolutionising Product development: Quantum Leaps in Speed, Efficiency and Quality.* Free Press, New York.
4. Booz Allen and Hamilton (cited in 1 above) concluded from their research that product development performance could improve by as much as 27% with each new product developed by a design team.
5. The Attwater quote was cited in Greunwald G. 1988, *New Product Development.* NTC Business Books, Lincolnwood, Illinois.
6. A good case study describing how the motor industry changed from 'over-the-wall' to a more functionally integrated product development procedure is given in Ingrassia P. and White J.B. 1994, *Comeback:The Fall and Rise of the American Automobile Industry.* Simon & Schuster, New York (e.g. p434 et seq.).
7. The basic principles of quality control are described and formalised in ISO 9000, available as *BS 5750, 1987* from the British Standards Institution, London. For a general introduction to quality management see Fox M.J. 1993 *Quality Assurance Management,* Chapman & Hall, London.
8. The survival curve for new products has been published in both Hollins B and Pugh S. 1990, *Successful Product Design*. Butterworth & Co. London and Cooper 1993 (cited in 1 above).
9. This data is compiled from Booz-Allen and Hamilton 1982 (see 1 above) and incorporating concepts from Cooper 1993 (see 1 above) and Wheelwright S.C. and Clark K.B. 1992 *Revolutionising Product Development: Quantum Leaps in Speed Efficiency and Quality.* Free Press, New York.

3 The principles of product styling

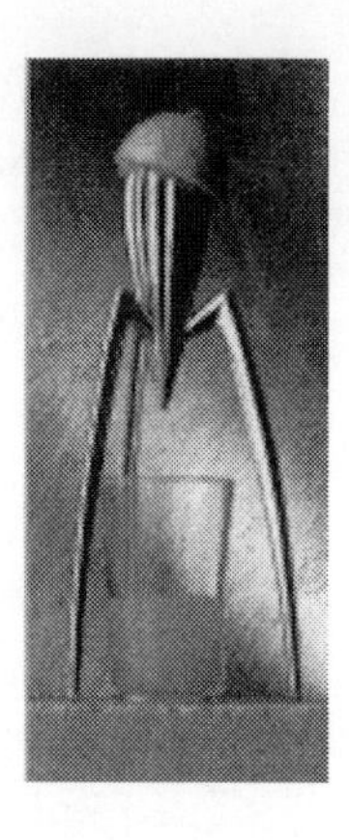

Designers cannot help but give their products visual form. That visual form may be non-descript, inelegant or just plain ugly. Or it can be transformed, by styling, into a thing of beauty, admired for how it looks rather than what it does. It is now accepted by everyone from consumers to national governments that styling is an important way of adding value to a product without changing its technical performance. This styling need not be flashy, elaborate or expensive. It certainly need not come affixed to a 'designer label'. Product styling is about creating visual attractiveness in every-day products.

Most design teaching emphasises the importance of skills development for product styling. This is usually achieved by means of studio exercises which allow designers to develop their styling skills through practice and tuition. This is vital and the ability to project onto a sketch, rendering or model whatever visual forms you can think of in your head is a basic skill all designers should have [1]. But an equally important part of product styling is having the inspiration to create the visual forms in your head in the first place. This requires an understanding of both the principles of product styling and the process by which good styling is most effectively achieved. It is here that I believe that a great deal of design teaching is lacking. Few courses offer any formal teaching on the principles of product styling. No structured process for getting from the start to the finish of a styling project is offered and rarely are relevant design tools and methods provided. Books on product design are little better. Processes, tools and methods on the technological aspects of design abound. But rarely do authors venture into the hallowed ground of product styling. My approach in this book is to try to treat styling as thoroughly as functional aspects of design. This does not mean that styling is reduced to a prescriptive and creatively constrained activity. Good styling will always be an art but it does not have to be a black art!

This chapter describes the underlying principles of product styling. In subsequent chapters the product styling process is explored further and translated into workable procedures. This is done in a way that provides a framework for both thinking about styling and organising the relevant design activities.

Visual perception of products [2]

When we talk of a product being attractive we rarely refer to its sound, feel or smell. This is a striking reminder that human perception is dominated by vision and that product style is usually an abbreviation for visual style. The attractiveness of products is, therefore, deeply rooted in visual perception and it is here we must start our exploration of the subject.

We see things when light enters our eyes and causes nerve impulses to be sent to our brain. The patterned images which we call vision are the outputs from arrays of sensor cells in the retina and subsequent arrays of nerve cells en route to the brain. Together, these cells divide visual images into their component parts, distinguishing for example, lines, colours and movement. Reduced to these component parts, visual signals are transmitted to the brain where they are examined for content and meaning, used to guide movement and then stored away in memory for future use.

Clearly, the brain does some clever processing of the fragmented visual signals it receives. The most obvious evidence of this is that the visual images we 'see' are perfectly coherent and whole. We are never actually aware of the jumble of lines, dots, colours and movements contained in the nerve impulses. As we will see, the way the brain manages this conjuring trick is a central issue for product styling.

Let us start by considering some of the individual pieces of the jigsaw of visual perception [3].

1. **Two stage visual processing.** Our analysis of visual information takes place in two discrete ways. Firstly the overall image is scanned to look for patterns and shapes. This is a very rapid process, requiring no deliberate

effort on the part of the viewer and is therefore described as pre-attentive. The second, attentive mode of visual processing involves deliberate focusing on details of the image to examine its component parts. To illustrate this look at Figure 3.1. At a glance, you should see that there is a patch in the upper right hand area of the figure which is different in some way. This is the result of pre-attentive processing. You will not, initially, know what is different nor perhaps where the exact boundary of the patch lies. Moving into attentive and detailed analysis of the image reveals that a square of 9 x 6 letters is of a different size and weight compared to the rest.

AAAAAAAAAAAAAAAAAAA
AAAAAAAAAAAAAAAAAAA
AAAAAAAAAAAAAAAAAAA
AAAAAAAAAAAAAAAAAAA
AAAAAAAAAAAAAAAAAAA
AAAAAAAAAAAAAAAAAAA
AAAAAAAAAAAAAAAAAAA
AAAAAAAAAAAAAAAAAAA
AAAAAAAAAAAAAAAAAAA
AAAAAAAAAAAAAAAAAAA
AAAAAAAAAAAAAAAAAAA
AAAAAAAAAAAAAAAAAAA
AAAAAAAAAAAAAAAAAAA
AAAAAAAAAAAAAAAAAAA

Figure 3.1
We become aware of the square of letters because of primary global precedence

2. **Primary global precedence.** The pre-attentive stage of visual processing has a property known as 'primary global precedence'. This term refers to (i) the fact that pre-attentive processing is global, in other words it looks at the entire image rather than its component parts, (ii) it occurs first and (iii) it has precedence over the subsequent attentive stage of visual processing. Saying it has precedence means that the impressions gained during pre-attentive processing will dominate and partly determine the subsequent attentive processing. In Figure 3.1 above, pre-attentive processing identified an area of interest in the upper right of the pattern and focused the attentive processing on that square. Did you happen to also notice that the letter in the lower left corner was also of the different size and weight? If not, it was because the precedence of primary global processing directed your attention to the square, to the exclusion of other areas. For another example consider the image shown in Figure 3.2 This is an example of an ambiguous image, showing either the head and shoulders of a young woman with her face turned away from view or a close-up profile of an old woman's face. It is impossible to see both simultaneously and this, again, is because of primary global precedence. Once your mind fixates on one of the two alternatives, global precedence is set. You can then go on to examine individual details: in the young woman image – the strong line of her chin, the elegant necklace, the flowing head-dress and luxurious coat; in the old woman image – the hook nose, the protruding chin, the thin lips and sad eyes. Generally, the image you see first will be held in your mind whilst you examine the details. To see the alternative image you will have to look away, blink or hold the image away from you. Again, once the new image is fixed in your mind then attentive processing can go on to study the details.

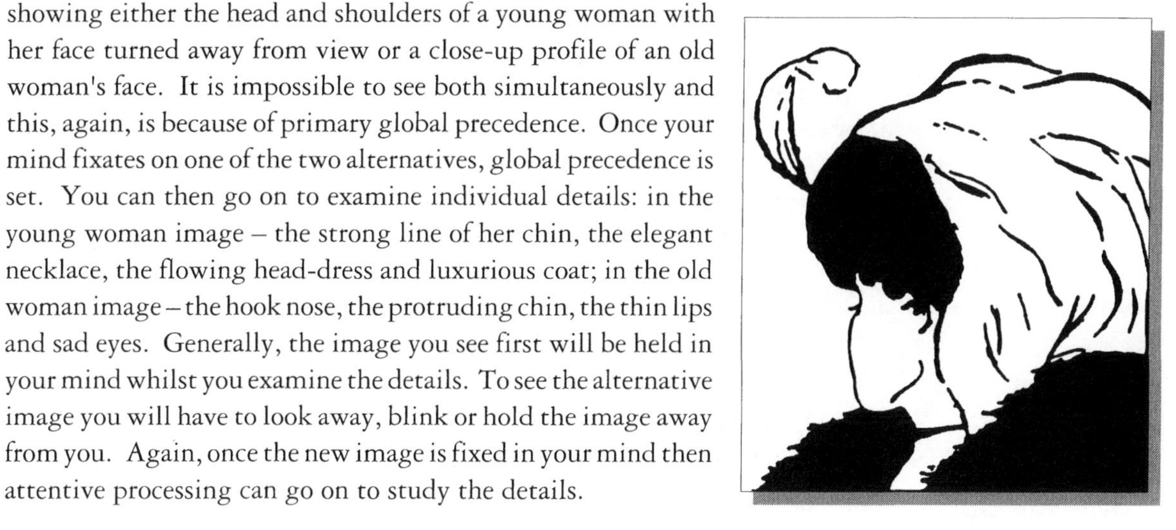

Figure 3.2
Young woman or old woman?

3. **The primal sketch** In cases of ambiguous or incomplete information we construct visual hypotheses in our minds and project them mentally on to the visual images in front of us. Figure 3.3 'seethes with activity' as your mind hunts through possible visual hypotheses (mostly circles) on what is the nature of the underlying pattern. Partially complete images of complex shapes are dealt with in a similar way; the mind rapidly forms what is called a 'primal sketch'. This is an extraction of key pieces of the visual image and their assembly into a recognisable pattern. Figure 3.4 consists of a series of blobs but for most viewers a dalmation dog quickly emerges from the blobs (the dog has its head down sniffing the ground and is facing the upper left corner of the figure). Once identified, it becomes a strong feature of the visual image. This is after your mind has formed the primal sketch of the dog.

Taken together, these findings provide a remarkable and completely

Figure 3.3
Seething circles [4]

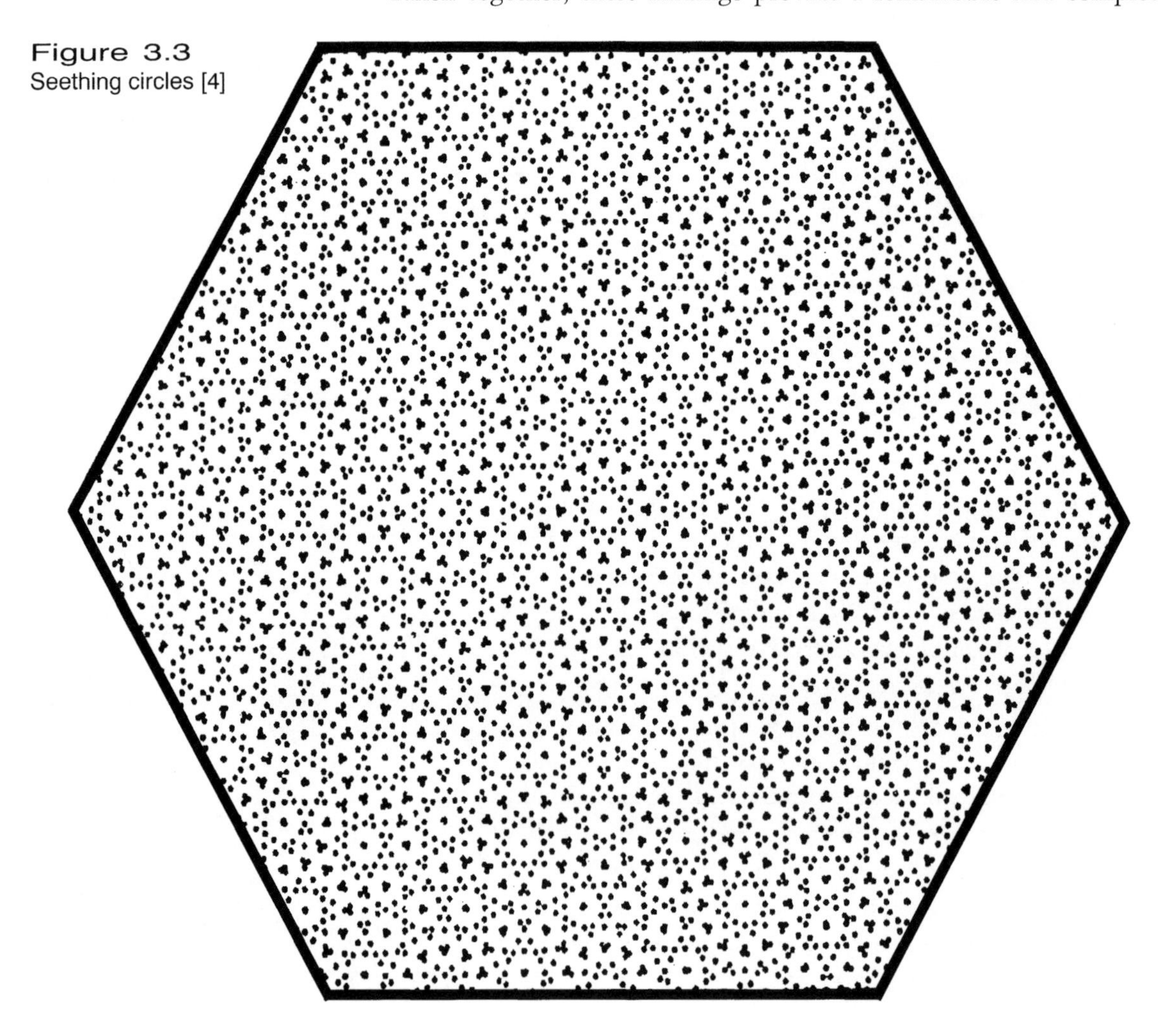

counter-intuitive understanding of how we see. Intuitively, we believe that our eyes are a window on the world. But this is not so. We actually see what we think we see. We look at an image and without thinking, we extract its main features. From these features, our brain works out what it thinks it contains by forming a 'sketch'. Then attentive processing, guided by this pre-conception, studies the component parts of the image and extracts the detail. Look at Figure 3.5 and you see a white triangle. No triangle actually exists; the image consists of three circles with segments missing and lines arranged to make three points. Our brain conjures up the image of a triangle from what is missing in the other parts of the image. And, as if to emphasise its existence, the triangle seems to stand out in brighter white than the rest of the page. Visual illusions, such as this, expose the lies that our eyes tell us about the world.

Figure 3.4
Spotty picture of a spotty dog [5]

Perception of product style

So, what does this tell us about product styling? The way we talk about styling suggests that much of our judgement of style is determined by pre-attentive global processing. We say that products have immediate appeal, they are striking, eye-catching or attention grabbing. These judgements are instant and pre-attentive. They require no careful deliberation and do not seem to be based on the attentive processing of the product's component parts. When we talk about the overall form or image of a product, we are referring to our global perception of it. Style is, therefore, at least partly judged on the basis of pre-attentive global processing. The beauty of a product is, consequently, more to do with the properties of our visual system than anything fundamentally beautiful about the product. If an extra-terrestrial alien was to look at the products we consider sublimely beautiful it might find them completely unattractive because

Figure 3.5
Non-existent triangle [6]

its visual system sees them differently. Beauty truly is in the eye (and brain) of the beholder. When we design an object to be beautiful, we must design it to correspond with the perceptual properties of human vision. Understanding vision, therefore, becomes the key to creating beauty and the rules of pre-attentive visual perception should translate into principles of product styling.

The rules of visual perception

The way our visual system works is a relic of our evolutionary past. As mentioned earlier, humans have evolved to be predominantly visual animals. In other words we use vision more than we use hearing or smell for the important tasks in our daily lives. Vision may even have had a profound influence on human evolution. It has been suggested that the reason humans first stood upright was to be able to see further than when on all fours. With vision playing such an important role in the lives of early man, it is obviously important that our visual system evolved to work well at the tasks which were important to survival. Amongst these was the ability to spot things that are dangerous (snakes, predators, other marauding humans) and to distinguish between things that were edible and those that were not. Being highly gregarious, early humans had many social skills to acquire. Being able to recognise other individuals and, more subtly, being able to read the mood of other people in their facial expressions all made for a more harmonious life. Our evolutionary heritage from this, is our present-day human vision. As we shall see, the visual rules by which we judge the beauty or otherwise of contemporary products were moulded from the survival pressures on primaeval man. We can examine these resulting rules of visual perception at two levels. Firstly, there are general rules of visual perception that allow us to extract visual information from any scene. Secondly, there are specific rules which make us particularly good at certain visual tasks, for specific survival reasons.

General rules

The 'gestalt' rules of visual perception are named after a group of German psychologists working in the 1920's, 30's and 40's [7]. Gestalt is the German word for pattern and the Gestalt psychologists suggested that human vision is somehow predisposed to see certain types of pattern. How this happened in the eye and brain was not the strong point of the gestalt psychologists.

Indeed, their fanciful proposals caused their theory to fall into disrepute. Then, more modern research discovered some of the real mechanisms and the former sceptics had to face up to the embarrassing fact that gestalt theory had been correct all the time, even if the reasons were wrong.

When we first see an image our brain is 'programmed' to extract certain types of visual patterns and these are then constructed into a meaningful image. This programme is not entirely hard-wired into our brain at birth but develops according to the visual stimuli we are exposed to during development. If, for example we were raised in an artificial environment consisting entirely of vertical stripes, our mature visual system would be blind to horizontal stripes. The gestalt rules of visual perception [8] are the operational rules for this programme in our brain.

Probably the strongest gestalt rule is the rule of symmetry [9]. We have a remarkable ability to detect symmetry, in complex forms, natural forms with incomplete symmetry and even in objects which have had their symmetry substantially distorted (Figure 3.6). Viewing objects at an angle, for example, distorts their appearance but we still find it relatively easy to see whether or not they are symmetrical. Related to the symmetry rule is the geometric rule, whereby we detect simple geometric forms easier than irregular forms or highly complex geometric forms. This may simply be due to our innate ability to detect symmetry (all simple geometric forms have strong symmetrical features), enhanced by our up-bringing in a world dominated by simple geometric forms.

We also have a remarkable ability to detect regular patterns and the gestalt psychologists break this down to three rules: the rule of proximity, the rule of similarity and the rule of good continuation [8]. The rule of proximity proposes that objects or features which are in close proximity will tend to be seen as a pattern. This is illustrated in Figure 3.7a. The dots on the left are seen as a horizontal pattern because they are closer horizontally than they are vertically. The dots on the right make up a vertical pattern because of their

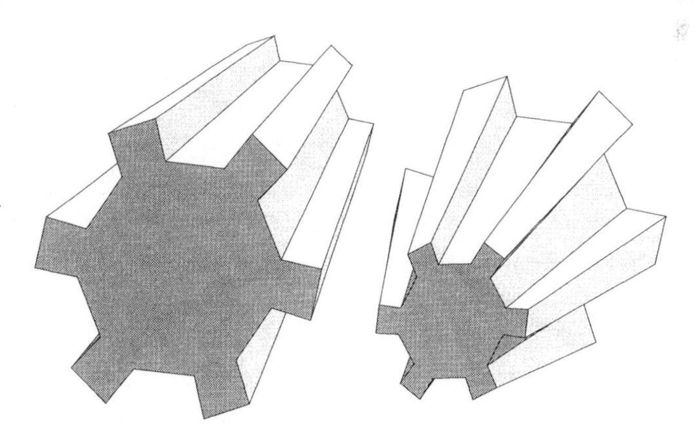

Figure 3.6
We have a remarkable ability to detect symmetry [9]

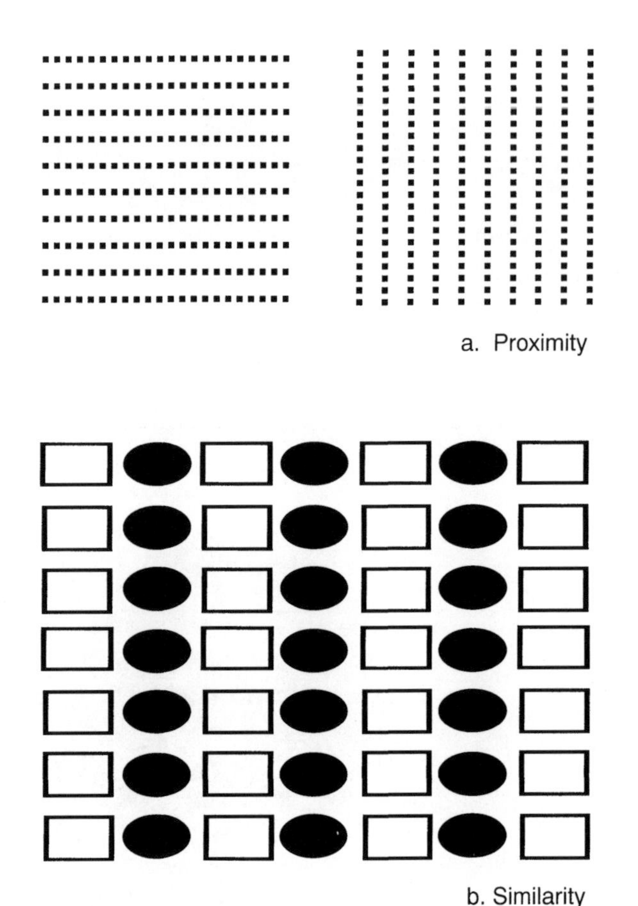

a. Proximity

b. Similarity

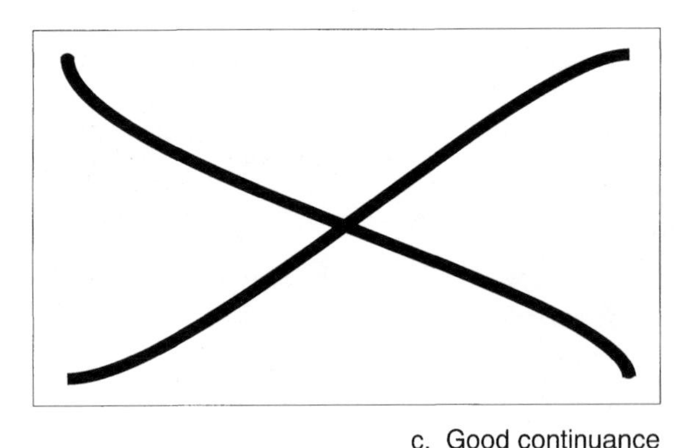

c. Good continuance

Figure 3.7
The gestalt pattern rules

vertical proximity. The rule of similarity proposes that objects or features which are of similar shape or form will tend to be seen as a pattern. Figure 3.7b illustrates this. Despite being closer together horizontally than they are vertically, the elements of this figure are seen to make up a series of columns because of their similar shape. The rule of good continuance proposes that patterns are perceived due to the continuity, trajectory or vector of their component parts. Figure 3.7c could be seen as two 'vee' shapes meeting at a point. We, however, tend to see two lines crossing and this is because each line has a continuity or trajectory which we interpret as continuing across the point of intersection between the lines.

Another general perceptual ability which is comprised of several gestalt rules is our interpretation of figure and ground in an image [8]. This is where we distinguish part of an image to be the figure or object and the remainder of the image to be the background to that image. This is best illustrated by another ambiguous visual illusion (Figure 3.8). This image is seen either as a white vase on a black background or two black faces on a white background. Just like the young woman/old woman image (Figure 3.2) it is impossible to see both simultaneously. Our visual system jumps between these two alternatives as it moves from one hypothesis to the next as to which is the figure and which is the ground. Figure/ground judgements are made on the basis of four gestalt rules: the rules of symmetry, relative size, surrounded-ness and orientation. The more one part of an image is symmetrical, relatively small, surrounded and oriented in the vertical or horizontal axes, the more likely it is to be seen as the figure. The three parts of Figure 3.8 illustrate these rules. The upper image shows the classic faces/vase illusion.

This has all the ingredients for figure/ground ambiguity. Both the vase and the faces are symmetrical, both are of approximately equal size, neither surrounds the other and both are vertically oriented. In other words there are no clues as to which is the figure and which the ground. The middle image shows how the vase can become the slightly preferred figure by orienting the faces at 45 degrees to the vertical. The lower image, however makes the vase far more prominent than the faces because it is smaller and surrounded.

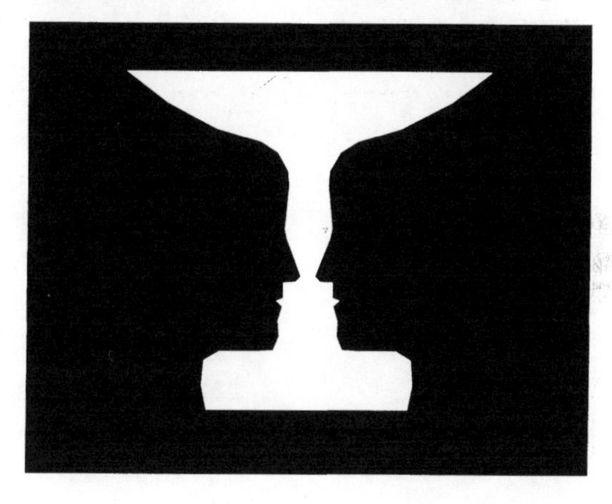

Figure 3.8
The face/vase illusion illustrating the gestalt rules of figure/ground

What do the rules mean for product styling?

The implications of the gestalt rules for product styling are profound and range from the specific to the sweeping. Let us begin with the specific. The effective integration of product components or product features can be derived from the gestalt rules of patterning. Product features which are related functionally, can be made to appear grouped together using these rules. The upper part of Figure 3.9 shows a mobile phone currently on the market. Its function keys are not in any way visually unpleasant but they do not appear to have been designed to take any account of the gestalt rules of patterning. The lower part of Figure 3.9 shows how these function keys could be redesigned to incorporate the gestalt pattern rules. Firstly, the rule of proximity could be used to make functionally related features appear visually associated, simply by placing them close together. So, for example, the power button switches on the screen. By placing them close together, this functional association would make the location of the power button more intuitively obvious and hence easier to find. The gestalt rule of continuity could be used to embody key-press sequences in the styling of other buttons. Store (STO), recall (RCL) and function (FCN), for example, are used immediately before the digit buttons. They are therefore placed above the digits with arrows pointing down. The send (SND) and

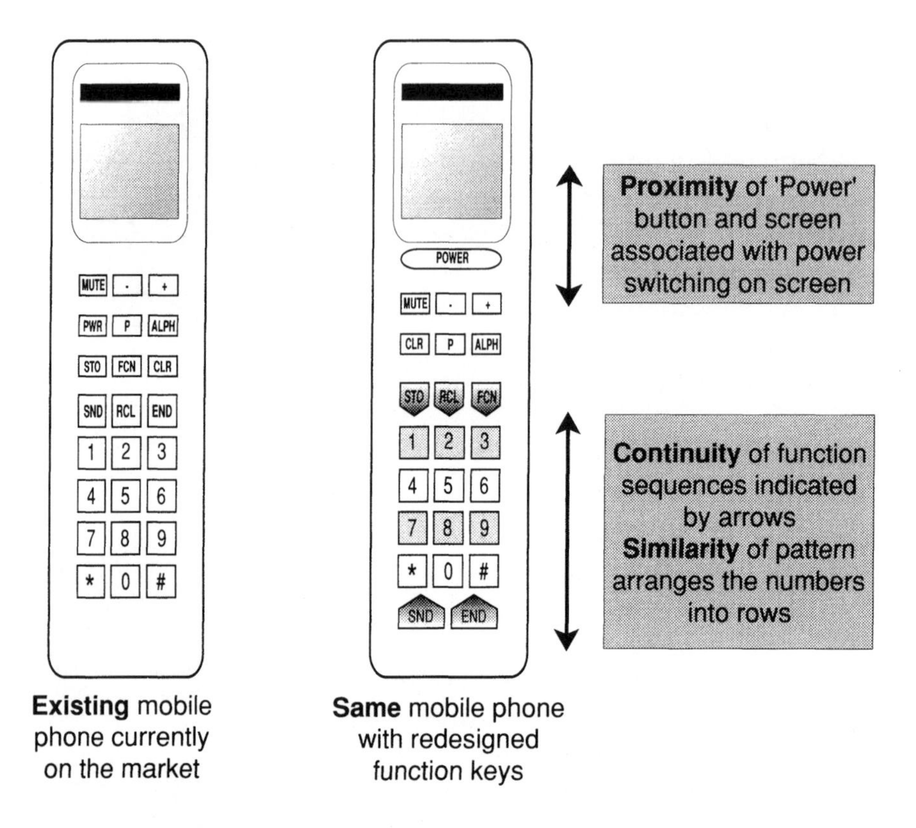

Figure 3.9
Redesigning phone function keys according to gestalt pattern rules

end (END) buttons, on the other hand are used after the digit buttons and hence are placed underneath with upward arrows. Finally, the gestalt rule of similarity could be used to indicate visually that the number buttons are arranged in horizontal sequence by having the buttons in each row of a similar pattern and in contrast with buttons in adjacent rows

The notion of harmony in visual forms has also been linked to gestalt rules [10]. Strictly speaking, harmony was never formulated as a rule by the gestalt psychologists. But the rules relating to visual simplicity combined with the rules regarding visual patterns suggest that visual harmony must be important. Let us imagine that our visual system has detected a particular type of geometric form in a product. If that geometric form is repeated in the product we will link them because of the rule of similarity. Remember how the dots and squares in Figure 3.7 formed a pattern of vertical columns. Intuitively we know that repetition of the same shape give rise to a sense of greater visual coherence or harmony than the repetition of different shapes. This is almost certainly a consequence of the pattern detecting abilities of our visual system. From this we can, therefore, talk of a 'derived' gestalt rule of visual harmony.

Figure 3.10
Harmony in visual styling can be achieved by repeating geometric forms

Breaking this gestalt rules can easily give rise to a sense of visual incoherence in products. In Figure 3.10, the left hand cup repeats a single geometric form and thereby gives a sense of visual harmony. In contrast, the right hand cup mixes a wide variety of geometric forms and consequently lacks visual harmony.

The issue of visual simplicity

The most sweeping implication of gestalt theory for product styling concerns the visual simplicity of products. To comply with the most powerful gestalt rules, products should be symmetrical and comprise clean lines which go together to make up simple, geometric forms. To take these rules, on their own, as the rules of product styling would amount to a manifesto for minimalism. In contemporary product styling minimalism is a dominant theme and elegant simplicity is the aspiration for most modern designers (see margin quotes on next page). But is this just a current fashion or has simplicity always been sought after? There is certainly no doubt that earlier generations of product designers produced a much higher degree of visual complexity. Figure 3.11 shows how typewriter design has changed over the past century. From a highly complex visual form in 1910, the typewriter's visual complexity has progressively reduced to the smoother, more enclosed 1947 version and then to the clean-lined, simple geometric form of 1970. Similar trends towards visual simplicity can be found in many different products, from cars to crockery. This, then begs two questions: firstly, why were products more visually complex in earlier eras and secondly, where is this leading us in the future of product styling.

A century ago, people had the same eyes and these eyes must have

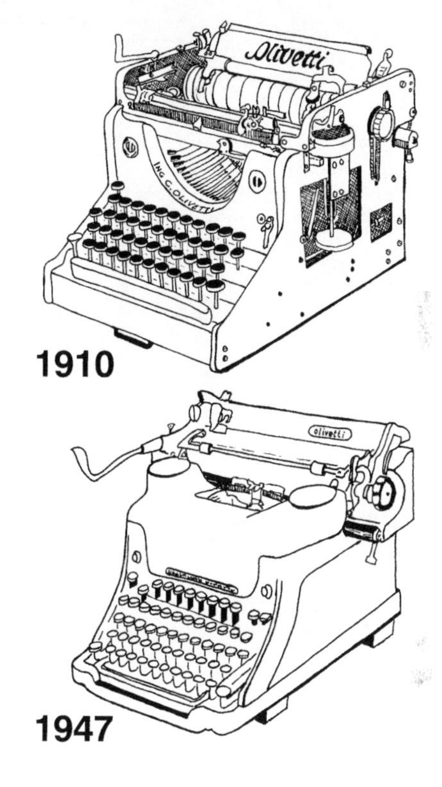

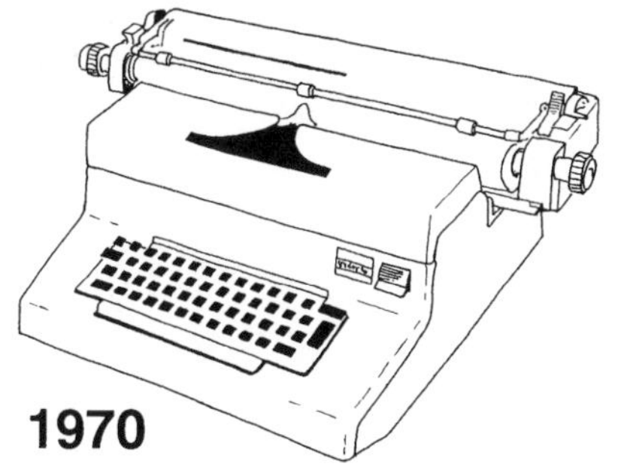

Figure 3.11
The trend towards visual simplicity in typewriter design [11]

> 'Complicated, unnecessary forms are nothing more than designer's escapades ... (in)... self expression'
> Dieter Rams 1984 [12]

> 'I have a taste for austerity and utility, but that's certainly not to say I have no appetite for pleasure. Quite the contrary. I firmly believe that plain simple things are superior to flashy complicated ones, precisely because they are more pleasurable.'
> Terance Conran 1985 [13]

perceived the world according to the same gestalt rules. It is faintly possible that they would actually have preferred simple visual forms but that technological constraints on visual form forced them to accept visual complexity. Because of these constraints, form was obliged to follow function to a much greater extent than it does currently. With more advanced technology, functional components have been miniaturised and are often packaged in compact prefabricated components. Visual form is, therefore, less constrained [14]. An alternative argument, which seems a lot more compelling, is that customers had different cultural values and hence different preferences for product styling. At a time when technology was novel and exciting, many early twentieth century designs made a visual feature of the technological complexity of the product. We will explore cultural influences on product style at greater length below.

As for where this is leading us in terms of future product styling, this is a much more profound and revealing question. Ultimately, according to the most radical interpretation of gestalt theory, we might end up living in a world of circular or spherical products – the simplest and purest visual form which all products might eventually assume. Yet we know intuitively that this would never happen. Rebellion against visual monotony would be a much stronger influence than the guiding principles of gestalt psychology. This intuition turns out to have a strong foundation in scientific research, championed by Daniel Berlyne, a Canadian psychologist. Reduced to its most basic form, Berlyne's research [15] revealed a preference curve for visual complexity (Figure 3.12). Products which are either too simple or too complex give rise to less feelings of attractiveness than those which are intermediate in visual complexity. Berlyne's research actually delved

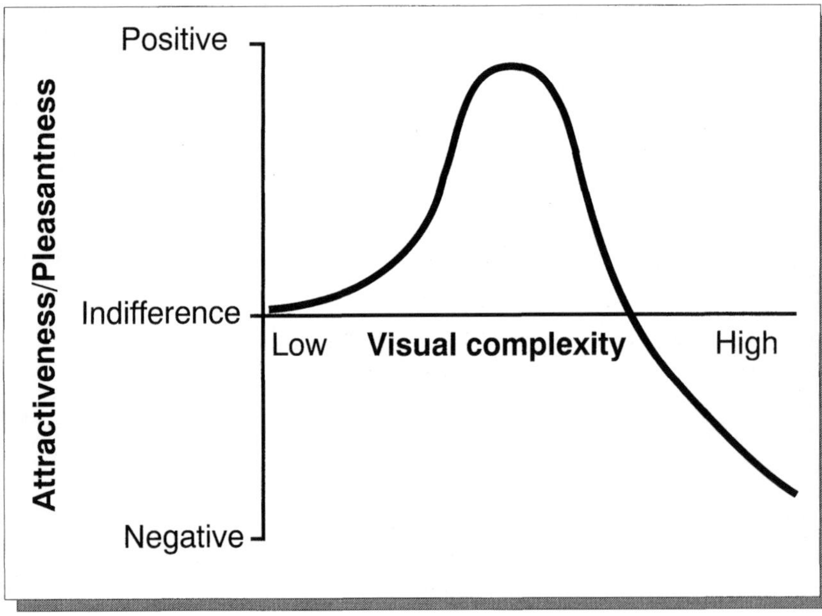

Figure 3.12
Berlyne's model of how people rate the pleasantness of objects depending upon their arousal potential [15].

deeper into the subject than this over-simplistic summary suggests. We can draw 4 main conclusions from Berlyne's research and the subsequent research which it spawned.

1. The main determinant of visual attractiveness is not the intrinsic complexity of an object but its perceived complexity to the observer. Thus, a product which is actually very complex may be perceived to be more simple and hence more attractive to some people because they are more familiar with it. To a chemist, the chemical structure of a benzene ring provides a shining example of elegant simplicity, yet to me, it looks quite complex and hence not especially attractive.

2. This interaction between complexity and familiarity can produce time-dependent changes in the attractiveness of objects. At first, a visually complex object may be unattractive but with repeated exposure, it becomes increasingly attractive as it becomes more familiar. This has been found to hold true for sounds as well as visual images. A complex piece of music may not be immediately attractive but its attractiveness grows with repeated listenings. Conversely, a very simple melody often has immediate attractiveness but this attraction wanes after a while because it becomes unattractively over-simple. Such changes in the familiarity of objects or sounds can also be used to explain why some things seem to came back into fashion after a period of being considered unattractive. After 10 years of mini-skirts in the 1960's, the familiarity factor appears to have stimulated fashion changes towards different skirt styles. By the mid 80's, however, mini-skirts were new and exciting again and came back into fashion.

3. Before an object is judged attractive, it is often said to be interesting. If it is interesting enough, it will hold the observer's attention for long enough for the object to become familiar and hence attractive. The recipe for interesting-ness appears to be to combine familiar and unfamiliar features within the product's styling. The familiar features provide the observer with known reference points and thereby prevent the object being seen as overly complex and unattractive. The unfamiliar features, however, demand the observer's attention if the unfamiliarity is to be overcome through exploration and understanding.

4. An important determinant of how familiar an object is seen to be is its symbolic meaning. An object could be of a visual form never seen before yet it may look familiar because it symbolises something which is familiar. The symbolism of an object allows the observer to pigeonhole it as 'this' or 'that' type of object even before you know what the object itself does. Exploring the symbolism of objects broadens our understanding of product styling considerably. We are no longer thinking about behaviourist determinants

of product style - the particular form of an object causes our brain to automatically and instinctively categorise it as attractive or unattractive. Now we have moved on to cognitive determinants of product style. Our perception of an object depends upon how it relates to our memories, emotions and feelings of other comparable objects.

So, where does all of this get us in terms of a better understanding of product styling. Firstly, the gestalt rules tell us how we form immediate visual impressions of objects. This determines the immediate visual impact an object has when we first see it and, as far as possible, we should try to embody these rules in the styling of objects. How attractive an object is judged to be is, however, determined by more than simply how well it complies with gestalt rules. Attractiveness is a function of both complexity and familiarity, with moderate complexity/familiarity being seen as most attractive. Products which have too simple a visual form will be seen as uninteresting and hence unattractive (at least after having been exposed to it for some time). The ideal mixture seems to be to combine familiar and unfamiliar features within the same product. This, if done well, will engage the observer's interest in the product allowing it to become more familiar and hence attractive. What makes an object familiar is determined by the symbolic meaning that object conjures in the mind of the observer. Product symbolism is a complex issue and will be returned to later.

Figure 3.13 Cute faces all have certain features in common [17]

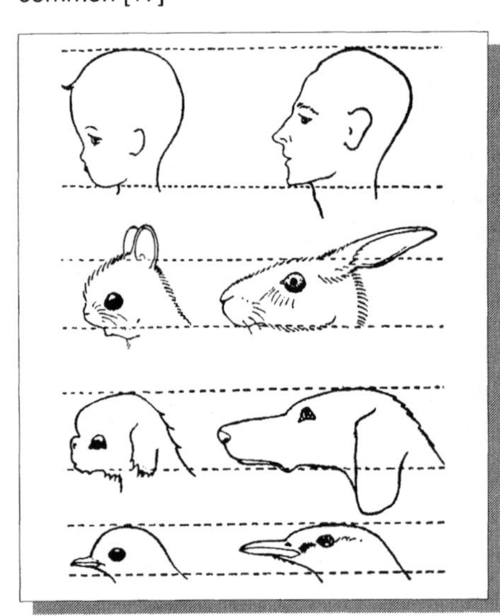

Specific rules

The gestalt rules are general rules of visual perception. We use them all the time, in the analysis of all types of visual images. Other aspects of visual processing are more specific to the visual processing of particular types of visual images. The perception of faces, for example, is a particularly well developed ability in humans [16]. Experimental studies have shown that it is an ability we are born with. Infants (of an average age of 9 minutes!) will watch and visually track patterns with face-like features more than any other types of pattern. By one month of age infants can recognise the facial features of their 'care-giver' (psychologists' political correctness for mother). As adults, our facial recognition abilities rise to dizzy heights. Eyebrows tell us about moods and emotions, forehead shape tells us about age, nose shape gives clues to masculinity/femininity and lip movements enhance speech perception. This expertise in face perception has some profound effects when it comes to products with facial characteristics. Research has shown how the facial features of two human creations have evolved to match our in-built facial preferences. People from all cultures are known to prefer certain facial features over others [17]. Figure 3.13 shows faces with preferred features on the left hand side and less preferred features on the right. The preferred faces all share features which psychologists call neoteny - the characteristics of neonates. Due to our developmental biology, our brains develop more rapidly than the rest of our bodies with the result that neonates have large heads with pronounced foreheads. The eyes of neonates are also large relative to head size. Our preference for these facial characteristics is probably nature's way of ensuring strong parental bonding to new-born children.

As a result of these preferences, both Mickey Mouse and Teddy Bear have undergone pronounced evolutionary changes since their first introduction. When first drawn in the 1930's, Mickey Mouse had relatively rodent-like facial features; a long snout, sloping forehead and small eyes. In a series of progressive changes, Mickey transformed into his modern version [18]. The size of his eyes nearly doubled (relative to head size), his head grew by 15 per cent relative to his body and his forehead bulged by nearly 20 per cent (measured by the length of the arc from nose to front ear, relative to distance from nose to back ear). Teddy bears followed an almost identical trend [19]. Since their introduction in 1900 the shape of teddy bear's head has changed profoundly due to consumer preferences (Figure 3.14).

Facial features are endowed upon a variety of products. The faces of cars recently entered the spotlight of public discussion when Japanese clients rejected an American design for a new car because it was not 'smiling'. The front grill was subsequently redesigned with an upturned mouth to the clients' reported delight. A more recent casualty was Ford whose new Scorpio (Figure 3.15), launched in the UK in 1994

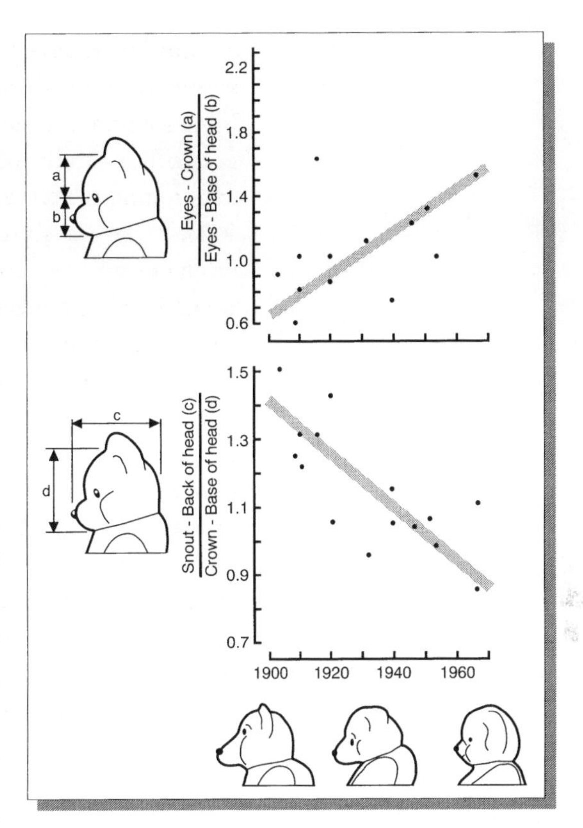

Figure 3.14
Evolution of facial characteristics in Teddy Bear

Figure 3.15 The Ford Scorpio '..... looks like someone's just rammed a banana up its bottom'. [20]

was said to have a facial expression which 'looks like someone's just rammed a banana up its bottom' [21].

There are two messages we can draw from this for product styling. The first is that our perceptual abilities are particularly well refined in the detection of certain types of visual forms, such as faces. This allows product styling to exploit these abilities and to bestow certain facial values upon products. Thus products can be made to smile, frown or look like they have had bananas stuck up their bottoms. Which brings us to the second message, that great care must be taken when using any visual theme which has deep rooted meaning in human values. Any such symbolism must be thoroughly explored and its basis in human psychology must be understood before it is used.

The fabulous Fibonaccis

A long standing, but still highly contentious, issue in product styling is whether we have any innate predisposition towards particular types of natural or organic forms. And if we do, which, of the myriad of organic forms, do we prefer?

The design of both plants and animals follows some remarkable mathematical consistencies [22]. If we examine how leaves are arranged in a spiral around the stems of plants, some consistent numbers begin to emerge. Take hold of a leaf and then follow the leaf pattern up the stem until you find another leaf in exactly the same position on the stem as the one you are holding. Irrespective of what type of plant you are holding, the number of times you encircled the stem and the number of leaves you counted on the way will be a number in the mathematical series 1, 1, 2, 3, 5, 8, 13, 21, 34, 55, 89, 144 etc. If, for example, it was an oak tree or one of the common fruit trees, you would have counted five leaves arranged round two full circumferences of the stem; both of which are numbers in the series. If it was the humble leek it would have 13 leaves arranged around 5 full circumferences of the stem. As a mathematical series, these numbers are unique. Each number is the sum of the previous two. Together they are called the Fibonacci series after the

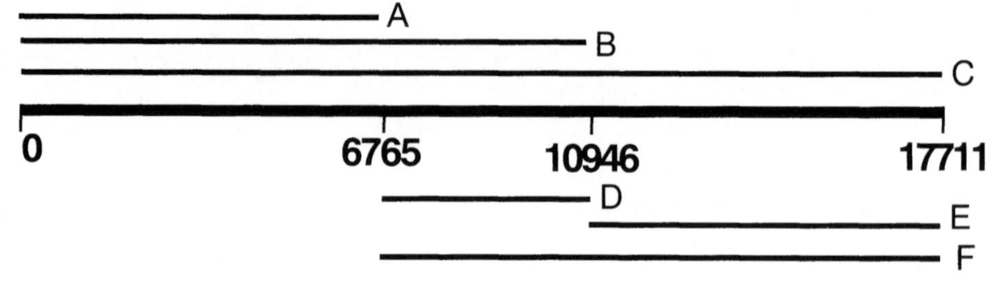

Figure 3.16
Consistent ratios in the Fibonacci numbers

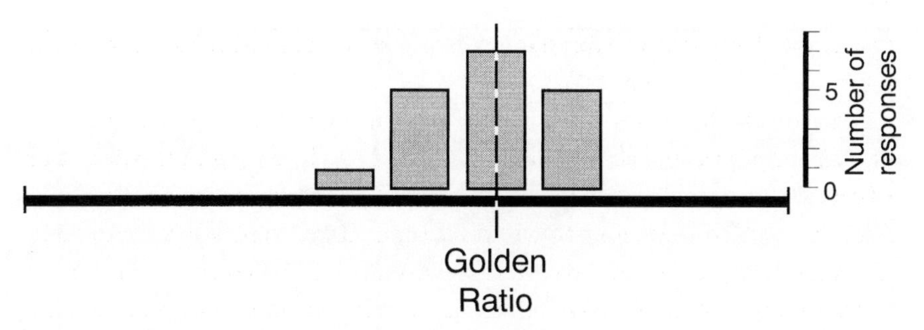
Golden
Ratio
0
5
Number of
responses

Figure 3.17
Preference for the Golden Ratio [26]

Italian mathematician who discovered them in the 13th century [22].

The series does not, of course, stop at 144 but continues to infinity. As the numbers reach into the thousands, consistent patterns between the numbers begin to emerge. Each number becomes a consistent fraction of its neighbouring numbers. Let us take, for example, three numbers high up in the Fibonacci series; 6765, 10946 and 17711. Looking at Figure 3.16 we can work out that the length of line A is 0.618 of the length of line B, line B is 0.618 of line C, line D is 0.618 of line E and line E is 0.618 of line F. So, what is the significance of 0.618? It is called the Golden ratio and it describes the mathematically perfect way to divide a line into two portions. If you take a line of any length and chop off 0.618 of its length, the bit you are left with is 0.618 of the length of the bit you chopped off (take line F, for example and remove 0.618 of its length, which is the length of line E. The bit you are left with is line D and this is 0.618 of the length of the bit you chopped off, line E). When asked to divide a line into two parts with the most pleasing proportions, people tend to do so at the Golden ratio. Figure 3.17 shows the result of one experimental study in which all but one response was within 15% of the Golden ratio [23]. The reason for this is not known; it just seems to look right. But it begins to suggest that Fibonacci numbers lead to more interesting conclusions than simply a mathematical curiosity. Extending this to two dimensions gives us the Golden Section. Draw a square of any size. Add to one of its sides a rectangle as long as the original square and 0.618 times as wide (Figure 3.18). You now have a large rectangle made up of a square and a smaller rectangle and the proportions of this large rectangle are known as the Golden Section [22]. Ratios of 0.618 start to appear everywhere! The length to

Figure 3.18
Construction of the Golden Section

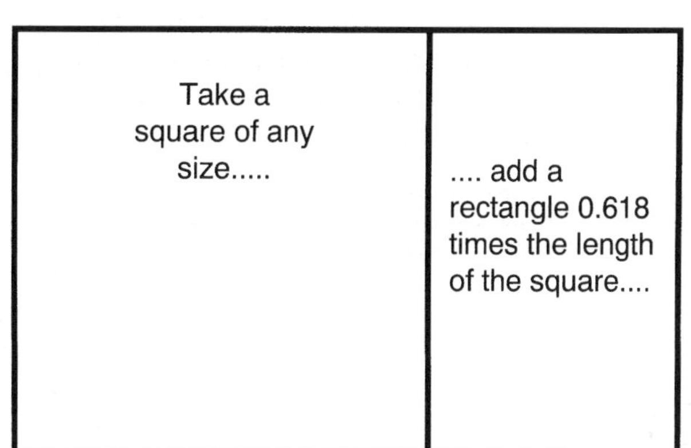
Take a
square of any
size.....
.... add a
rectangle 0.618
times the length
of the square....

.... and you get a rectangle in the proportion of the Golden Section.

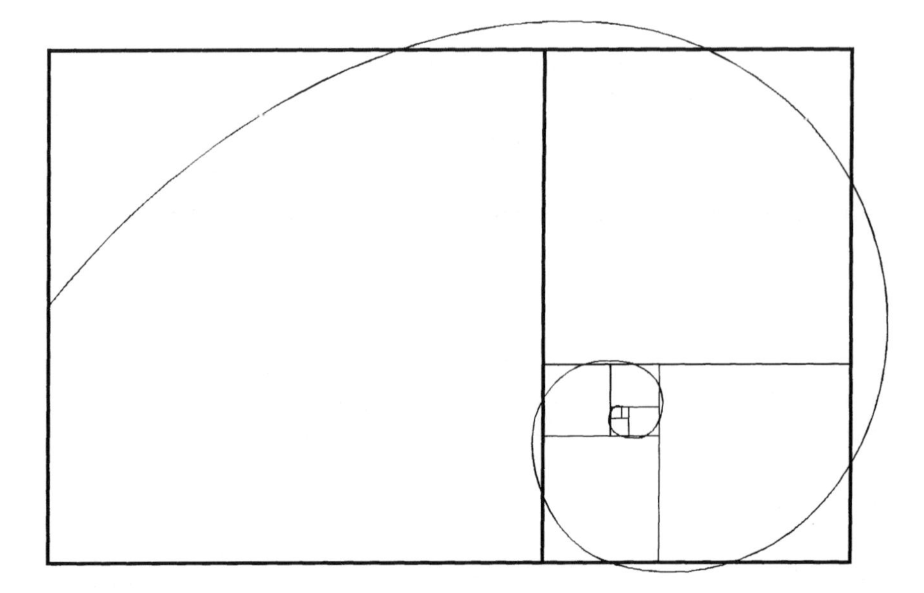

Figure 3.19
The Golden Section and the equi-angular spiral [22].

breadth ratio of the large rectangle is 0.618. The length to breadth ratio of the small rectangle is 0.618. The length of the square is 0.618 of the longer side of the large rectangle. And the length of the small side of the small rectangle is 0.618 of the length of the square. Take the smaller rectangle and divide it in two portions at a point 0.618 along its longer side and guess what you get? Yes, it is an even smaller rectangle with a length to breadth ratio of 0.618. Keep going with smaller and smaller rectangles and they will eventually spiral down to a point (Figure 3.19). This spiral, which is called an equiangular spiral, brings us back to very close to our starting point with this discussion. Leaves which spiral around the stem of a plant do so in an equiangular spiral. The fabulous Fibonacci numbers describe not only how many leaves there are and how tightly they spiral round the stem, they also tell us how big each leaf is relative to its neighbours. The shells grown by snails follow exactly the same rules, as do the horns grown by antelope, sheep and goats. Fibonacci numbers, golden ratios and equiangular spirals are to be found everywhere in nature [22]. If life on earth was created according to some grand plan, it is beginning to sound like the chief design consultant was a 13th century Italian mathematician!

Now let's get back to the significance of all of this for product styling. It would make sense for humans to be particularly good at recognising the natural shapes and patterns of plants and animals. Some patterns, we would learn, signify things that are good to eat, whereas others are dangerous and must be avoided. If the ability to recognise natural patterns was an innate ability, just as our ability to recognise faces, it might have a profound effect on our judgement of product styling. The perfect shapes for products might be golden section rectangles or equiangular spirals; their perfection arising from our innate ability to recognise their natural shapes based on Fibonacci numbers. The obvious question is 'Are they perfect?'. Are they even preferred over other shapes and forms? The ancient Greeks believed that the golden ratio rectangle was the perfect proportion and the front elevation of the Parthenon in Athens is surviving testimony to that belief. Michaelangelo was also a fan of golden sections, as his painting of Adam's Creation in the Sistine Chapel proves (Figure 3.20). An intriguing modern day application of the Golden Section is found in the proportions of cars. The cars themselves never reach golden section proportions, presumably because of the inconvenient requirement of having to fit people inside them. But despite this failure to meet perfect proportions, pictures of cars in advertisements are stretched to precisely golden section proportions. Figure 3.21 shows the real Nissan QX and underneath the advert of the QX as it appeared in UK newspapers.

Experimental studies of the golden section, have repeatedly shown that it is actually preferred over rectangles of other proportions, although the strength of that preference is not great. In one classic study, for example, ten rectangles of different proportions were presented to 500 people. When asked which was of the most pleasing proportions, the one chosen the most, by 33% of the people, was the golden section [24]. Another study used a rectangle which was drawn over two sheets of paper. In the study, people were asked to move the sheets to produce a rectangle which in their view had the most pleasing proportions. Of the people studied, those that were good at

Figure 3.20
Golden sections galore! The Parthenon in Athens (top), and Michelangelo's Adam's Creation in the Sistine Chapel (bottom)

Figure 3.21
The Nissan QX. In its true proportions and as it appeared in adverts (a perfect Golden Section)

judging differences in the length, ratio or volume of a rectangle anyway (those with good visual skills) were remarkably accurate at positioning their preferred rectangle in the Golden Section proportion [25]. To my mind these studies are particularly fascinating since rectangles are the least likely of the Fibonacci-based shapes to be innately preferred. Our argument so far has been that certain mathematically natural shapes will be preferred because of the evolutionary advantage of being able to detect natural objects. How many rectangles do you find in nature? The golden rectangle is a human abstraction based upon the mathematical basis of organic design. It is much less likely to have significance than curves formed from equiangular spirals or parts or features of an object positioned in golden ratio to each other. But it **is** preferred and this suggests that shapes more commonly found in nature may be even more preferred.

What, then can we conclude about using these principles in product styling? Clearly, there are mathematical rules which determine a great deal of the shape and pattern of plants and animals. Designers who seek to style their products according to organic forms cannot do so faithfully without taking these rules into account. For other designers, simply seeking the perfect visual form for their product, these rules are relatively simple and straightforward to apply. They provide nourishing food for thought and they might just trigger a response in the primeval depths of the mind of customers.

Bisociative attraction

Bisociation is a term coined by Arthur Koestler [26] to describe the nature of humour. Humour, he argues, is funny because it leads to an unexpected and often absurd or ridiculous conclusion from an otherwise commonplace train of thought. He relates the tale of a man who comes home unexpectedly to find his wife in a passionate embrace with the Bishop. The man turns quietly, walks into the street and starts blessing everyone who walks by. 'What are you doing?' cries his distraught wife. 'The bishop is performing my functions, so I thought I would perform his.' The bisociative, and hence humorous part of this story is that we immediately expect certain consequences to be associated with the events described. The man should confront his wife, confront the Bishop or both with an emotional and possible violent outburst. But instead of this expected association we get bisociation – a reaction which, in other circumstances might be logical and reasonable human behaviour but which, in this context is unexpected, ridiculous and humorous. Bisociation, therefore, is where two trains of thought, each of which is independently sensible, collide in a way which jars or surprises our normal powers of association.

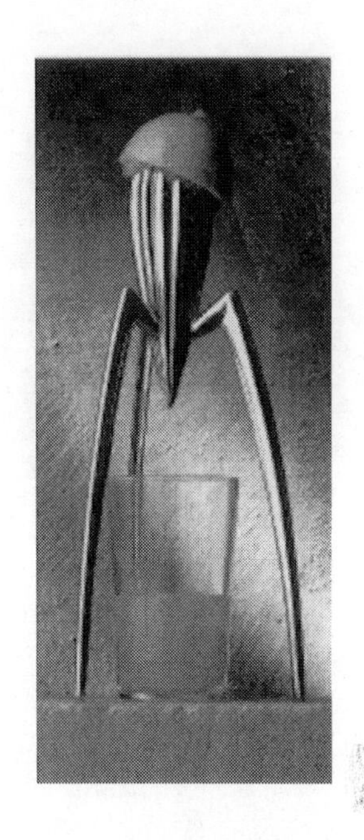

Figure 3.22 Bisociation gives products a humorous and hence attractive appearance [27]

Certain products which we perceive to be attractive have strong bisociative themes. Phillipe Starck's lemon squeezer (Figure 3.22) is one such example. The product's neat tapering body and elbowed spindly arms symbolise some sort of exotic insect or extra-terrestrial spacecraft. Yet the patterning on the body clearly reflects the typical pattern on a conventional lemon squeezer. The contrast between these two images is bisociative with the inevitable result of raising a smile when you see the product for the first time. Humour is a rewarding experience for people and as a result, the source of such humour becomes not only memorable but also valued and appreciated. Giving products bisociative visual themes is, therefore, a route to attractiveness in product styling. It does, however, require subtlety. Every joke-teller knows how fine the line can be between attractive, tasteful humour and crass vulgarity.

Social, cultural and business effects

So far, we have seen how certain fundamental aspects of visual perception determine the way we see the three dimensional form of products and how this, in turn, affects our judgement of their attractiveness. Clearly these are not the only determinants of product style. Social factors, cultural factors and commercial or business factors all play a role and, as we shall see, they sometimes play such a strong role that they may override the more basic perceptual factors.

We only have to look as far as the clothes we wear to see that product styling can be influenced by social trends. The fashion industry is structured in the most ruthlessly efficient way to ensure that the styling of clothes is led in slightly different directions every year by social trends. The trend is set with the annual round of fashion shows by the 'big names' in the fashion industry. These show the world what 'beautiful people' are wearing this year, a message

Figure 3.23
Product design in earlier times often featured great complexity

reinforced by the idols of contemporary society film stars, musicians and the occasional fashion-conscious Princess. These fashion trends then percolate down to the mass market in the form of 'this years spring, summer, autumn or winter collections' in the retail outlets. The media herald these arrivals with an advertising and editorial fanfare, encouraging consumers to keep up to date with the latest fashions. Implicit in all of this is that your social value will be somehow diminished, in the eyes of your peers, if you let yourself be seen wearing last year's style.

This, of course make the most wonderful business sense. Clothes with a useful life of several years are given an annual obsolescence, based on social trends in style. Dedicated followers of fashion are the fashion industry's greatest business asset.

Cultural influences on the perception of product style have a more long term effect. The cultural context within which a society operates can have a pronounced effect on the values and beliefs held by individuals within that society. This in turn causes certain aspects of product style to be valued and others to be disparaged. In Russia around the time of the revolution, for example, the lavish and luxurious lifestyle of the ruling classes became a focus for widespread public discontent. Stemming from this, and fuelled by the political principles of Communism, all symbols of conspicuous consumption became politically incorrect. This led to strong utilitarian and industrial themes which emerged in product design throughout the Soviet era. This contrasts sharply with the trends in Western design during the 1980's when Reaganomics and Thatcherite monetarism gave rise to a hedonistic and materialistic culture in which conspicuous consumption thrived.

Even longer term cultural influence on product styling can be seen operating over century-long time scales. As discussed previously, product design in earlier times often featured great complexity for cultural reasons. At a time when products were manufactured on a craft basis, their cost of manufacture was largely due to the amount of time spent by the craftsman or artisan in making the product. One important way of adding value to products was to decorate them ornately, making them more expensive to produce and, therefore, increasing their value. Visual complexity, thereby, became a matter of product prestige – a trend which appears to have had a greater influence on product styling than customers' innate preferences for visual simplicity. Contemporary mass manufacturing techniques can produce visual complexity at little extra cost. Visual complexity has, thus, lost its prestige value, allowing the more innate preferences for simple visual forms to prevail.

Business decisions about new products, as we have seen in Chapter 2, are all about risk management. This is no less true for product styling than it is for decisions about functional aspects of products. It is a much greater risk to try to pioneer a new style than it is to stick with existing styles. As a result, cautious business decision-making tends to channel product styling along visual themes. One of the most striking examples of style channelling, and perhaps the most ridiculous, was the addition of fins and wings to American cars in the 1930's. This began with Wayne Earl setting out to transfer some of the visual styling from aeroplanes to cars (Figure 3.24). This style was copied and developed until it became a visual theme and established itself as the norm in the US motor industry. Now, put yourself in the position of a motor industry executive facing a decision about a new car at that time. Virtually all your competitor's products have elaborate wing and fin designs. Your styling choices are simple: you either break ranks and try to pioneer a new style, risking, in the process, the failure of what might be a technically innovative car, just because it does not look right. Or, you follow the theme and perhaps develop it further so that your car is even more winged and finned than the others. This latter, more cautious styling strategy prevailed for many years, with the wing and fin theme becoming more pronounced and exaggerated all the time. By 1959 it was claimed that, as a result of this style channelling cars were carrying an unnecessary half ton of metal! [28]

There inevitably comes a time when a visual theme has run its course. Most of the styling possibilities have been explored and the visual theme has been developed as far as it can go. At this point, the time is ripe for styling innovation. The rewards for pioneering the next visual theme are sufficiently great that the risks become worthwhile. A good example of this happened again in the US auto industry two decades later. American cars of the 1970's were predominantly boxy and angular. Concerns over fuel saving were,

Figure 3.24
Style channelling created the elaborate fin designs on American cars in the 1950's [29]

Style channelling by customer demand

In the late 1980's Donald Petersen, then Chairman of Ford, decreed that vinyl roofs would be banned from all new Ford cars. Petersen, a man confident of his own good taste and sense of style, hated garish ornaments on cars. He, however, was forced to lift his ban after car dealers around the USA inundated Ford's marketing department with customers' requests for vinyl roofs [30].

however, forcing engineers to look seriously at aerodynamics. It soon became obvious that smooth curves were more aerodynamic than sharp corners and edges. But that would mean a major departure from the prevailing visual theme. Ford were the first to take the plunge with the curved and sleek shape of the Taurus. The risk paid off handsomely. The Taurus quickly established itself as market leader and set the aerodynamic visual theme which remains dominant in the worldwide auto industry today [30].

The determinants of style

The perceptual determinants of product style can be thought of at several levels. Style is determined at its most basic level by what our visual system sees. The immediate visual impact of a product is determined by the way we visually process images at a pre-attentive level. This not only gives us our initial visual impression of the product but also determines what we focus on during more detailed attentive processing. There are well established rules governing pre-attentive perceptual processing – the gestalt rules – and these can be translated into product styling rules. At the next level, there are specific attributes of our visual processing which make us particularly good at perceiving and evaluating certain types of visual images. Faces are a well established example and natural shapes are also likely contenders. At the highest level, style is determined by social, cultural and business factors. The cultural values, prevailing at any time in history determine the importance of different aspects of product styling. Within this cultural context, social trends determine styling fashions. And fashions tend to become channelled into visual themes because of the business risks of departing from established styling norms. Product styling is a complicated process with a lot of complex factors to consider. Nobody ever said it was simple!

The creation of attractiveness in a product continues throughout the design process. It must not be seen as something that happens at a single point in time –a sort of injection of styling at an appropriate moment. Worst of all, it is not something added on at the end. Many manufacturing companies who use design cons-ultancies make this mistake. They complete the technical and functional aspects of the design and send the finished product off to be styled. By this stage so much of the design is constrained by other factors that styling can be no more than the most superficial cosmetic exercise; a smooth curve here, a pattern on the casing there and a choice of colour. Product styling must be an integral and harmonious part of the entire design process. Styling decisions need to be made at every stage from product planning to production engineering.

Attractiveness and product style

Attractiveness is itself an inelegant word. But it describes so perfectly what products should be in the eyes of a customer that its inelegance is forgiven. Products should be attractive to customers in three subtly different ways:

- Firstly, an object can be said to be attractive if it grabs your attention (by being visually pleasing, of course, rather than outstandingly hideous!). Architects refer to houses as having 'kerb appeal'. When you walk or drive past a well-designed house, its visual appearance, from the kerb, should be immediately attractive. Similarly, products should have their own version of kerb appeal. They should immediately grab your attention as you walk past them in a shop or as you flick over their photograph in a leaflet.
- Secondly, an attractive object is a desirable object. Making customers want to own a product is more than half the battle in product marketing. If a product can be made desirable to customers simply by its visual appearance, then this is powerful product marketing, indeed.
- Taking these two meanings together – a product which is both attention grabbing and desirable – means that customers are 'drawn towards' the product; the literal meaning of the word attractive.

The four faces of attractiveness

In order to make products attractive, we need to understand the way in which products can be seen as attractive by customers. There are four main ways.

1. **Prior knowledge attractiveness** The most obvious reason that a that, like me, most people have suffered the frustration of buying a novel only to find that it is one you had bought and read years ago when it had a completely different front cover design. This make business sense in the publishing business but it can be a disaster in product design. Many types of products

Figure 3.25
The four faces of attractiveness

are dependent upon repeat sales for their commercial success. If, during a redesign of such a product, its visual appearance is so radically changed that customers no longer recognise it as the product they previously used and liked then the prospects for repeat business are destroyed. So, if the product that you are designing is an update of an existing product it is important to maintain the visual identity of its predecessor. To fail to do so could jeopardise prospects for repeat purchases by existing customers. If the product is part of a product range or is sold under a brand or company identity, that must be made obvious to potential customers from its visual appearance. Even if it is unique and completely new product, customers must be able to recognise what kind of product it is, from its visual appearance. There is no point in producing the most innovative and imaginatively designed new lawn mower, for example, if customers are going to walk past it in the shop, not realising it is a lawn mower. This an important aspect of product styling which is often overlooked by designers or resented as an undue constraint on styling innovation. From a marketing and business point of view it is vitally important and it must be acknowledged and respected during the styling process.

Styling is not injected at any single point in time, nor is it something added on at the end. Styling continues throughout the design process.

2. **Functional attractiveness** For customers with no prior knowledge of a product, visual appearance must somehow inspire the confidence that prior use would otherwise provide. For products in which functional value is of importance, this is achieved by making the product look like it will perform its function well. This is different from actually making the product work well. Very often customers will not have the opportunity to thoroughly test a product before buying it. A great deal of their judgement on performance is, therefore, based on how the product **looks** like it will function. A whole branch of product design has developed around the principle of making products look like they perform their intended purpose well. It is known as product semantics: literally product meanings.

3. **Symbolic attractiveness.** Where appearance-value is an important part (or all of) the reason for purchasing a product, a different approach to its styling is required. Here the product's symbolism is important. Purchasing

confidence is inspired by the extent to which the product reflects the customer's self image and the statement that they wish the product to make in the eyes of others. Designers can embody these in the appearance of the product by having it make some sort of symbolic statement. Examples might include, 'This is a fun product and I am a party animal', 'This is a refined, traditional product and I am a pillar of the village community' or 'This is a very respectable but slightly risky product and I am not as old as my daughter thinks I am'. These statements and the images they conjure are developed and articulated during the design process by the use of lifestyle, mood and theme boards – of which more later.

4. **Inherent attractiveness of visual form.** At the root of visual appearance, for a product of any sort, is that most elusive and intangible quality: its elegance, its beauty, its intrinsic aesthetic appeal. This is the embodiment of the perceptual, social and cultural determinants of the attractiveness of products, as described above.

The styling process

This brief tour of product styling has come full circle. We began by exploring the perceptual determinants of style, considered how these are distorted by social cultural and business factors and ended up considering the practicalities of creating attractiveness in products. The last of these attractiveness factors is the intrinsic attractiveness of products – the attractiveness which is determined by their basic perceptual 'eye-appeal'. The question remains as to whether we can systematically and methodically style products to be visually attractive. The answer, in a book entitled 'systematic methods for new product design and development' must be Yes!

- Chapter 6 on product planning will look at how styling objectives can be researched and specified.
- Chapter 7 on concept design will examine how these styling objectives can be interpreted into visual themes by the use of mood and theme boards and then translated into the first styling concepts for the new product.
- Chapter 9 will then show how to convert these visual concepts into a physical embodiment which can be modelled, market tested and ultimately manufactured as the finished product.

Key Concepts

1. What we see is what we like!

The properties of our visual system determine, to a large extent, what we see as attractive in products.

- The Gestalt rules show how products which are symmetrical, have clean lines making simple geometric forms and give a sense of visual harmony will be easier to process visually and hence will have more immediate visual appeal. Berlyne's theories show how products with moderate complexity and familiarity have an over-riding effect over the gestalt rules.
- Specific visual abilities make us perceptually attuned to patterns such as faces and natural organic forms.

2. Social, cultural and business effects

Our basic, perceptually-based, preferences for simplicity in product styling are often over-ridden by social, cultural or business influences.

- Social trends, such as those dominating the fashion industry can change our styling preferences form one year to the next.
- The cultural environment in which we live affects our personal and social values, thereby changing the types of visual symbolism which we like and dislike. This generally changes over time periods of decades or centuries.
- The business decision-making by the managers of manufacturing companies can channel product styling along particular visual themes.

3. The attractiveness of products

The attractiveness of products can be categorised in four ways:

- Prior-knowledge attractiveness – recognition of a product previously used and liked
- Semantic attractiveness – the product looks like it works well
- Symbolic attractiveness – the product appeals to the personal and social values of the customer
- Intrinsic attractiveness – the inherent beauty of the product's form.

Notes on Chapter 3

1. This wonderfully simple objective, that design students should be able to draw or model any visual form that they can imagine in their minds, was first suggested to me by Prof. Ronald Hill of the Art Centre, College of Design at Pasadena, California.
2. An excellent review of the current scientific understanding of visual perception is Bruce V. and Green P 1990, *Visual Perception: Physiology, Psychology and Ecology* (2nd Edition). Lawrence Erlbaum Assoc. London.
3. Uttal W.R. 1988 *On Seeing Forms*. Lawrence Erlbaum Assoc, London gives a comprehensive review of how we perceive visual forms.
4. The seething circles illusion was first produced in Marroquin, J.L. 1976. *Human visual perception of structure*. MSc thesis, Massachusetts Institute of Technology, Boston.
5. Adapted from Thurston, J.B. and Carrahar, R.G. 1986. *Optical illusions and the visual arts*. Van Nostrand Reinhold, New York.
6. The non-existent triangle was first developed by Kanizsa G. 1979, *Organisation in Vision: Essays on Gestalt Perception*. Preager, New York.
7. The two classic texts on Gestalt psychology are: i) Koffka, K. 1935. *Principles of Gestalt psychology.* Harcourt Brace, New York. ii) Kohler, W. 1947. *Gestalt psychology: an introduction to new concepts in modern psychology.* Liveright Publishing Company, New York.
8. The Gestalt rules are summarised in Bruce and Green 1990 (see 2 above) pages 110 to 115 and Uttal 1988 (see 3 above) pages 153 to 155.
9. Symmetry detection abilities are described in Uttal (see 3 above) pages 144 to 146.
10. Lewalski, Z.M. 1988 *Product Esthetics: An Interpretation for Designers.* Design & Development Engineering Press, Carson City, Nevada.
11. Beeching W.A. 1974. *Century of the Typewriter.* Heinemann Ltd. London.
12. Deiter Rams quote is cited in Greenhalgh P. 1993, *Quotations and Sources on Design and the Decorative Arts*. Manchester University Press, Manchester, UK. p 235.
13. Terence Conran's quote is cited in Greenhalgh 1993 (see 12 above) p 235.
14. Krippendorf K. and Butter R. 1993. Where meanings escape functions. *Design Management Journal*, Spring 1993, 29-37.
15. Daniel Berlyne's work and related studies are summarised in Crozier R 1994, *Manufactured Pleasures: Psychological Responses to Design*. Manchester University Press, Manchester UK.
16. Bruce and Green, 1990 (see 3 above) pages 360 to 374 for a review of human face recognition.
17. Lorenz, K. 1971, *Studies in Human and Animal Behaviour*, Vol 2. Methuen & Co. London.
18. Gould, S.J. 1980, *The Panda's Thumb: More Reflections in Natural History.* Penguin Books, London. pp 81-91.
19. Hinde, R. A. 1987, The Evolution of the Teddy Bear. Animal Behaviour.
20. Picture from Ford (UK) Ltd, reproduced with permission.
21. Clarkson, J. 1994. *Not a pretty face.* Sunday Times.
22. For a simple and largely pictorial introduction to the Fibonacci series, the Golden Ratio and the Golden Section see Hargittai I. and Hargittai M. 1994. *Symmetry: A Unifying Concept*. Shelter Publications Inc, Bolinas, California,

pp 154-164. For more academic treatments see Cook T.A. 1914. *The Curves of Life.* Constable & Co. London. or Huntley H.E. 1970, *The Divine Proportion: A Study in Mathematical Beauty.* Dover Publications Ltd, London.

23. Angier, R.P. 1903. The aesthetics of unequal division. *Psychological Monographs*, Supplements 4: 541-561. See Pickford, R.W. 1972. *Psychology and visual aesthetics*. Hutchinson Educational Ltd, London. Discussion of the Golden Section on pp 27 to 30.
24. This original Golden Section experiment was conducted by Fechner G.T. 1876. *Vorschule der Aesthetik*. Breitkopf & Hartel, Leipzig. See Pickford 1972 (see 23 above) and, more recently, Crozier 1994 (see 15 above) for critical discussions of the current Golden Section evidence.
25. McCulloch W.S. 1960. *Embodiments of Mind.* MIT Press, Cambridge, Mass. See 24 above.
26. Koestler, A. 1964. *The Act of Creation*. Hutchinson & Co., London.
27. Starck P. 1991 (Text by Olivier Boissiere) *Starck*. Benedict Taschen & Co. Koln, Germany. Photograph reprinted with permission.
28. Mingo J. 1994. *How the Cadillac Got its Fins.* Harper Business, New York.
29. Photograph from General Motors (US), reproduced with permission.
30. Ingrassia, P. and White, J.B. 1994. *Comeback: the fall and rise of the American automobile industry.* Simon and Shuster, New York. p127.

4 The Principles of Creativity

Creativity is one of the most mysterious of human abilities. It has been written about by everyone from problem page columnists to the great creative scientists and artists themselves. Psychologists and philosophers have dedicated their lives to its study. And a whole new branch of consultancy has emerged, offering to unblock the most creatively constipated organisation.

But are we really any the wiser? Are there ways of stimulating creativity or is it something you are born with? The psychologist's answer is that yes we are wiser, yes you can stimulate creativity and no, people are not to any significant extent born with in-built creativity. So, everyone can be creative, if they try hard enough.

The importance of creativity

Creativity is at the heart of design, at all stages throughout the design process. The most exciting and challenging design is that which is truly innovative; the creation of a radical departure from anything currently on the market. Unfortunately, most of the working lives of designers is taken up with more pedestrian design projects. The improvement of an existing product, the extension of an existing product line or the development of a product to catch up with a competitor. But this does not diminish the importance of creativity. In most markets today there is sufficient competition to leave little room for price elasticity. Competition purely on the basis of price is, consequently, severely limited and, therefore, product differentiation remains the main competitive weapon. Securing a competitive advantage through product differentiation means creating differences between your product and those of your competitors. These do not need to be radical differences. Indeed, many companies believe that they cannot afford the investment and risk involved in radical innovation. There do, however, need to be differences which customers perceive to be significant. And this requires you to be creative at every stage in the product development process from spotting an opportunity to engineering the product.

The psychology of creativity

Like many great wonders of human intellect, creativity is not claimed to be completely understood by psychologists. But there is enough known to signpost the route to creativity and to describe some characteristics of the main landmarks along that route. It is a tantalising sort of understanding, so common in psychology: the steps taken to be creative are known, but taking them is no guarantee of success. Despite this, it is well established that you are more likely to be creative if you understand and follow these steps than if you do not.

First insight

The steps, which I have described as the stairway to creativity (Figure 4.1) are a structured way of presenting a common sense understanding of creativity. The first insight is the way you frame in your mind the need for some creative discovery. Modern folklore conjures up an image of inventors beavering away in garden sheds being inspired at random with a new mousetrap one minute and a perpetual motion machine the next. In truth, most inventors, and nearly all the great scientists, artists and technologists are highly focused on one type of problem. Their first insight frames the

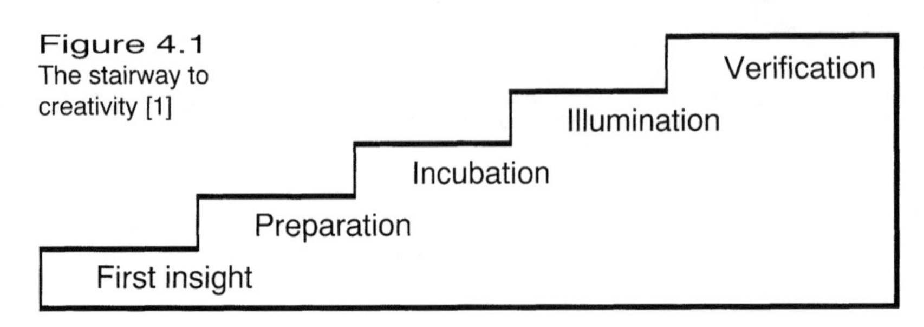

Figure 4.1 The stairway to creativity [1]

problem in one area and focuses their mind on that problem until they solve it.

Archimedes' Eureka

As Archimedes climbed into his bath on that fateful day in 230 BC he did so having wrestled with a particular problem over several months [2]. His protector and patron Heiro, the tyrannical ruler of Syracus had been given a crown, allegedly of pure gold, but which Heiro suspected had been adulterated with silver. Archimedes was given the task of testing it. The specific weight of gold was well known at the time and was used to determine whether an ingot of a known size was of the correct weight to be gold. Archimedes' problem was that the crown was a highly complex shape with intricate filigree decoration. It would not have gone down well to suggest melting it into a block, measuring it and weighing it. He had to find out the volume of the crown in some other way. Having puzzled over this at length and unsuccessfully thought through as many different ways of solving the problem as he could imagine, he got into his bath. There before his eyes, the solution became obvious. As he got into the bath, his body displaced the water, making the level of the bath water rise. To measure the volume of his complex crown, all he had to do was immerse it in water and measure the volume of water it displaced. With a shout of 'Eureka!' he ran off down the street in the nude and straight into history books. Archimedes was a brilliant man and he lived at a time when the need to measure the volume of objects had become important. He was presented with a specific problem and the incentive to solve it. His first insight framed the problem and focused his attention upon ways of measuring the volume of a complex object. The solution of this problem had a value much greater and more general than determining the purity of a tyrant's crown. But without the tyrant and the consequent need to focus on a particular problem, the principles of volumetric displacement may have been a long time in coming. Inspiration is rarely a 'bolt from the blue'. The majority of great discoveries arise from the need to solve a specific problem. And the need to solve that particular problem is the first insight which shapes the entire problem-solving activity.

Preparation

Thomas Edison's suggestion that 'Creativity is 1 per cent inspiration and 99 per cent perspiration' has a lot of truth to it. From research, from accounts of great discoveries and from the practical experience of more ordinary mortals it has become clear that the key to creative thinking is preparation. A creative idea is generally a connection, an expansion or a perception in a new light, of a set of existing ideas. Preparation is the process by which the mind becomes immersed in these existing ideas. Immersion in these ideas then fuels the creative breakthrough. Not even the greatest of creative breakthroughs occurs in a vacuum.

Faraday's electricity

The developed world seems to have an almost addictive dependence upon electrical devices. In the average Western home and car there are an average of 100 motors [3]. To power them, electricity flows through every home and car, generated by electric dynamos. That both the electric motor and the electric dynamo were both invented by the same man, Michael Faraday is remarkable. That this man had no formal education in either physics or engineering suggests a genius of epic proportions.

There can be no doubt that Faraday made more progress of both academic and technological significance than any other scholar who studied in his field. But the common belief that Faraday single-handedly discovered electricity, and then created the devices for its generation and application, is far from reality. The importance of mental preparation and the building upon the work of others is perfectly exemplified by Faraday's discoveries [4]. Faraday's preparation began when he was apprenticed to a bookbinder and bookseller at the age of 14. As well as learning his new trade, Faraday began reading all the books he came across on physics and chemistry. By 19 years of age he started attending lectures and discussions at the City Philosophical Society in London and at 20 he attended a series of Royal Institution Lectures by Sir Humphry Davy. The following year he managed to secure a position at Davy's laboratory as an assistant and one of his first tasks was to accompany Davy on a tour of Europe visiting the great laboratories of the time, including Volta's and Ampere's. From his reading, his discussions with Davy and his European tour, Faraday would have found out that Volta had invented the electric battery and Schweigger had invented the galvanometer for measuring electric current. The Dutch scientist Oersted had found that a wire carrying an electric current will make a compass needle move and Ampere went on to find that two wires, each carrying an electric current behave as if they are magnets, attracting or repelling each other. Before Faraday's historic discoveries, the nature of electricity was known, as was the principle of electro-magnetic induction. What Faraday managed to do was develop these ideas one small but highly significant step further to discover electro-magnetic rotation. By passing an electric current through an apparatus, he made a freely suspended wire rotate around a fixed magnet. Although only a small step from Oersted's movement of a compass needle, this was the first time anyone had used electricity to produce continuous motion. More importantly, it was the first, albeit modest, electric motor. His subsequent discovery of the electric dynamo was similarly incremental in nature but it was an equally true breakthroughs in science and massive technological invention. So, Faraday **was** a brilliant scholar and the most creative of inventors. But he would have been neither if he had not immersed himself in the works and discoveries of others for the 16 years prior to his own breakthroughs. His discoveries testify to the truth of Isaac Newton's words 'If I have been able to see further than others, it was because I stood on the shoulders of giants.' The message for principles of creativity is encapsulated by Louis Pasteur, 'Fortune favours the prepared mind'.

Incubation and illumination

First insight and preparation are the logical and rational parts of creativity. It makes sense that a creative breakthrough needs some first insight to frame the problem and establish a goal. It then makes sense that an understanding of current knowledge is required in order to advance beyond that knowledge. The next stage of how to actually make that creative breakthrough seems more to do with luck, inspiration or, as many scientists and artists have suggested, divine intervention. It certainly seems to have little to do with logic and rationality. Here are two examples of great creative breakthroughs.

Buckyballs! [5]

That carbon exists in two crystalline forms, diamond and graphite, is a memory most people take away from their school chemistry classes. The possibility of a third crystalline form of carbon was something that most professional chemists would have bet against with a sizeable proportion of their salaries, as recently as the early 1980's. It is perhaps a face-saver for chemists that the new carbon was not discovered under their noses - it was first found in outer space. Several scientists from around the world shared a common research interest in dust that appeared to be generated by 'red giants', dying stars composed mostly of carbon. By vaporising graphite at around 10 000 degrees centigrade, they managed to produce samples of carbon dust for study, under conditions that they believed might be similar to a red giant in space. The carbon dust they collected had a number of odd features about it but the oddest by far was the predominance of C_{60}, a molecule containing 60 carbon atoms. Why 60? Why not 59, 61 or even 160? There had to be something special about 60 atoms that could arrange themselves into a structure which was more stable and hence more common than any other molecular structure. But what was it? Harry Kroto, a professor from Sussex University and his co-workers Bob Curl and Richard Smalley from Rice University in Texas, puzzled at length over how to arrange 60 carbon molecules into a uniquely stable shape. The breakthrough refused to come, until Kroto's thoughts drifted back to the famous geodesic dome that Buckminster Fuller had designed for EXPO '67 in Montreal. As soon as that came to mind, Kroto also remembered a three-dimensional map of the stars he had made years ago for his children. Both the dome and the map had been made up of pentagons and hexagons. It was Richard Smalley who cut out paper pentagons and hexagons and began to stick them together. Soon an almost perfect sphere emerged, made up of 12 pentagons and 20 hexagons. Upon asking the mathematics department if such a structure had a formal name, Smalley was told it was a soccer ball! In recognition of the part played by the geodesic dome in discovering its structure, Kroto and his co-workers named the new C_{60} molecule a Buckminsterfullerene or Buckyball for short (Figure 4.2).

Figure 4.2
Buckyball structure

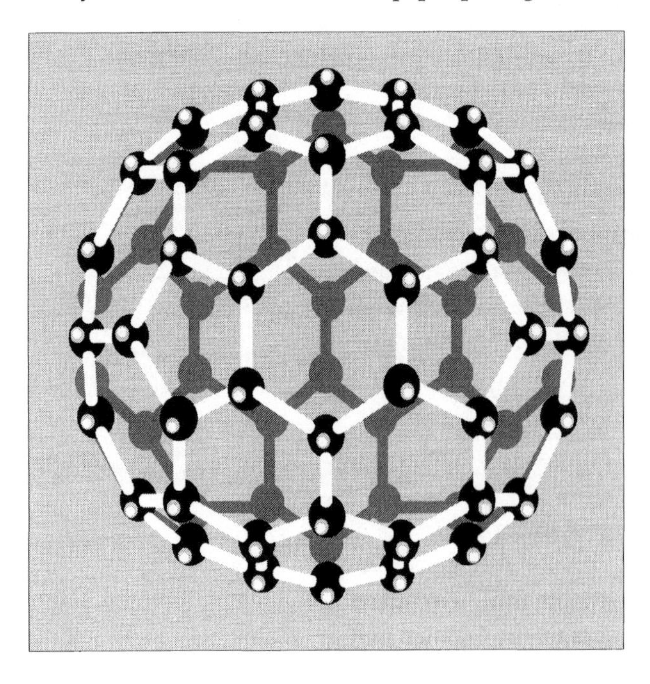

Since its discovery, a race has begun to explore commercial applications for the new molecule. At present they range from molecular ball-bearings to some of the neatest molecular packaging for the containment of other molecules.

From hymn books to Post-it notes [6]

A breakthrough of much less scientific prestige but still of great commercial value was the invention of the Post-it note by 3M. In 1968, Spencer Silver discovered a glue that wasn't very sticky. Within the adhesives department of 3M, a team with a track record for developing some of the strongest known adhesives, his discovery was hardly considered earth-shattering! But Silver persevered and felt there must be a useful application for his new glue. For 5 years he tried various possible applications but none provided the breakthrough. And then, one Sunday, a colleague, Arthur Fry went to church and found the inspiration. For his regular Sunday church service Fry always marked the hymns for that service with slips of paper in his hymn book. On this particular Sunday, whilst flicking from one hymn to the next, he dropped his hymn book and several slips of paper fell from their place. That was it! What he needed was slips of paper with a light adhesive to temporarily hold them in place. The 3M Post-it note was born. The next breakthrough came when a process was discovered by which the unsticky glue could be made to stick much more firmly to one surface than another. Thus, when the glue was applied to a slip of paper using this process it stuck more firmly to that paper than to any surface that the slip of paper was later stuck to. For the first time, a slip of paper could be stuck to a surface and then removed without leaving a trace of adhesive.

The nature of incubation and illumination

My request to a former boss, many years ago, to install a bath in our design offices received a less than enthusiastic reception. Upon adding that the bath was not for the purposes of personal hygiene but rather to help liberate the creative talents of myself and my fellow designers, the response deteriorated to intolerant disbelief. Although never intended to be entirely serious, my request did have some degree of genuine purpose. Lamentably, I cannot claim to have had any creative thoughts of Darwinian or Newtonian magnitude, but by far the majority of my more modest inspirations have happened in the bath! I am convinced (or deluded?) that I am not peculiar in this respect. Several acquaintances have admitted to similar balneary inspirations, whilst others say that lying in bed, in that half-awake, semi-conscious state has been their source of inspiration. One scientist described the 3 B's of his major discoveries; the Bus, the Bath and the Bed. These share a common theme that has been widely reported in studies of creativity.

Inspiration that has been struggled over for weeks and sometimes years of deliberate and painstaking effort sometimes appears from nowhere in a moment of quiet relaxation. The ancient Alchemist's Rosarium suggests 'Thou seekest hard and findest not. Seek not and thou whilst find.' The psychologist Lloyd Morgan advises 'Saturate yourself through and through with your subject .... and wait'. Souriau's advice is a little more oblique but closer to the truth of these type of inspirations 'Pour inventer il faut penser a cote' [to invent you must think aside].

To incubate an idea, you need to let the idea 'settle' in your mind. It is believed that constantly thinking hard about a problem just forces you up against the same creative wall that probably produced the problem in the first place. By relaxing (in a bath or in bed) and letting your mind wander, more diverse thoughts come to mind and some of these diverse thoughts just might make an unusual or unorthodox connection that breaks through the creative wall. There is even a theory that, during sleep, the previous day's memories are sorted and filed away in the appropriate parts of your memory. Some sleep theorists suggest that at least part of dreaming is our semi-conscious awareness of this memory-sorting and filing process. Occasionally, great thinkers have reported waking suddenly from deep sleep with the most intractable of problems completely solved [7]. Here the vital connection of ideas is likely to have been made in the process of filing away memories during sleep. To sleep on a problem may have more validity than you might imagine. For most everyday design problems, there is rarely enough time to sit back and wait for inspiration to strike. The illumination part of problem solving must be actively sought and is often needed by yesterday. Techniques are, therefore needed to accelerate the process that might otherwise take days or weeks of lying in bed and taking baths.

Figure 4.3
A ski jump inspired the design of the take-off platform on aircraft carriers

Bisociation and lateral thinking

Arthur Koestler's concept of bisociation was introduced in the last chapter to explain how associating two absurd or ridiculous ideas gives rise to humour. Koestler goes on to describe how bisociation may be the key to creativity. A creative breakthrough is very often the association of two known ideas or principles which have not been connected previously. A widely quoted example is the invention of the take-off platform on aircraft carriers which was inspired by the shape of a ski-jump (Figure 4.3). Edward de Bono has popularised this concept as lateral thinking [8]. The problem as far as creativity is concerned is that human thought is too logical and too constrained to conventional frames of reference. When we think of ways of getting aircraft to take off our thinking is usually constrained to the world of aircraft; could we make them take off vertically (like the Harrier Jump Jet), can we make them go faster at take off (by using a catapult). Rarely, in the course of deliberate and concentrated effort on making aircraft take off would the idea of a ski jumper come to mind. Except, of course, if you are lying in bed or in the bath and not thinking about the problem at all.

We can think of our brains as devices for making associations. When we see an object we associate it with words (what is the object called), with memories (when did we see it last), with emotions (how did we feel when we saw it last) and we associate it with the object's basic significance to our lives (can we eat it, have sex with it or should we run away in terror from it). When we learn, we make new associations (the number 4 becomes associated with 2+2). And when we sit staring into space, daydreaming, we associate a stream of past images and thoughts in our memories. Our brains are, in most respects, the best association devices in existence but they are constrained in the way they work. All of our perceptions, thoughts, emotions and memories are filed away in little networks designed to provide the most useful associations for going about our daily lives. Knowledge about aircraft will reside in one network and knowledge about ski-jumping in another. For these networks to be closely connected would not be useful. I have enough difficulty getting myself organised to get on the correct aircraft from the correct terminal on the correct day without having my brain cluttered up with images of ski-jumpers! Our brains are wired up to stop such irrelevant and unnecessary association being made. Where, then, does this leave the poor aircraft carrier designer whose place in history is dependent upon making an association which his brain is designed to prevent? There are two possible solutions. The first is to take lots of baths and lie in bed for long periods each morning and hope that it happens spontaneously. More usefully, the mind can be prepared, the problem can be allowed to incubate and then various thinking methods and creative tools can be used to force the mind into bisociative or lateral thinking.

Preparation involves absorbing all the relevant facts and ideas to fuel the creative breakthrough. Incubation involves digesting, assimilating and sorting them in your mind. The barrier to progress must be clearly defined and understood. Logical solutions must be explored and, if inappropriate,

Key issues in the practicalities of creative thinking and the corresponding design toolkits

1. Preparation:
- Explore, expand and define the problem
- Exhaust the possibility of existing solutions

Design Toolkit

Parametric analysis
Problem abstraction

2. Idea-generation:
- Think only of ideas in principle; leave the practicalities to later
- Seek ideas from outside the problem's normal frame of reference
- Use techniques for:
  - Problem reduction
  - Problem expansion
  - Problem digression

Design Toolkit

Procedures:
Collective notebook
Group-stimulated brainwriting

Techniques:
Product function analysis
Product feature analysis
SCAMPER
Orthographic analysis
Analogies & Metaphors
Cliches and proverbs

3. Idea selection:
- Consider both the good and bad aspects of all ideas
- Hybridise ideas which have complimentary good and bad points

Design Toolkit

Evaluation matrix
Dot sticking

4. Reviewing the creative thinking process

Design Toolkit

Phases of Integrated Problem Solving

Figure 4.4
Key issues in practical creative thinking

rejected. All the information relevant to the problem must be sorted, understood and filed away for easy access. You must become immersed in the problem and totally familiar with it. At this point your mind is prepared for the moment of illumination. The time has come to start forcing your brain to think laterally about the problem. And there is a substantial armoury of weapons to help you fight this particular battle with your brain. We will return shortly to the systematic methods available for stimulating creative thinking. But first, let us look more systematically at the entire process of creativity, from first insight to illumination. How, practically, can we control, manage and enhance creative thinking?

Creative thinking in practice

The practicalities of creative thinking have been written about extensively. In one book [9] alone 105 different techniques are described! In fact, so much has been written on the subject that it is difficult to see the wood for the trees; there are creative thinking techniques, variations on those techniques and variations on the variations. From all that has been written, however, several key issues emerge and these are mapped out in (Figure 4.4) alongside the relevant design toolkits.

Preparation

Design problems are usually complex, in as much as they have several goals, many constraints and an even greater number of possible solutions. In designing a new product, you will be trying to satisfy the needs of a wide range of customers, exploit to the full the abilities of sales, marketing and distribution channels, fit in with existing manufacturing facilities and suppliers and end up making a profit for the company. Defining a design problem to take account of all of this takes a lot of preparation.

Exploring, expanding and defining a problem seeks answers to several questions. Exactly what problem is it that you are trying to solve? Why is it a problem? Is it a part of a bigger or more general problem? Would solving this more general problem also solve the immediate problem? Is, therefore, the immediate problem the best problem to be tackling? What is the ideal solution to the problem? What is it about this solution that makes it ideal? Is this solution ideal only in particular circumstances and if so, what are these circumstances? The answers to these questions should establish three key

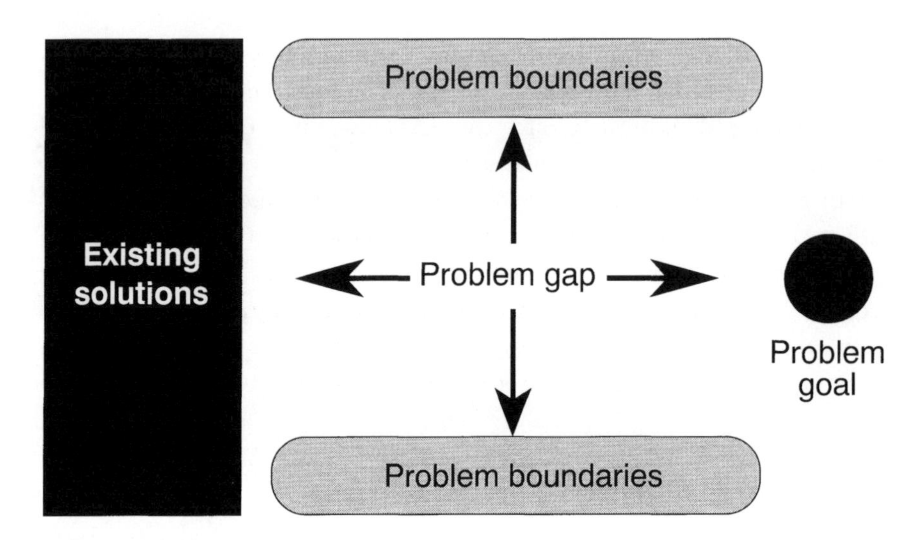

Figure 4.5
Visualising the problem gap [10]

characteristics of the problem: the problem goal, the problem boundaries and the problem gap (Figure 4.5). This exploration of the problem aims to reach a simple, concise and workable definition. It should specify the problem goal sufficiently so that you know when a solution has been found. It should also allow potential solutions to be compared and ranked according to how well they solve the problem. The problem definition must also define the problem boundaries. These define the limits of acceptability for potential problem solutions. The problem gap is the space between existing solutions and the problem goal, and is limited by the problem boundaries. It is this gap which needs to be bridged by creative thinking in order to solve the problem. Of course, both the problem goal and the problem boundaries will be different depending upon how widely or narrowly the problem is defined. The key to effective preparation for problem solving is to challenge the problem thoroughly and not simply take it as stated or grab at the first problem definition that comes to mind.

Preparation toolkit

Two different methods for exploring, expanding and defining problems are presented in the design toolkits .

- Parametric analysis explores the quantitative, qualitative and categorical measurements of the problem (toolkit pp 82-83)
- Problem abstraction tries to reduce the problem to progressively more

abstract concepts. Given a problem as initially stated, problem abstraction asks why you would want to solve that problem in the first place. This is a good way of trying to get to the root of a problem (toolkit pp 84-86).

By means of these techniques, problems can be thoroughly explored. Goals are most often derived from problem abstraction, although problem mapping can also be highly effective (see opposite page) . Boundaries can also be set from problem mapping but parametric analysis usually plays a key role.

Preparation for problem solving involves divergent thinking and ends up throwing a wide net to catch all the possible aspects of the problem and all possible angles to its solution. Next comes the convergent phase in which you reduce this myriad of possibilities to a single, manageable problem definition. This does not mean reducing the problem to a single product with known features or attributes. A problem definition can be quite broad but it needs to have a clear goal and clear boundaries.

The process of preparing for problem solving inevitably raises a great deal of complexity. Nearly all problems which require a concerted effort to solve are complex and revealing that complexity is the job of good preparation. It is important, however, for the next stages of problem solving to define the problem in a simple and concise way which is easy for the participants to grasp. With a complex problem, this will require distilling the problem down to its core elements and simplifying the complexity. The person leading the idea generating and idea selection sessions must be aware of this complexity and should try to lead the discussions in ways which takes account of that complexity. But no problem is so complex that it cannot be broken down to its roots (using problem abstraction) and expressed simply.

Idea generation

Idea generation is at the heart of creative thinking. The ideas produced are the lifeblood of the creative process. They are what puts the 'creative' into creative thinking. We have already seen that truly creative inspiration is often the result of bisociative thinking, the joining together of previously unrelated ideas. Many idea generating techniques try to force participants to make these unlikely connections. These methods are criticised by more rationally and literally minded people as being 'wacky' and 'grasping at straws'. This is true and often they do not work. But bisociative thinking is intrinsically difficult and we have yet to discover sure-fire ways of achieving it. Most idea generating techniques, to their credit, require little time and effort. A typical idea generation session might take an hour and involve less than ten people. Consequently, the possible benefit of coming up with an innovative solution is usually considered justification of the cost.

Several techniques are, however, more conservative. They try to find solutions by rearranging, refining or developing ideas already related to the problem. What is important about idea generation is to try all the different techniques and become familiar with their strengths and weaknesses. Having an armoury of problem solving techniques means that different techniques can be applied to different problems.

There are three main categories of techniques for the creative generation of ideas (Figure 4.6):

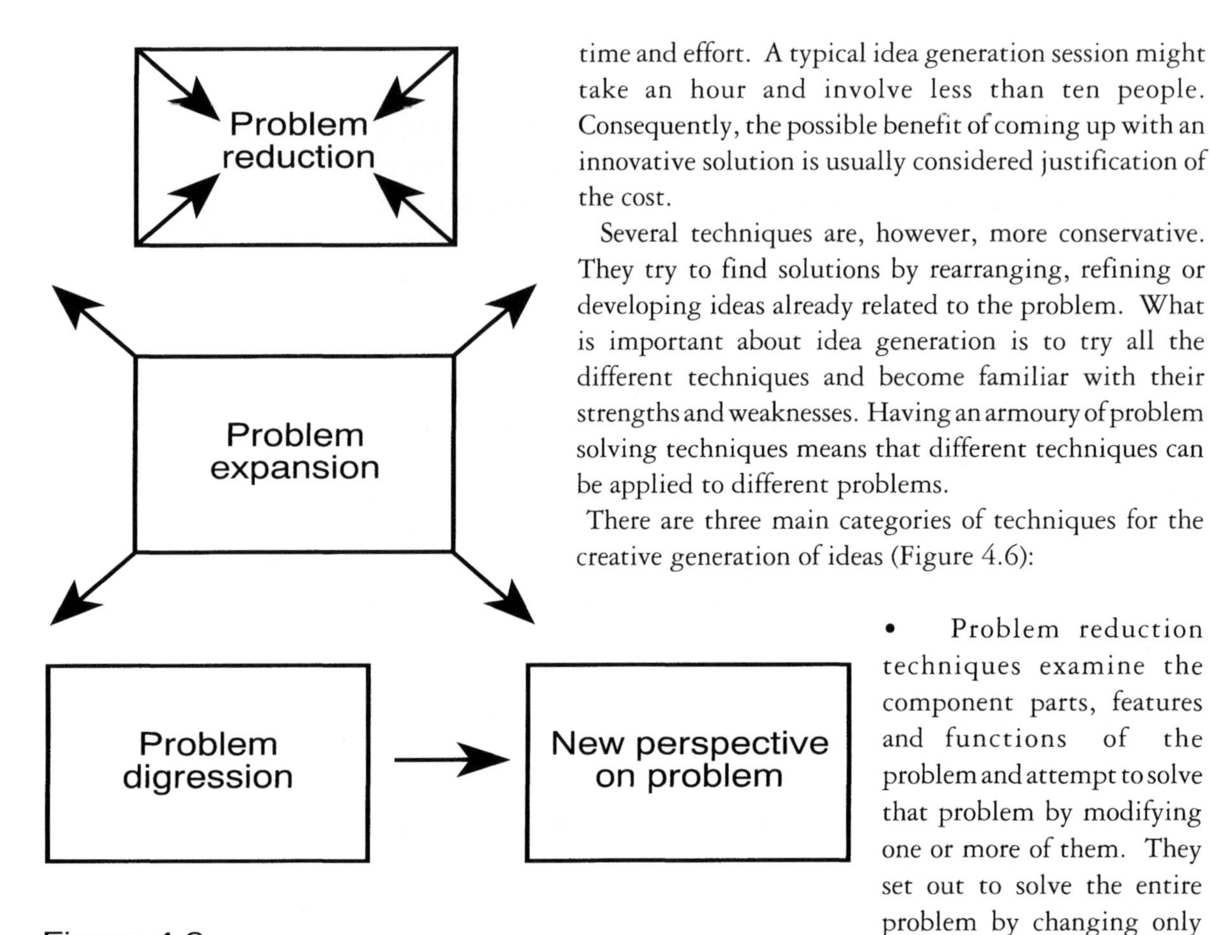

Figure 4.6
The main categories of creative idea generation techniques

- Problem reduction techniques examine the component parts, features and functions of the problem and attempt to solve that problem by modifying one or more of them. They set out to solve the entire problem by changing only part of it. The techniques are reductionist because they look only at the existing characteristics of a product and do not look beyond the product for its solution.
- Problem expansion techniques look beyond the immediate domain of the problem for inspiration. Problem expansion is primarily a way of broadening the perspective on a problem and thereby opening up a wider range of potential solutions.
- Problem digression moves away from the problem domain for inspiration and is totally concerned with lateral thinking. Some techniques begin with the original problem and use digressive methods to stimulate lateral thinking outwards from the problem. Others start with a totally unrelated stimulus and try to work from there back to the problem, thereby stimulating lateral thinking in the process.

Idea generating procedures

A key issue in idea generation is how to go about organising and managing the process. For many people, and most companies, idea generation means brainstorming. Ironically, brainstorming, in its classical or traditional form has been shown to be relatively inefficient [11]. A classical brainstorming session involves a group of people sitting round a table and coming up with new ideas for solving a stated problem. A key aspect of brainstorming is that the ideas from each person are meant to fuel ideas from the other people in the group. The problem with this is that all the ideas from the group tend to get channelled into a limited number of lines of thinking. One person will raise a particular idea and all the others round the table immediately develop and expand upon this idea, constraining the group's output to that one limited line of thinking. Occasionally, different themes will emerge during the discussion and these will be explored. But the clear conclusion from research is that the number and quality of new ideas produced from classical brainstorming is generally less than the members of the group would have produced, working individually. This does not mean that it is a bad idea to have groups of people involved in the generation of ideas. Everyone who has been in a group idea generation session knows how ideas are stimulated in your own mind by listening to other people's ideas. It just means that precautions must be taken to avoid the channelling of ideas into limited lines of thinking. The technique 'brainwriting' (see toolkit p 81) takes such precautions.

Another procedural issue is the extent to which judgement should play a part in the idea generation process. We have already seen, in Chapter 2, the logic of separating idea generation from idea selection. In product design, thoroughly evaluating an idea requires it to be developed through concept design, embodiment design and then prototyped and tested. To develop and evaluate every single idea in this way before generating the next idea, would waste an enormous amount of time. To this extent, therefore, ideas should not be judged and evaluated in-depth as they are generated. At the other extreme, however, there is an argument that absolutely no judgement at all should occur during idea generation. The logic of this is that even the most absurd and ridiculous idea could stimulate a more reasonable and practical solution later in the problem solving process. There is some degree of truth in this and, indeed as we will see below, some creative thinking techniques deliberately set out to generate 'way-out' ideas as a creative stimulus.

Research has shown that deferring judgement does actually increase the number of ideas generated in a creative brainstorming session. What is gained in quantity, however, is lost in quality [12]. Deferring judgement during idea generation results in ideas of lower quality than when judgement is encouraged. To some extent, the issue of whether or not to defer judgement during idea generation depends on what you are trying to achieve. If the problem is highly intractable and the goal is to generate highly original approaches to its solution then judgement should be deferred to a great extent, resulting in many ideas of variable quality. This variable quality will necessitate further work after the idea generation process is complete to develop, refine and select those which can be made workable. If, on the other hand, the problem is quite well defined and circumscribed and a more directly applicable solution is sought, then a greater degree of judgement is called for during idea generation.

Another aspect of the judgement issue is that ideas should only be outlined in principle during idea generation sessions. The practicalities should be left for later idea selection or as an integral part of the more general product development process.

As a general rule, the best approach is to have all the participants in the idea generation process well prepared in advance. This preparation will inevitably cause them to conduct a certain amount of judgement during idea generation. But, during the idea generation session itself, the participants should be encouraged to suspend judgement as far as possible. Their preparation and in-depth understanding of the problem will prevent them from doing so totally, thereby achieving a balance between being overly judgemental and not judgemental at all. Certainly all criticism of other people's ideas should be forbidden because this will tend to inhibit the less confident participants from putting forward ideas.

One final procedural issue is that most people imagine creative thinking taking place in a group session around a table. This, of course need not be the case. A technique which encourages participants to integrate creative idea generation into their normal work routine is the collective notebook. Here each participant is given a notebook in which to collect together ideas over a prolonged period of time. At the end of that period, the notebooks are exchanged to fuel further ideas or a group session is convened to compare notes. The technique is described in a toolkit on p 87.

Idea generation toolkit

Six different methods for generating ideas are presented in the design toolkits.

Synectics – a critique

Next to brainstorming, the problem solving technique that comes to mind is probably synectics [13]. If my business acquaintances are representative, few people understand it and even fewer are able to use it. This is a situation I see little reason to change and synectics would have happily been omitted from these pages had it not been such an integral part of the jargon of creative thinking. Let me explain my feelings.

My main complaint about synectics is that it is portrayed as complex, highly structured and inflexible. It is conducted in 8 stages, uses 4 types of analogy and is intended to engender 5 different psychological states! A great deal of obscure, but very technical-sounding claims have also been made for the technique. 'An operational theory for the conscious use of the pre-conscious psychological mechanisms present in man's creative ability', was how the inventor of synectics, William Gordon described it [13]. There is no doubt that, hidden amongst all of this, there are many general rules of creative problem solving, developed by William Gordon and his colleagues under the banner of synectics. These are extremely valuable but these are not specific to synectics. They should be applied to all problem-solving and have been described in the section on general rules on page 73 onwards.

At the heart of synectics is the use of personal analogy, direct analogy, symbolic analogy and fantasy analogy. Of these, direct analogy is the method nearly always used spontaneously by people [14]. According to William Gordon direct analogies are the also the most productive. Stripped of the complexity and all the psycho-babble, synectics is, therefore, little more than a structured way of using the simple analogy technique described in the design toolkit on page 91. Personal analogies require participants to imagine being the product and think what it would be like to do whatever it is that the product does. These can be fun to liven up an otherwise dry problem solving session and can occasionally be useful. But, again, their use is not limited to synectics. Personal analogies should be used at any time and as a part of any problem solving technique. Symbolic analogies are the most obscure of the synectic techniques. Gordon describes them as a poetic response, which although technologically inaccurate, are aesthetically pleasing. Fantasy can be a useful way of stimulating unusual ideas but it is an unnecessary constraint to limit its use to analogous thinking.

William Gordon and his colleagues were great pioneers in the development of creative problem solving techniques. Their legacy, however, is the individual parts of synectics which should be applied wherever they appear useful. Synectics as a whole only serves to wrap them up in complex, confusing and unnecessary baggage.

- Product function analysis starts with an existing product and systematically arranges its functions in a hierarchy. This forces you to work out what is the primary function of that product and how the subsidiary functions contribute to that main function. This is a technique for problem reduction. The technique plays a key role in concept design and hence the toolkit describing it is found on page 236 in the concept design chapter.
- Product feature permutation also starts with an existing product and explores all the different ways in which its components or elements can be arranged. This technique is also for problem reduction, and it plays a key role in embodiment design. It is described and exemplified in the embodiment design chapter (See p 275)
- Orthographic analysis presents two or three attributes of a problem in a graphical two or three dimensional array. This allows possible solutions to be explored by means of combination, permutation, interpolation or extrapolation. This technique can be used for both problem reduction and problem expansion and its toolkit is described on pages 88-89.
- SCAMPER is an acronym for 'substitute, combine, adapt, magnify or minify, eliminate or elaborate and rearrange or reverse'. It is intended to stimulate thinking about different ways an existing product could be changed. This technique is for problem reduction and its toolkit is on page 90.
- Analogies and metaphors of problems can be used to stimulate lateral thinking. Synectics is a specific technique, involving the use of analogies. These are problem digression techniques and the toolkit is on pages 91-92.
- Clichés and proverbs are used to give a new perspective upon a problem and, thereby, fuel lateral thinking. This is a problem digression technique and the toolkit is described on pages 94-95.

Idea selection

An essential feature of problem solving, and indeed the whole of product design, is to think of all possible solutions and pick the best It is the role of idea generation to come up with all the possible solutions and it is the role of idea selection to pick the best. To be able to do so requires a problem definition by which the best is judged. This is why the preparation stage of problem solving is so important. A common misunderstanding is that the creative part of problem solving ends with idea generation and that idea selection is a rather mindless routine procedure. In fact, a great deal of creativity is often required during idea screening. It is at this stage that ideas must be expanded, developed and even hybridised to get closer and closer to an ideal solution.

Idea selection toolkit

Two different methods for selecting ideas are presented in the design toolkits.

- Dot sticking is the simplest form of idea selection. Several judges are given a number of sticky dots which they use to vote for potential solutions by sticking them to the sheets of paper describing the solutions. The toolkit is on page 93.
- An evaluation matrix arranges potential solutions along one side of a matrix and selection criteria along the other. Each solution is then scored against each of the selection criteria or, alternatively they are judged better or worse than a reference solution. This is a key technique in product planning, concept design and embodiment design . The technique is introduced on p 169 and further examples of its use appear on p 229.

Reviewing and improving creative thinking procedures

It is clearly important to continuously improve and refine the methods used for creative thinking. A structured way of doing so, called Phases of Integrated Problem Solving (PIPS), was devised by Morris and Sashkin [15]. This offers a checklist for the evaluation of different stages of the problem solving process and allows you to rate several aspects of each stage to identify areas needing improvement. This technique is described in the toolkit on pages 96-99.

Key Concepts

1. Step by step creativity

Creative thinking is most effective when i) a period of preparation absorbs and digests the available information then ii) ideas are generated with as much imagination and creativity as possible and finally iii) the best ideas are selected, against criteria originally laid down for the creative thinking process.

2. Preparation

Key aspects of preparation are to familiarise yourself with all the available information relevant to the problem. This, in itself can be a highly creative process. It requires the problem to be manipulated and abstracted in a wide variety of ways in order to maximise the 'food for thought' in creative idea generation. As a result, many of the techniques used for idea generation can also be used to explore the problem initially.

3. Idea generation

It is an unfortunate but unavoidable fact that ours brains do not work in anything like an ideal way for creative idea generation. To be creative we need to escape from the sensible and rational associations that we commonly make in our everyday lives. We need to come up with lateral thoughts (bisociative thinking) in order to break new creative boundaries. Many techniques for doing so are given in the design toolkits

4. Idea selection

Idea selection is a more rigorous, systematic and disciplined procedure than idea generation. Idea selection aims to identify, from the wide and (hopefully) imaginative range of ideas created, those which best solve the problem originally tackled. A great deal of creativity can be required in idea selection to mix and match useful elements of different ideas.

5. Practice makes perfect

Creative thinking should be an on-going activity for most designers. To make sure that your creative abilities continually improve, you need to regularly evaluate your creative abilities. A method for doing so is given in the design toolkits.

Brainwriting

Brainwriting is a procedure which has all the advantages of brainstorming without the disadvantages. It is similar to brainstorming in so far as a small group of participants gather together but instead of talking about their ideas, they begin by writing them down. Every individual writes down their ideas and, initially at least, keeps them to themselves to avoid everyone's ideas getting channelled into one train of thought. This continues for a period of time until participants begin to run out of ideas. Then, when anyone wants the stimulus of fresh ideas they look at what another participant has been writing.

Brainwriting procedures vary considerably in the way the ideas are recorded and communicated to others within the group. Some suggest using small index cards, slips of paper or post-it notes each with only one idea written on it and which each individual keeps hidden from everyone else until the time comes to share ideas. Others suggest using large sheets of paper which can be pinned up on the wall for everyone to look at after a period of time. A key variable is the amount of time that individuals are left to think of ideas on their own before these ideas are shared between the group. Clearly, this should be long enough to allow most individuals to have nearly exhausted their own ideas. This, in turn will depend upon the complexity of the problem being tackled. Some authorities advocate as little as 5 or 6 minutes. This seems too short a period for anything but the simplest of problems. Ten to fifteen minutes is probably better as a general rule. An alternative method is to let individuals decide for themselves. If each idea is written on a single slip of paper and placed face down on the table, they can be looked at by another participant whenever they feel in need of inspiration. The slips are then returned to wherever they came from for others to use.

A more structured version of brainwriting is for each person to write a limited number of ideas (usually no more than 10) on a single sheet of paper, either in columns or rows. Each sheet is then handed to someone else in the group and they have to try to improve or develop all of the ideas a step further by adding a new row or column. This can be repeated several times until the ideas have been exhausted or until each sheet has been round every group member.

Once the ideas have been generated on paper, a final session of conventional brainstorming is used to bring out any completely new ideas not on any of the sheets but stimulated during the brainwriting process.

Design Toolkit

Parametric analysis

A parameter is literally something that can be measured and usually refers to dimensional measurements (e.g. metres, kilogrammes, Newtons). The parametric analysis of a problem or product, however, usually covers quantitative, qualitative and categorical features.

- Quantitative. What is the size, weight, power, speed, strength or price of the product? What are the quantifiable measurements of its efficiency? What is its durability?
- Qualitative. Qualitative parameters are those which can be ranked or scaled relative to other products but which do not have conventional dimensions. Do these scissors cut better or worse than other scissors? Which scissors are the most comfortable to use? Is the calculator easy to learn to use? Does the car door close smoothly and solidly or does it need to be slammed shut?
- Categorical. Categorical parameters state which of several alternative categories the product belongs to. Does the potato peeler have a fixed or swivel blade? Does the portable computer have a built-in or separate mouse? Is the lawn mower electrical or petrol driven? Categorical parameters can also refer to the presence or absence of something. Does the product come supplied with a built-in circuit breaker? Does the music player have a remote control?

To conduct a parametric analysis, it is usually easiest to pick an existing product which comes closest to solving the problem. A parametric analysis is then completed on this product, with particular attention being paid to the parameters in which the product fails to provide a complete solution. It is then indicated how these parameters would have to be different to fully solve the problem.

Parametric analysis example A cosmetics manufacturer wishes to make their packaging more environmentally-friendly. A competitor has, what they describe as environmentally-friendly packaging, but it does not go as far as the manufacturer would like. The problem they have set themselves is to develop a bottle for shampoos, liquid soaps, lotions and oils which is fully recyclable. The parametric analysis shown opposite illustrates how the company analyses 6 different parameters, quantifies the competing product on these parameters and then establishes targets for its own new product in comparison.

Parametric analysis (Cont'd)

Parameter	Dimension	Competing product variable	Comments	Ideal product variable
Bottle	Material	HDPE*	Good for re-cycling	HDPE*
Bottle	% from recycled material	0%	Must use more	Minimum target of 40%
Bottle	Mass (g)	105	Could use less material	Less than 100
Cap	Material	Polypropelene	Different material**	HDPE
Label	Material	Paper	Different material**	Direct printing on to bottle?
Label ink	No. of dyes	4	Synthetic dyes	2

* High density polyethylene
** Needs to be separated during recycling process

Design Toolkit

Problem abstraction

Problem abstraction sets out to reduce problems to their most basic elements and then set problem goals and boundaries. The procedure for problem abstraction begins with a simple statement of the problem. The problem statement is then challenged by asking 'why' you want to solve this particular problem. The answer is then challenged with further 'why' questions until the company's ultimate objective is reached. For most companies this is to make a profit.

The value of this process for exploring and expanding a problem is that each level of abstraction reveals a new set of potential solutions. Let us imagine that, in implementing its innovation strategy, a company is considering improving an existing product. It, however, wants to make sure that this is the best way to proceed. The reason for improving its product, revealed at the first level of problem abstraction (Figure 4.7), is to catch up with a competitor who has recently introduced a better product. An alternative way to catch up with this competitor is to develop a totally new product, rather than to improve an existing product. Realising this challenges the company to question what problem they really want to solve. Subsequent abstraction of the problem reveals that the company wants to avoid losing market share, prevent any drop in annual turnover, maintain its existing distribution of overhead costs and thereby sustain its current level of profit margin. This reveals a range of options for the company, from expanding into a different market, increasing advertising, reducing overheads or freezing

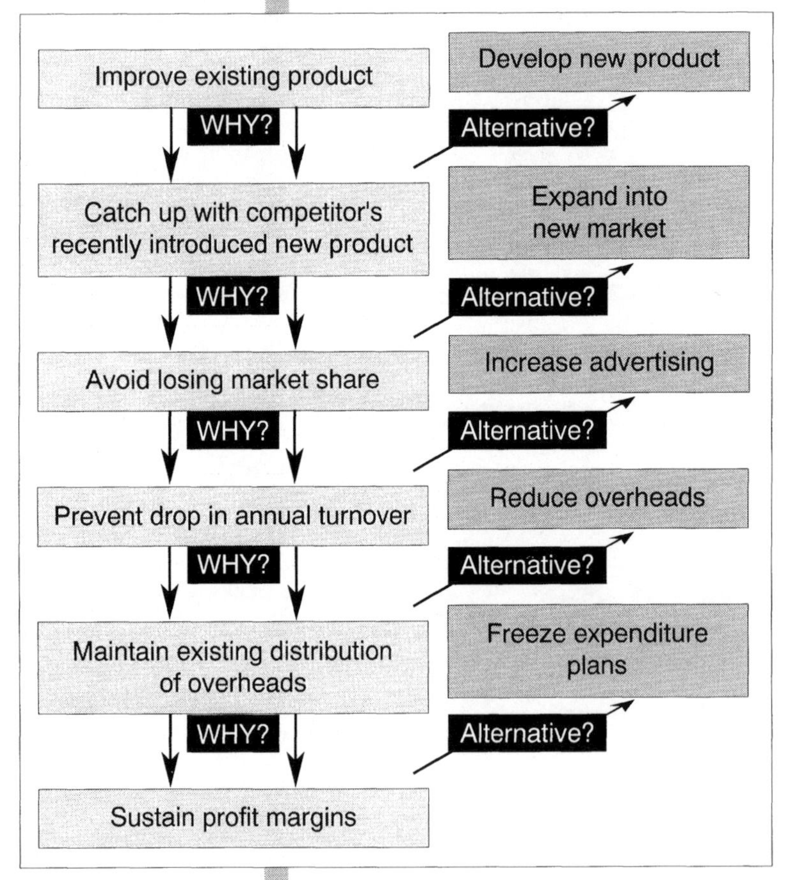

Figure 4.7 Problem abstraction reveals a range of alternative options to the original problem

Design Toolkit

Problem abstraction (Cont'd)

capital expenditure plans to preserve profit levels. At the end of all of this, the company may still decide to develop an improved version of the existing product. However, after having abstracted the problem they will, at least, have explored all the alternatives and decided that this is the best option rather than just the first one they thought of.

Problem abstraction example The aim in problem abstraction is to establish the problem goal and the problem boundaries. To illustrate how this is done, let us consider the example of a garden equipment manufacturer who needs to decide what type of new product to develop next. The company's marketing department has suggested that a lawn mower is an important omission in their current product range. Before commencing development, however, the company wants to make sure that this product is most suited to customer needs. So, why do customers buy a lawn mower? To cut grass. Why do they want to cut grass? To maintain a neat lawn. Why do they want a neat lawn? To maintain a neat garden. At each of these levels of problem abstraction a different set of problem goals and problem boundaries can be established (Figure 4.8). The goal for cutting grass is to design a lawn mower. This could be a mower of conventional design (either cylinder or hover) or it could be of an innovative design; perhaps a scissor mechanism, a rotating drum of abrasive (sandpaper?), a hot surface to singe the top of grass or a cutting device based on a laser or high-pressure-water cutting system. The problem boundaries are that the product must be able to compete with well-established market leaders. To do so, the new design must offer significant customer benefits over existing products. The next level of problem abstraction is to maintain a neat lawn. Not all lawns are made of grass. In Victorian times, before lawn mowers existed,

Figure 4.8 Problem goals and boundaries at different levels of problem abstraction

	Level of problem abstraction		
	Cut grass	Maintain neat lawn	Maintain neat garden
Problem goal	Design innovative lawn mower	Design innovative product for maintaining moss lawns	Design innovative product for maintaining paved gardens
Problem boundaries	Must offer advantages over established market leaders	Is there a market for such a product?	Benefit for cost

Design Toolkit

Problem abstraction (Cont'd)

gardeners made lawns from mosses, a practice which is becoming popular again in modern gardens. Main-taining this type of lawn would require a very different product which would feed the lawn, weed it and irrigate the soil underneath. The problem boundary here is that there may not be a market of sufficient size for the product. This would need to be explored through market research before proceeding any further with the idea. The next level of problem abstraction is maintaining a neat garden. Many modern gardens, particularly in towns are paved. How about a paving maintenance product which weeds and cleans the paved surface? The boundary to this problem is whether a product can be designed which offers sufficient benefit to the customer to persuade them to pay the price to buy it. Some initial design work followed by market research would be needed to resolve that issue. For the company wanting to develop a new gardening product, problem abstraction challenges assumptions and preconceptions. It forces them to decide what problem they really want to solve by making them think of all the alternatives and pick the best.

Collective note book

The collective note book is a very general procedure, originally developed by the Proctor and Gamble Company. A group of participants are selected and each is given a common statement of a problem written at the front of an otherwise blank note book. They are then given a period of time (usually one month) in which to record their ideas for solving the problem in their individual note books. Usually, they are told that at least one new idea per day is expected from them. At the end of the agreed time period, the note books are collected and summarised into a single document (the collective note book which gives the technique its name). The results are then either circulated back to the participants or a group brainstorming session is convened to discuss the proposed solutions. This brainstorming session can often be an excellent way of stimulating further new ideas or hybrids of two or more individual ideas.

The collective note book has the great advantage that it can involve virtually any number of people. It can also involve people based at different sites. It can be a useful way of tackling a wide variety of problems

A variation on the collective note book procedure is to have participants exchange note books after two or three weeks of recording their own ideas. The ideas from another participant is an excellent way of fuelling further idea generation.

Another variation is to provide the participants with idea generating techniques, written or printed at the front of the note book.

These could include:

- Product function analysis
- Product feature analysis
- SCAMPER
- Orthographic analysis
- Force fit tables
- Product hybridisation
- Analogies and Metaphors
- Problem exaggeration
- Cliches and proverbs

Design Toolkit

Orthographic analysis

Orthographic analysis is a way of arranging one, two or three attributes of a problem in a graphical representation of, correspondingly one, two or three dimensions. The simple presentation of problem attributes in this way allows solutions to be visualised by different combinations of attributes, or by expanding the problem in any of the given dimensions. A general purpose version of orthographic analysis takes a product and represents its material, its manufacturing process and its market along the three orthographic axes. A schematic diagram and an example for a potato chip is shown in Figure 4.9.

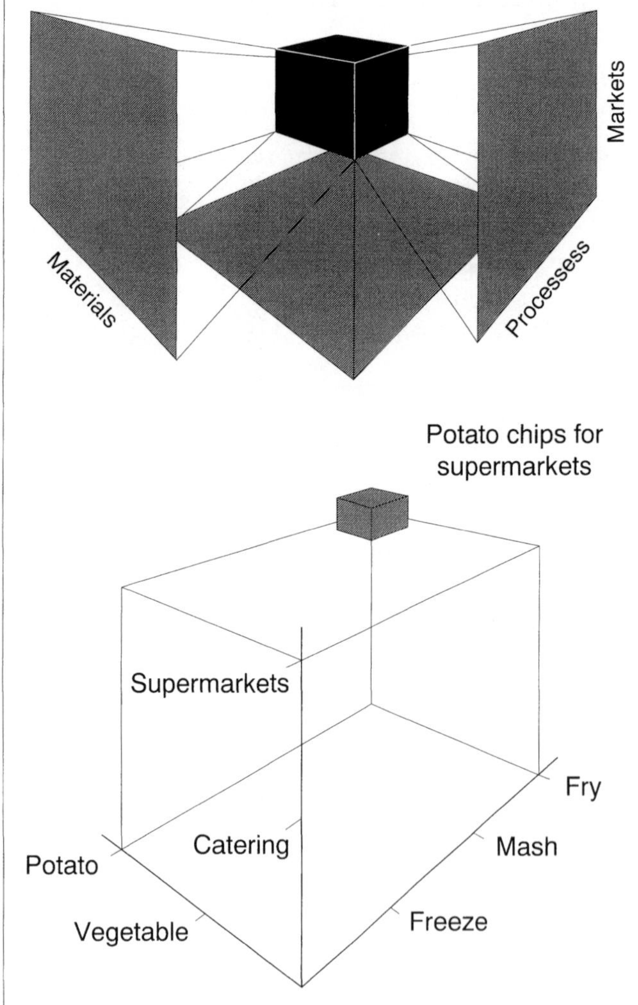

Figure 4.9 Orthographic analysis for problem expansion

The starting point for orthographic analysis is to pick as many as three attributes of a problem and use these attributes to label the axes of a one, two or three dimensional array. Identify where existing solutions fit within the array and then explore potential new solutions by means of:

- Combination – are there any new solutions which combine one or more existing solutions.
- Permutation – can the attributes of existing solutions be inter-changed to give rise to new solutions
- Interpolation – is it possible to have a new solution which is intermediate between two existing solutions, or
- Extrapolation – are there any new solutions which lie outside the range of existing solutions.

Orthographic analysis (Cont'd)

Orthographic analysis example

A major manufacturer in the food industry seeks to develop an innovative new product. Currently, they produce frozen vegetables and frozen mashed potatoes which are sold to both supermarkets chains and the catering industry. In another plant they produce potato crisps for supermarket chains. Looking at this in terms of an orthographic analysis we could present raw materials, manufacturing processes and market outlets as the three orthogonal axes. The materials are potatoes and vegetables, the processes are boiling, mashing, freezing and frying and the markets are supermarkets and the catering trade.

By combination, permutation, interpolation and extrapolation we get the following possible new product ideas:

- Combination of processes – combine 'fry' with 'boil and mash' to produce frozen potato croquettes.
- Combination of materials – combine 'potato' and 'vegetable' to produce a vegetarian patty.
- Permutation of products and markets – sell crisps to the catering trade.
- Permutation of materials and processes – fry vegetables to produce vegetable-based crisps.
- Interpolation/extrapolation of materials – produce pasta-based crisps or frozen pasta.
- Interpolation/extrapolation of processes – produce part cooked, frozen potato crisps, for final cooking at home.
- Interpolation/extrapolation of processes and markets – use only organic produce and sell into health-food outlets.

Design Toolkit

SCAMPER

SCAMPER is an acronym and stands for 'substitute, combine, adapt, magnify or minify, put to other uses, eliminate or elaborate and rearrange or reverse'. Together, these headings make up a checklist of possible product modifications for stimulating ideas. When thinking about possible product modifications it is all too easy to concentrate only on the obvious ones and overlook others. So, for example, when trying to make a product less expensive to manufacture it is likely that you would think of making it smaller, eliminating some of its features or substituting expensive materials with cheaper ones. It may, however, be easy to overlook the possibility of rearranging the components to reduce assembly times or even making it bigger to relax manufacturing tolerances. The use of a checklist forces you to go through all the possible modifications to the product which might solve the problem.

Be warned, using product modification checklists can be a mind-numbing experience, especially with complex or multiple component products! But if it helps solve the problem, the gain is worth the pain. It might also save a lot of frustration and wasted time in the long run.

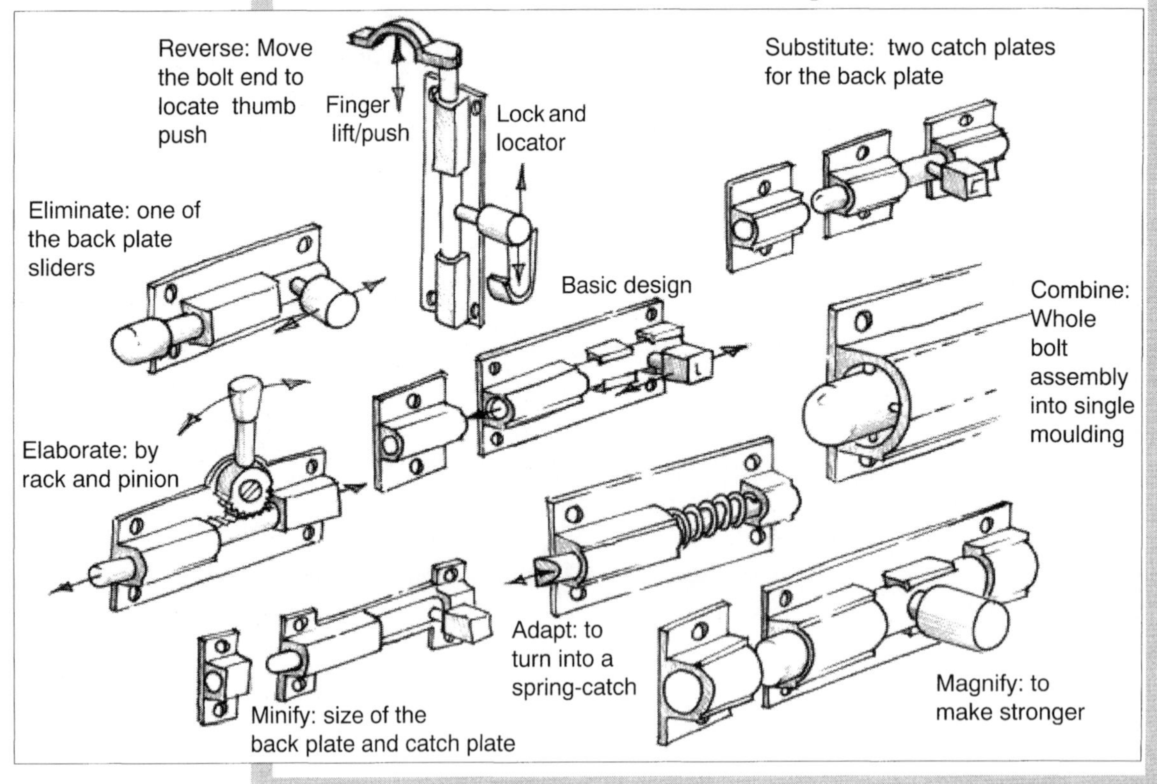

Analogies

Analogy *n.* an agreement or correspondence in certain respects between things otherwise different: agreement in function *(biol)* Chambers 20th Century Dictionary.

An analogy is a particular form of thinking or reasoning in which the properties of one object are thought of in terms of a second object which is different but has certain properties in common. So, a length of rope could be thought of as a rattlesnake when it is coiled up on the floor or as an escape chute when suspended from an upstairs window or a gripping hand when tied tightly round a bundle of sticks. There are several ways in which analogies can be used in creative thinking. They can be used simply to change your perspective on a problem and to 'free up' your thinking. More specifically, they can be used to explore new functions, new features and new applications for a product. If a piece of rope is like an escape chute when it is suspended from an upstairs window, how could we improve the piece of rope to make it a better escape chute. Escape chutes are generally smooth and certainly do not burn your hands when you slide down them. They often have a steep slope at the top but then run out gently at the bottom to slow you down before you reach the ground. Is there any way a rope could be designed with these properties to make it function better for escape. Finally, they can be used to create completely new solutions to problems by discovering how the problem is solved or partially solved in a totally different context. The ski jump inspiring the take-off platform on an aircraft carrier is an example.

When using analogies try to stick to the following rules:

- Think of the essence of the problem in abstract terms. A can-opener 'removes' part of a can, a cup is a 'container', a belt 'grips' or 'squeezes' the object it is tied around. Use this abstract description to stimulate analogies
- Find analogies which have an element of action or movement associated with them. Biological analogies such as the rattlesnake or gripping hand above are usually good in this respect.
- Do not force the analogy to fit the problem too quickly. Take your time and think in simple steps. Generate a list of analogies without thinking about the problem. Then write down a list of associations for each analogy. Again withhold judgement about their relevance to the problem. Now try to force a fit between each association and the problem to be solved.

Design Toolkit

Analogies (Cont'd)

Analogy example

A manufacturer of home security equipment is looking for innovative products to protect homes from burglary. The essence of this problem is seen as 'prevention'.

'Prevention' analogies	Associations	Stimulated ideas
Contraception	Pill	Locking device secured in 'mouth' of VCR or CD player.
	Vasectomy Withdrawal method	'Cut' power cables to equipment; make them inaccessible, withdraw them into locked box fixed to back of equipment.
Seat belt	Head hits windshield Inertia reel	Prevent windows being broken for entry; window shutters; perhaps automatically closed by motorised reel when front door is locked.
Peace negotiator	Face to face	Hidden flash camera at most likely point of entry to photograph burglar's face.
	Ultimatum	Large notices promising horrible consequences for uninvited visitors.
Lifejacket	Flare	Device which switches on all lights and turns stereo on to maximum volume when triggered.

Dot sticking

A simple but effective way to select ideas is to give all participants a number of votes, in the form of small sticky dots and get them to vote for the ideas presented by sticking dots beside them. Any number of dots can be given to each participant so they can give some weighting to their votes. As a rule, no more that 5 ideas should be voted for by any participant and each idea can be given up to five votes. An advantage to voting with sticky dots is that each individual can be given a different colour. This allows you to see whether any idea has been given most of its votes by one person (perhaps their own idea!). Alternatively, different colours can be given to people with different functions within the company. Thus it may emerge that some ideas are preferred by marketing staff, whereas others are preferred by the production engineers.

Another variation on the dot sticking technique is to run through the procedure twice. The first 'pass' is a coarse filter to cut the number of ideas down to a manageable number; say between 5 and 10. Then the second round of voting establishes the most popular 2 or 3 ideas, in order of preference.

A key part of the dot sticking technique is the discussion that accompanies the voting procedure. This should establish why some ideas are preferred over others and why. Short brainstorming sessions can then try to improve the less favoured ideas to make them match the best. In addition, the good ideas should be developed to try to incorporate any good features in the poorer ideas. Always remember that throwing away an idea with any advantage over the other ideas can be an expensive mistake. It is also a mistake that may come back to haunt you later in the product development procedure, when you realise that the discarded advantage is the vital missing feature in the product you have been developing!

Design Toolkit

Clichés and proverbs

An interesting way to jolt conventional thinking out of a rut is to use cliches and proverbs. The idea here is that most of these common sayings have sufficient generality to be relevant in just about any situation. Armed with a list such as the one provided below a problem can be explored by taking one interesting or relevant saying and examining how it might apply to the problem.

Most familiar

- Practice makes perfect
- Better late than never
- If at first you don't succeed, try, try, try again
- Like father, like son
- A place for everything and everything in its place
- Two wrongs do not make a right
- Two's company, three's a crowd
- Where there's a will there's a way
- Don't count your chickens before they are hatched
- Easier said than done
- All's well that ends well
- Practice what you preach
- You can't tell a book by its cover
- An apple a day keeps the doctor away
- A penny saved is a penny earned
- Cleanliness is next to godliness
- Mind your own business
- Beggars can't be choosers
- Easy come easy go
- Beauty is only skin deep
- Beauty is in the eye of the beholder
- You can't teach an old dog new tricks
- Better safe than sorry
- Two heads are better than one
- Actions speak louder than words

Most 'visual'

- When the cat's away the mice will play
- The early bird catches the worm
- Like father like son
- Kill two birds with one stone
- Don't count your chickens before they are hatched
- If the shoe fits wear it
- Monkey see, monkey do
- A man's home is his castle
- The bigger they are the harder they fall
- Birds of a feather flock together
- Two's company, three's a crowd
- You can lead a horse to water but you can't make it drink
- Don't cry over spilt milk
- Two heads are better than one
- We're all in the same boat
- Never bite off more than you can chew
- One bad apple spoils the barrel
- Put on your thinking cap
- You can't teach an old dog new tricks
- You can't tell a book by its cover
- When it rains it pours
- Don't rock the boat
- Too many cooks spoil the broth
- Look before you leap
- A penny saved is a penny earned

The above selection was found to be the top 25 sayings out of a list of over 200 presented to US undergraduate students. The column on the left gives the sayings that were judged most familiar. Those on the right were judged to evoke most visual images [16].

Clichés and proverbs (Cont'd)

Clichés and proverbs example

A pet food manufacturer wishes to develop a high-value, high-price cat food. Since the market is highly competitive, the company wishes to explore some radically different solutions and has decided to try to do so using the cliches and proverbs technique. Scanning through the list, several interesting possibilities emerge.

Like father like son. Let us imagine that the cat's owner is the father and the cat is the son (a relationship not too far from the truth in some cases). We could design a cat food that the owner would like to eat. Perhaps a ready-meal, served hot out of a microwave. Perhaps different types of food in the one packet (a cat TV dinner). Alternatively, we could interpret the expression more literally. Cats should be given natural food that their forefathers would have eaten. How about robin, rabbit and rat flavoured food?

Kill two birds with one stone. Maybe luxury cat food should provide food and a drink in the same packet; a bowl of food alongside a bowl of milk, all in the same carton with a foil-sealed lid.

You can't tell a book by its cover. Do customers believe that fancy packaging does not always contain luxury contents? Maybe it would be more honest and more appealing to customers if a clear see-through packaging was used so that the high quality of the contents could be seen clearly as it sat on the shelf in the shop.

If the shoe fits, wear it. Would customers pay extra for a diet specially formulated for their type of cat. This could be achieved by having an exceptionally wide range of luxury food, some for large and some for small cats, some for active and some for quiet cats, some for kittens, some for adults and some for elderly cats.

An apple a day keeps the doctor away. So what keeps the vet away and can we include it in the cat food in order to promote it as a super-healthy diet?

Evaluation - PIPS

Problem solving should never be a one-off activity within any organisation but rather, should be on-going, tackling new problems as they arise. To be most effective, they should get better with practice. Problem solving techniques need to be developed and customised to suit both the organisation and the people involved and to do so, some form of evaluation and feedback from each problem solving activity is needed. Often, this is judged on the basis of the success or failure of the result. This, however, is at best, a crude method of evaluation and at times can be highly misleading. Success or failure may be due to extraneous factors; an excellent idea, well translated into a manufactured product might fail because of a competitors even better idea or even because of an unforeseen change in regulations or statutory requirements. This may be no reflection on the problem-solving abilities of the organisation.

Some way is, therefore, needed of evaluating how well or badly the problem-solving activity was conducted independent of its commercial outcome. A technique for doing this is called PIPS (phases of integrated problem solving). As the name suggests, PIPS divides the problem solving process into stages and looks at each individually. It also divides the problem solving activities into the tasks and processes involves. By scoring, on a 1 to 5 scale, a series of tasks and processes within each phase of problem solving the relative success of the entire procedure can be evaluated. Several participants in the problem solving activity should be asked to complete the PIPS evaluation to see whether different people have different views on how successful it was. The structure of the PIPS evaluation is given in the next 3 pages.

Phases of integrated problem solving

To what extent has each step been accomplished?

Task-related issues in problem solving

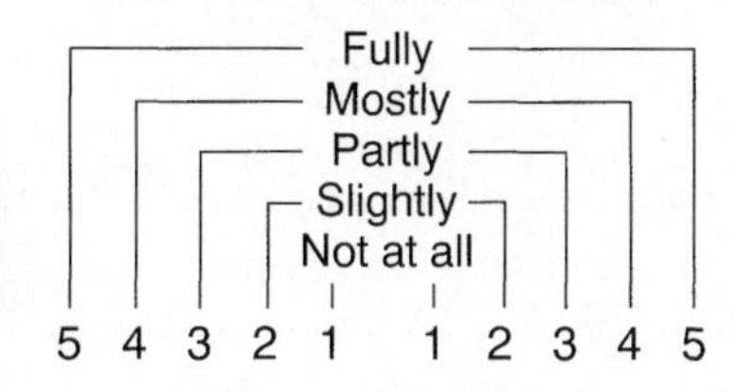

People-related issues in problem solving

Phase 1 Problem definition: Exploring, clarifying, specifying

Task-related issues		People-related issues
To what extent were information resources been sought out? Was everyone who might have relevant information present or represented?	5 4 3 2 1 1 2 3 4 5	Were those people with specialist knowledge of the problem encouraged to provide information?
Was all of the available information raised and discussed	5 4 3 2 1 1 2 3 4 5	Was there an 'open and sharing' atmosphere? Did all group members feel free to speak?
Was there any attempt to integrate information in order to clarify the problem definition?	5 4 3 2 1 1 2 3 4 5	Did group members keep the discussion problem-centred and avoid considering solutions?
Was the problem finally stated in a way that everyone understood and agreed to the problem definition?	5 4 3 2 1 1 2 3 4 5	Did everyone have ample opportunity to agree/disagree with the problem definition?

Phase 2 Idea generation: Creating, elaborating,

Task-related issues		People-related issues
Were the rules of the idea generation session reviewed and agreed at the start?	5 4 3 2 1 1 2 3 4 5	Were all ideas recognised and welcomed, regardless of their content?
Were everyone's idea generating capabilities exhausted at the end?	5 4 3 2 1 1 2 3 4 5	Were less forthcoming group members encouraged?
Once all ideas were generated, were they reviewed by the group for clarification, elaboration or addition?	5 4 3 2 1 1 2 3 4 5	Was criticism tactfully discouraged and evaluation effectively postponed?
Were the ideas clustered into sets with common features or attributes?	5 4 3 2 1 1 2 3 4 5	Was any one member prevented from dominating the discussion or imposing their ideas on the group?
Was a summary list of the most innovative, feasible or interesting ideas produced	5 4 3 2 1 1 2 3 4 5	Were the ideas finally presented or posted for all to see?

Phases of integrated problem solving

To what extent has each step been accomplished?

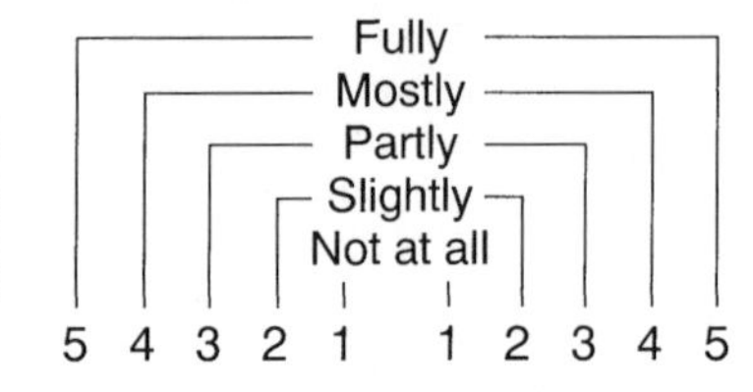

Task-related issues in problem solving		People-related issues in problem solving
Phase 3 Solution-finding: Evaluating, combining, selecting		
Was each idea discussed and weighed up against supporting and refuting evidence?	5 4 3 2 1 1 2 3 4 5	Did the group deal well with people whose ideas were being criticised or rejected?
Was the group systematic in its use of selection criteria?	5 4 3 2 1 1 2 3 4 5	Did the group concentrate on selecting the best idea rather than rejecting the poor ideas?
How well did the group modify and combine the initial ideas?	5 4 3 2 1 1 2 3 4 5	Were differences of opinion negotiated to a point of mutual satisfaction?
Was one idea (or set of ideas) finally selected for more thorough exploration and evaluation?	5 4 3 2 1 1 2 3 4 5	Was the solution chosen by consensus and if not, was the extent of the agreement within the group established?
Phase 4 Solution development: Planning, assigning, coordinating		
Did the group list the action needed and the people responsible for developing the solution?	5 4 3 2 1 1 2 3 4 5	Did all group members contribute to make sure no critical steps were left out?
Are all the necessary resources available?	5 4 3 2 1 1 2 3 4 5	Were group members willing to assume the necessary responsibility for solution development?
Were potential problems foreseen and planned for?	5 4 3 2 1 1 2 3 4 5	Were group members open and forthcoming about the internal obstacles and difficulties likely to be encountered?
Have the group's activities been effectively conveyed to those developing the solution?	5 4 3 2 1 1 2 3 4 5	Did all group members agree to the dissemination of the ideas generated?

Phases of integrated problem solving

To what extent has each step been accomplished?

Task-related issues in problem solving

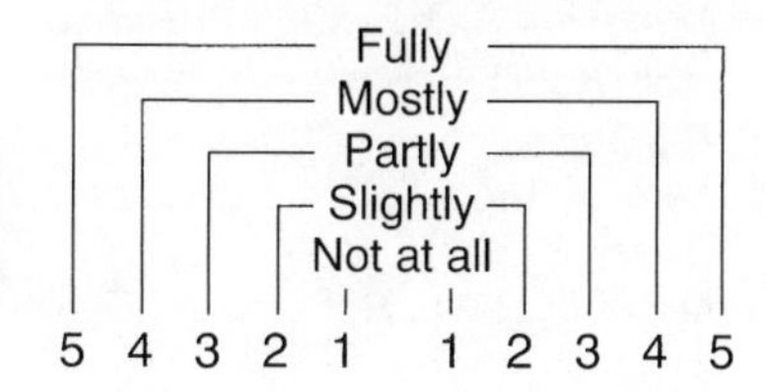

People-related issues in problem solving

Phase 5 Solution evaluation

Task-related issues		People-related issues
To what extent did the solution correspond to the original problem definition?	5 4 3 2 1 1 2 3 4 5	To what extent was every group member's knowledge and abilities exploited?
To what extent has the problem solving activity been seen within the organisation as creative, workable and useful?	5 4 3 2 1 1 2 3 4 5	To what extent did the group members make and keep to group commitments?
To what extent was the action to develop the solution well planned, managed and executed?	5 4 3 2 1 1 2 3 4 5	To what extent have communications been open, expressive and constructive?
Has the group met again to review the way the problem was tackled?	5 4 3 2 1 1 2 3 4 5	To what extent did the leader play a steering, peace-keeping and visionary role?

Notes on Chapter 4

1. Originally proposed by Wallas, G. 1926. *The Art of Thought*. Harcourt, New York. More recently described in Lawson 1992 *Creative Thinking for Designers* Architectural Press, London.
2. From Koestler, A. 1964. *The Act of Creation*. Hutchison & Co., London. pp 105-106.
3. Kenjo T. 1991. *Electric Motors and their Controls*. Oxford University press, Oxford.
6. Ashall F. 1994. *Remarkable Discoveries.* Cambridge University Press, Cambridge pp 1-15.
7. Ashall F. 1994. See 6 above pp 93-103.
8. Nayak P.R., Ketteringham J.M. and Little A.D. 1986 *Breakthroughs!* Mercury Business Books, Didcot, Oxfordshire, UK pp 29-46.
9. Van Gundy A. 1988 *Techniques of structured problem solving.* 2nd edition. Van Nostrand Reinhold, New York. This provides the most comprehensive treatment of creative thinking techniques. Other worthwhile books on the subject include Von Oech R. 1990, *A Whack on the Side of the Head: How you can be More Creative.* Thorsons Publishing Group, Wellingborough, UK; Davis G.A. and Scott J.A. 1971, *Training Creative Thinking.* Holt, Rinehart and Winston Inc, New York; De Bono E. 1991, *Serious Creativity.* Harper Collins Publishers, London.
10. Van Gundy 1988 See 9 above.
11. Van Gundy 1988 See 9 above.
12. Van Gundy 1988 See 9 above.
13. Gordon, W.J.J. 1961. *Synectics.* Harper & Row, New York. An objective review of the value of synectics was conducted by Thamia, S. and Woods, M.F. 1984. A systematic small group approach to creativity and innovation. *R & D Management* 14:25-35. This study asked long-term users of different problem-solving techniques to score the techniques for usefulness. Giving a 1 for excellent and 6 for failure, synectics was given an average score of 3 (good).
14. Khatena, J. 1972. The use of analogy in the production of original verbal images. *Journal of Creative Behaviour* 6: 209-213. This experimental study looked at spontaneous use of analogies by 141 college students who had been pre-selected for their high scores on a test of originality. They were asked to produce unusual verbal responses to the 'sounds and connotations of words'. A total of 5640 analogies were produced of which 5545 (98%) were direct analogies.
15. Morris, W.C. and Sashkin, M. (1978) Phases of integrated problem solving (PIPS). In Pfeiffer, J.W. and Jones, J.E. (eds) The 1978 *Handbook for Group Facilitators.* University Associates Inc. La Jolla, California. pp 105-116.
16. Higbee, K.L. and Millard, R.J. (1983) Visual imagery and familiarity ratings for 203 sayings. *American Journal of Psychology* 96: 211-222.

5 The innovative company

Being innovative doesn't just happen. It needs to be nurtured and developed within a business environment. Many authorities talk about an innovative culture within a company. But what does this mean? Is it a tangible and quantifiable set of company characteristics? More importantly, can it be brought about by systematic methods or is it one of those 'either-you-have-it-or-you-don't' bits of business magic.

An innovative business culture is mostly to do with the attitudes of people within the company and, as such, will be ingrained in the company's management style and the responses it engenders in company staff. It will, therefore, be difficult to change and very difficult to change radically. It can be as big a mistake to try to make a company too innovative as it can be to make it not innovative enough. The round peg of innovation simply cannot be forced into the square hole of a traditional company.

This chapter attempts to provide a recipe for creating that rare delicacy, the innovative company. Like all good recipes it describes both the ingredients and the procedure. That procedure involves advance preparation, careful execution and delicate presentation to those partaking the feast; in this case the company staff. Like all good food, it will be easy to swallow for enthusiastic gourmets with a healthy appetite. Those with no hunger for change and with the expectation of being force fed will enjoy the food little and benefit from it even less. Like all good cooks, the company managers will find the recipe improves with practice. Some ingredients will be found to mix better than others. New procedures will be discovered to improve upon the old.

This recipe for the innovative company begins by thinking what it about a company that needs to be innovative. The interaction between management and the process of innovation in a company is shown in Figure 5.1. The process of innovation has inputs, which are transformed to create outputs; new products. Inputs to the innovation process are, to a large extent the creative ideas of individuals within the company. Effective innovation management encourages these ideas and provides individuals with the freedom to develop them. They, of course, have to be assessed and screened to ensure that only viable ideas are pursued. This is best done by a team containing representatives with different functions within the company. Marketing, product development and production engineering must all be happy that an idea is viable and worth pursuing before resources are invested in its development. If it is, indeed, considered a viable idea, a multi-disciplinary team should also have responsibility for developing a design specification to guide and quality-control its development. Given appropriate strategic guidance from a corporate level within the company, this team should be able to make decisions as to the acceptability of new product ideas. The alternative is to waste time seeking corporate approval for every idea as it arises.

The actual development process which transforms ideas into manufacturable products is a task also best tackled by a product development team. To work

Figure 5.1
The innovation management matrix

		Innovation		
		Inputs	Transformation	Outputs
		←	Investment in innovation	→
Management	Corporate	• Screening procedures and acceptance criteria	• Use of formal product development procedures	• Strategy: what new products are wanted
	Team	• Multi-disciplinary teams for early assessment and specification	• Team composition, and organisation • Responsibility and accountability for development decisions	• Hand-over responsibility **or** • Continued involvement throughout product lifecycle
	Individual	• Creative freedom • Access to decision-makers	• Involvement, commitment and 'ownership/championing' of new products	• Recognition and reward for success

effectively the team must comprise individuals committed to the success of the new product. The product development team need not necessarily be the same team as did the idea screening. If, however, they are, it often generates a higher degree of ownership and product championing since they will feel some responsibility for having deemed the idea worthwhile in the first place. Although the team should, to a large extent, be left to decide on their own working procedures, corporate decisions may deem certain development procedures important for quality control and reporting progress as the new product develops.

The output, in the form of the finished product, is, of course, the important part of innovation management. The only way to make sure that the correct sort of products get developed is to have an innovation strategy which makes it clear to everyone in the company what new products are sought, consistent with the company's overall corporate strategy. In terms of delivering these new products, the development team must understand where their responsibilities lie. Some companies are taking the radical step of extending the principle of product ownership to the point that the development team continues to be responsible for 'their' product throughout its entire sales lifecycle. Other companies consider this a poor use of staff expertise. People who are good at new product development may not be so good at production, marketing and sales. If responsibility is to be handed over, it must be managed carefully. It is easier to overlook potential problems if you, yourself, are not going to have to sort them out. For individual members of a product development team, handing over 'their' creation may not be easy. This is especially so if, as often happens, they feel that the sales and marketing team are going to reap the benefits of its success. The entire principle of encouraging new product ideas and delegating responsibility for product development can be destroyed if individuals fail to receive recognition and reward for the success of the finished product.

....ideas are best screened by a multi-disciplinary team from marketing, product development and production engineering .... who then go on to develop a design specification to guide and quality control the product's development.

Key measures of product development effectiveness

Commercial success for a new product means that the product must sell to customers in sufficient numbers and at a sufficient price:

- To cover its full cost of sales,
- To cover its full development costs and
- still return a sufficient profit to the company.

A great many things can go wrong, making new products commercially unsuccessful. In Chapter 2 we reviewed the general success and failure factors in new product development. The most important of these were:

- Market orientation. Successful products were seen by customers as substantially better than competing products and better in ways which they valued highly.
- Early feasibility assessment and specification. Successful products were (i) thoroughly and stringently assessed and (ii) sharply and well defined in a design specification before development.
- Quality of the development process. Successful products were developed by teams (i) with skills well matched to the needs of the product, (ii) with harmony between technical and marketing staff and (iii) with design activities conducted to a consistently high standard.

Now we need to look in greater depth at the more proximate causes of product failure in order to find out how companies can improve their product development management.

When a company makes the decision to develop a new product, it does so on the basis of certain projections. The development work is projected to take a particular period of time. This leads the company into expectations about its development costs and about when the product will be on the market to generate its first return upon investment. Clearly, if these estimates of development time are over-run, development costs will rise and this may turn a projected profit over the product's lifetime into a loss. In addition, the late launch of a new product disrupts cash-flow projections. The greatest danger from the late launch of a product, although the most difficult to quantify, is the lost market opportunity. A competitor might introduce their own new product and steal the opportunity you had planned to exploit. There may be a fixed window of opportunity for your new product (e.g. until the next technological advance renders your product obsolete). Every month of delay caused by development over-run may, therefore, be a month of lost sales.

Another important projection is the projected cost of the new product. In a competitive market, there is likely to be little price elasticity and, therefore the only way to achieve the projected level of profit is to meet the projected sales cost for the new product. Again failure to meet target costs may turn a projected profit into a loss.

The actual cost implications of these development problems have been calculated [1] as shown in Figure 5.2. For a product in a highly competitive and growing market but with a limited product life, the consequences of launching the product six months late are severe, a loss of one third of total profits. Being forced to reduce prices will take 15 per cent off profits, whereas exceeding production cost targets or development costs, are more minor, a loss of only 3.5 per cent profits. This shows how important it can be to invest extra resources in product development in order to get the product developed on time and at the right cost.

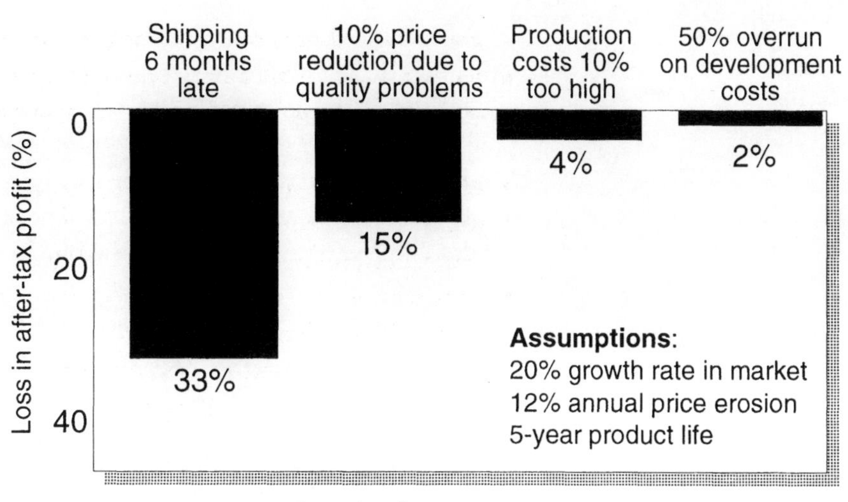

Figure 5.2
The costs of product development problems

How well do companies do?

A recent study by the Design Council in the UK [2] surveyed over 500 companies to find out how well they were performing in key measures of product development. Less than half of all products were delivered to target production costs and, similarly, less than half were delivered on time. Compared to development targets, the products were an average of 13 per cent more expensive and they were launched six months late. The survey revealed several reasons for this. Functional aspects of the design specification were changed an average of 12 times during product development. Once development was supposedly completed, major engineering changes still had to be undertaken. Of all the engineering changes, as recorded in formal 'engineering change orders', an average of 13 per cent occurred after product development was 'signed off'. As a consequence, companies required an average of 16 weeks of production 'ramp-up time' to resolve the problems that should have been sorted out during development. In short, companies were failing to get

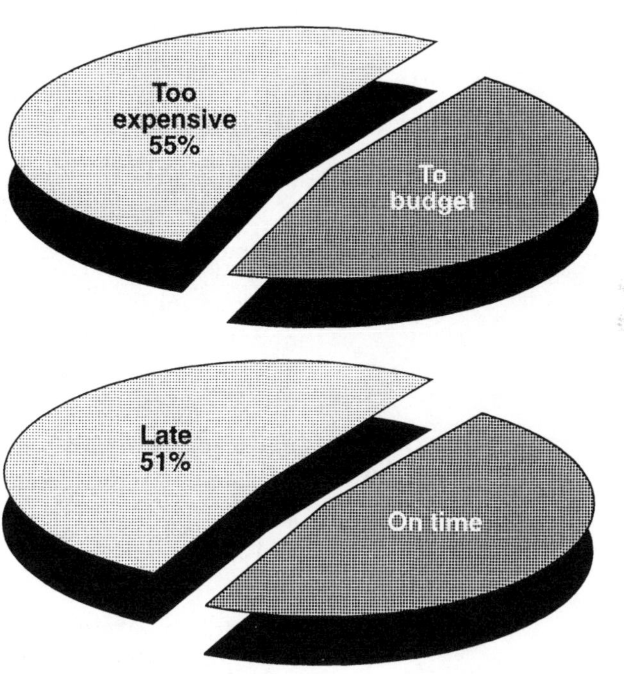

Figure 5.3
Typical product development performance in companies [2]

products 'right first time' at the end of the development process. Products which reach the end of the development process with functional failures, reliability problems and incomplete engineering solutions can cause enormous problems. The effort needed to solve problems at this stage is nearly always disproportionate to the nature of the problem. This is because the simplest of changes to a finished design can have knock-on effects to other aspects of the design. A simple problem can very quickly escalate into a significant re-design.

A Japanese shipbuilder was once asked to comment on the biggest difference he noticed between the Japanese and British approach to shipbuilding. In Japan, he said, we take four years to design a ship and one year to build it. In Britain, however, you take only one year to design it and it then takes you four years to build it.

Strategy for product development

Very often, companies say they are opportunistic with regard to innovation. If they spot an opportunity to be innovative, they will seize upon it and try to be the first to exploit it. If, however, their competitors get one step ahead of them, they will be forced into responding with a catch-up product in order to defend their share of the market. To an extent, saying that a your company is going to be opportunistic is missing the point of strategic planning. Being opportunistic is what you do when you are unable to control or at least predict the pressures and influences which affect you. To have a company strategy of opportunism is virtually a self fulfilling prophecy. If no attempt is made to seek particular types of opportunity in particular ways, the only possible outcome is to have to deal with the opportunities when you stumble across them. Opportunism is a form of 'fire brigade' management in which the manager's main task is to steer the company through a succession of crises.

Strategic planning, on the other hand looks for the goal or mission which the company should strive to achieve, establishes a strategic plan for reaching that goal and then implements that strategy by investment, recruitment, business organisation and management. Different strategies require different investments of people and resources in different functions within a company. Strategies with regard to innovation can be categorised as one of four basic types [3].

Pioneering strategies aim to give technical and market leadership. They depend heavily upon research and development to introduce both radical and incremental innovation to the market ahead of their competitors.

A pioneering strategy is pro-active and has a long-term perspective on return on investment. It is highly dependent upon an effective innovation culture within the company. Considerable importance is attached to patent protection, since substantial monopoly profits will generally be required to pay for investment in product development and the inevitable product failures.

Responsive strategies aim to respond to pioneering competitors but deliberately let others take the new product development risks and open up new markets. Depend upon fast times to market and often strive to improve upon pioneering products (a 2nd-but-better strategy).

Traditional strategies are adopted by companies operating in a largely static market with a largely static range of products where there is little or no market demand for change. Innovation normally limited to minor product changes in order to reduce costs, ease production or increase product reliability. Traditional companies are usually poorly equipped to innovate when forced to do so by competitive pressures.

Dependent strategies are the result of companies depending upon their parent company or their customers for innovation. In-house innovation in dependent companies is usually limited to process innovation.

As far as organising and managing a company to implement these different strategies is concerned, Figure 5.4 shows how the importance of different company functions can be ranked. A dependent strategy, for example, will generally require high standards of production engineering to enable the company to be responsive to a wide variety of requests from clients or the parent company.

Figure 5.4
Different strategies give rise to different priorities and resource requirements [3]

Company priorities for different innovation strategies

	R&D	Innovative Design	Time to market	Production Engineering	Technical Marketing	Patents
Pioneering	***	***	**	**	***	***
Responsive	*	***	***	**		*
Traditional				***		
Dependant				***		

Traditional companies also need strong production engineering skills, although for different reasons. Typically, traditional strategies are only able to survive when their product(s) has reached a mature stage in the market. During this stage, competition on the basis of price usually becomes fierce. Development costs have been recouped, marketing costs are minimal, sales and distribution costs should been honed to a minimum and hence retail price will depend mostly upon how lean and efficient the production engineering has become. Pioneering companies need a wide range of skills and expertise. They need strong R&D in order to discover opportunities with pioneering potential. They then need to invest heavily in patenting to protect their new products against copying by competitors. They need strong design skills to be able to translate innovative ideas into commercially successful products. They also need strong marketing skills to be able to persuade customers that a product they may never have heard of before is what they always wanted. By contrast the responsive company does not need such strong marketing (the market will already have been opened up by the pioneers), patenting (although expertise in avoiding patent infringement is important) or R&D (although some research is needed to make sure the company does not fall too far behind the pioneers). They need particularly strong design skills, combined with a very fast product development procedure. Time to market is often the most important single determinant of commercial success for responsive companies.

In principle, then, we can distinguish between different types of strategy and see how adopting different strategies causes companies to evolve in different directions. The company must become organised, managed and resourced to suit its business strategy. Because of this, strategic planning must inevitably be long term. It takes time to get the right people with the right abilities working well together to develop the right kind of products. Strategic planning, therefore, is visionary and goal-oriented. Strategy forms the basis for the entire structure and management of the company and provides the guiding light to determine the type of new products the company seeks to develop. A company's mission can, and often should, be idealistic. Ideals can be inspiring and motivating to staff. But the way in which the strategic goals are pursued must be realistic. Every step towards the ultimate strategic plan must be carefully considered and commercially justified. This applies equally to the planning of innovation. Just because a company seeks to be pioneering does not mean it turns its nose up at an opportunity to respond to a competitor's new product. In fact, in tracking the historical development of a new type of product it is sometimes difficult to decide who were the pioneers and who did the responding (See ballpoint pens example on next page). Of course, companies will always have to be

The ballpoint pen - blurred distinctions between pioneers and responders [4]

In 1944, two Hungarian brothers Ladislao and Georg Biro produced the first commercial ballpoint pen. The US distribution rights were bought by an established fountain pen manufacturer called Eversharp, but before Eversharp received their first delivery, a Chicago businessman, Milton Reynolds who had seen Biro's pen overseas, had a copy product on sale in a New York department store. The ballpoint pen which actually pioneered the US market was, therefore, the copy of an original not yet delivered. Much legal wrangling then ensued, until it emerged that the Biro brothers' patent, filed in 1938, had been preceded, exactly 50 years earlier, by another patent for a ball-writing pen. Reynolds enjoyed considerable commercial success with his copy product and, in a curious reversal of roles, the product copier then became a product pioneer. Reynolds developed the first retractable ballpoint with a point which clicked in and out of the pen barrel.

Consumers' initial enthusiasm for ballpoint pens quickly cooled, mostly because of quality problems. Too many pens leaked, blotted or stopped writing long before the ink ran dry. Sales of both Eversharp and Reynolds pens dropped dramatically and both companies were eventually forced out of business. Parker was the next to introduce a ballpoint pen. Although still essentially a copy product, the Parker pen was re-engineered to overcome the reliability problems suffered by both Eversharp and Reynolds. It rejuvenated sales of ballpoints and gave Parker significant commercial success for several years.

Even with Parker's success the ballpoint pen market, at that time remained a drop in the ocean compared to the market today. The true pioneer, in terms of market success was Bic, the French company who took the revolutionary step of making the whole pen disposable, rather than just the ink refill. Bic pens sold for around 50 cents, at a time when other ballpoints sold for several dollars and fountain pens cost even more. Within a few years the fountain pen was all but obsolete. Bic was classically responsive in its business strategy. They copied the basic ballpoint technology at a time when a market for ballpoints had been proven and Parker had stabilised that market by overcoming the initial reliability problems. Bic, however, were sufficiently pioneering in both product design and production engineering to transform the ballpoint pen from a consumer durable to a consumer disposable and to thereby completely transformed its market potential.

This simple example shows how difficult it can be to decide who really pioneered any particular product. There are three messages we can take from this regarding innovation strategy. Firstly, pioneering, in the strictest sense is very rare. The Biro brothers discovered that their pioneering efforts had been foreseen 50 years earlier. Secondly, therefore, a pioneering strategy does not necessarily only refer to the original inventor of an idea. A pioneering strategy is one which seeks to make radical innovations and pioneer significant improvements in existing products, based on new designs, new materials or new technology. Such a strategy requires significantly more pro-active effort and the resources to make that effort than other product development strategies.

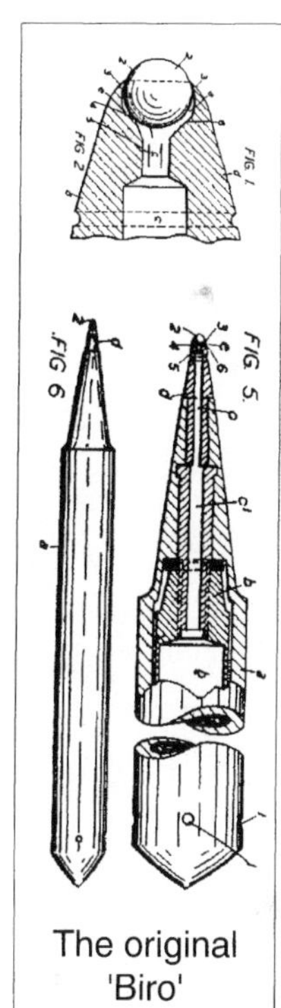

The original 'Biro'

opportunistic, at times. The events which occur in a competitive market rarely fall neatly into place alongside a company's innovation strategy, no matter how carefully planned. But companies should not depend upon opportunism for commercial success. They should not be waiting for others to dictate how they, themselves, run their business. Opportunism should happen despite company strategy, not because of it!

So, what can we learn from all this about how to make an innovative company? First and foremost, there are vast differences between those companies that seek to develop new products and those that do not. Companies with either a pioneering or responsive strategy need a broad range of skills ranging through marketing, opportunity identification, design and development, production engineering and manufacturing. Companies with either a traditional or dependant strategy, on the other hand, need to focus their skills in manufacturing (mostly production engineering). It is important to bear in mind that developing new products is not a pre-requisite for commercial success in a manufacturing company. Companies can make lots of money and be highly successful, manufacturing products that they have had little or nothing to do with designing.

Secondly, for companies seeking to develop new products, there are many ways of doing so, ranging from the most radically inventive pioneer to the most plagiaristic responder. Deciding where to position your company between these extremes is largely a matter of risk management. At one extreme, pioneering an entire new branch of technology carries the ultimate risk - that the technology fails or does not gain market acceptance; the VHS format became the standard for videocassettes, after Sony spent millions developing the Betamax format [5]. At the other extreme, producing identical clones of another manufacturer's products runs the risk of prosecution (under copyright, design or patent legislation) or achieving no product differentiation. There is no advantage in producing a copy product if the product which you copied is every bit as good and already accepted in the market.

Elements of strategy [6]

The origin of business strategy is the company's mission statement. This is a statement of the company's vision of the future from which all strategic planning is derived. From this, corporate objectives are developed. These relate to the most general measures of business performance (e.g. turnover, profits, costs, markets, staffing levels) but describe specific targets for change, so that the company can realise its mission. Corporate strategies are the ways in which these corporate objectives are to be achieved. Again they relate to general descriptions of business activities (e.g. targeting new markets, developing new products) but indicate the type of action to be pursued. Implementation of corporate strategies requires specific actions by nominated individuals with agreed targets and timescales.

Subsidiary to the planning of corporate strategy, is the planning of product development strategy. This follows an almost identical procedure, defining the product development objectives, specifying strategies and then identifying actions for individuals to carry out with targets and timescales. Product development strategy should, of course, be contributing to (or even better, be an integral part of) corporate strategy. Equally, the delivery of the product development objectives should deliver part of the corporate objectives. This strategic planning procedure is shown in Figure 5.5, with each stage of the planning process exemplified for a company seeking market leadership in the luxury end of the leisure yacht market.

There is no single correct solution to corporate and product development planning. A company could be successful by choosing any one of a number of missions and seeking to reach that mission by any one of a number of strategies. The key to strategic planning is to choose a direction for the company, and ensure that business decisions are made consistently to facilitate progress in that direction. So, for example, a company which manufactures boats for the leisure market could aim for market leadership, either in terms of producing the most boats, the best boats or the cheapest boats. Alternatively ,if it was a smaller company needing to focus on a market niche because of its inability to compete head-on with its larger competitors, it could specialise in making a specific type of boat (e.g. traditional timber-hull) or could offer a unique service to attract customers (e.g. a free mooring and service contract). The implications of such different corporate strategies on product development strategy is shown in Figure 5.6. The company featured in Figure 5.5 is shown as the 'leadership - best' category in Figure 5.6.

Steps towards strategy development

What does the company do well and what does it do badly? What do customers like and dislike about the company? What changes have taken place in the company's recent past, what stimulated them, who championed them and what brought them to fruition? What sort of decisions have led to success and failure in the past? What could happen beyond the company's control to threaten its survival? These are the sort of questions that every

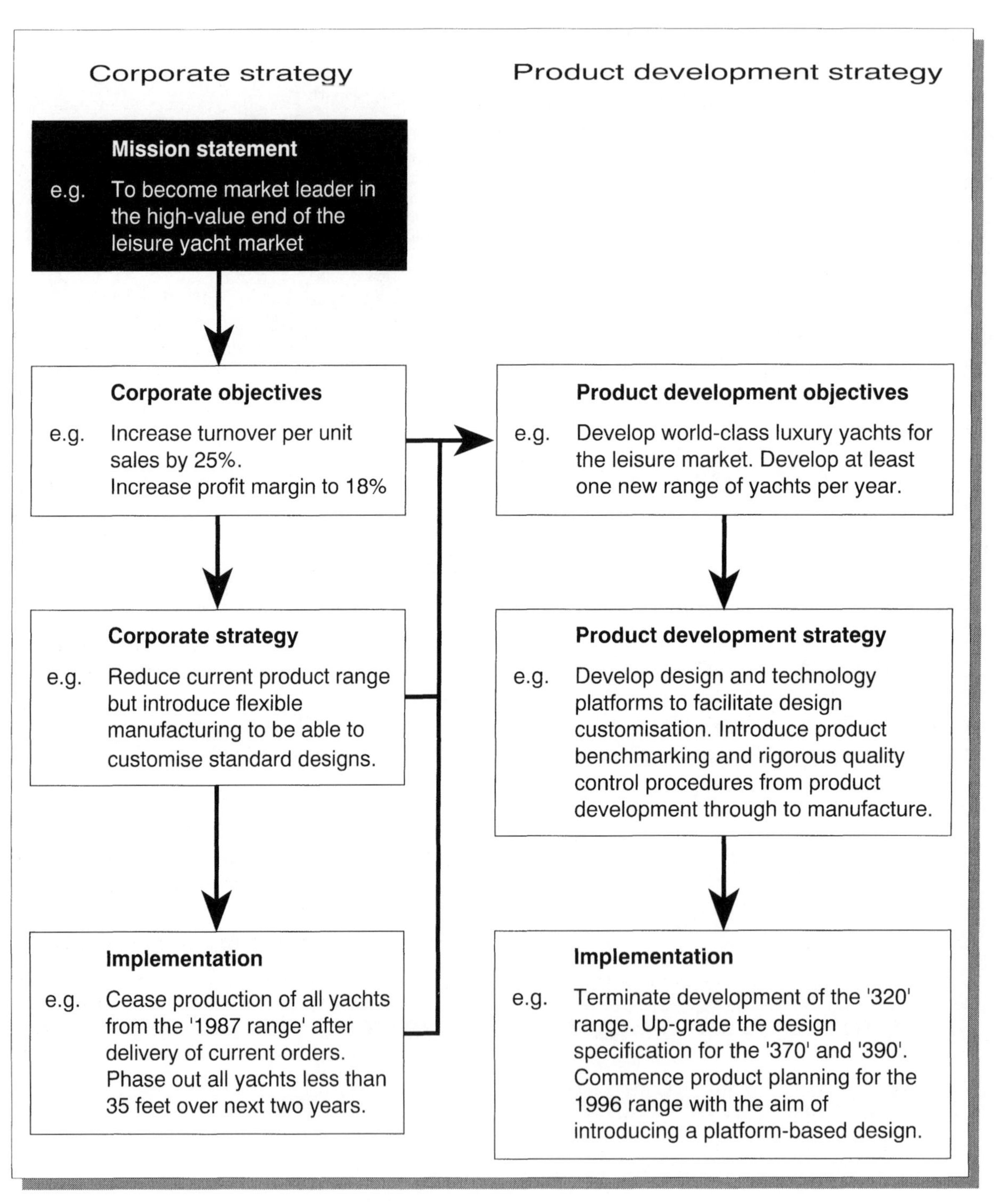

Figure 5.5
The strategic planning process

Mission			Product development strategy
Leadership	Biggest	To acquire (or retain) the largest share of the leisure boat market.	To **pioneer** developments in the leisure boat market by means of market research and pro-active product development.
	Best	To become (or remain) the market leader for luxury leisure yachts.	To **pioneer** high-value developments in the leisure yacht market and **respond,** selectively, to high-value innovations by competitors.
	Cheapest	To deliver the lowest price leisure boats.	To **respond** to innovations by competitors and develop low-cost versions by means of value analysis.
Niche	Specialist market	To produce traditional timber-hull boats.	To refine **traditional** timber boat building and **depend** upon customer specifications for innovation.
	Product + service	To offer leisure boats with a free 5-year mooring and service contract.	To maintain **traditional** boat designs and **pioneer** added-value services to augment the product package.

Figure 5.6 Business strategies in the leisure boat market

business manager thinks they can answer. Most, however, admit to some surprises when they tackle the issue systematically. There are four key steps in developing both a corporate and a product development strategy (Figure 5.7).

- Where are we?
- Where are we going?
- How are we going to get there?
- How do we know when we have got there?

Let us consider this process firstly in relation to corporate planning.

Corporate planning

Reviewing the current status of a company is partly an internally-focused task and, therefore, reasonably self contained. It can, however, be difficult for company staff to do themselves. They require an overview of the company and this is something that only the most senior, and often the busiest, staff have. To be done effectively, they need an element of detachment from the immediate day to day problems and challenges facing the company. Most importantly, they need to be self-critical; to generate a falsely reassuring sense of confidence can be worse than not doing any strategic review at all. As a result, many of the larger companies employ management consultants for this task. This overcomes some of the above problems but will rarely be as penetrating because the consultant cannot possibly know as much about the company as the company knows itself.

The externally-focused part of the review is fundamentally more difficult. Reviewing the external forces for change is necessarily more open-ended and based on less certain information.

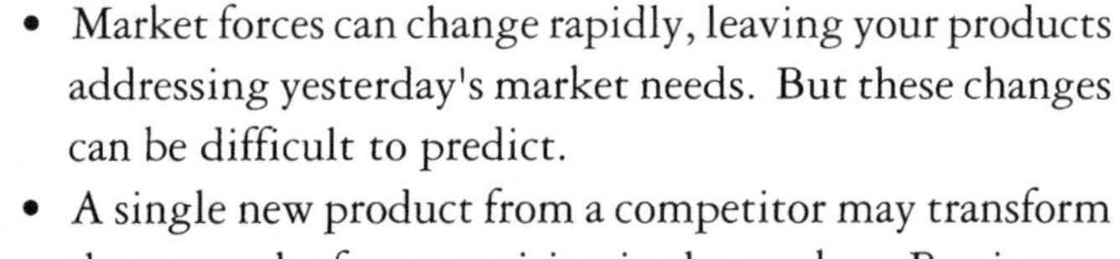

- Market forces can change rapidly, leaving your products addressing yesterday's market needs. But these changes can be difficult to predict.
- A single new product from a competitor may transform the strength of your position in the market. But it may have been under development for years and it might take many months before you could respond.
- New technology is always being developed, and being left behind in a technology race is a common cause of business failure. But emerging technology is difficult to keep track of and almost impossible to anticipate.

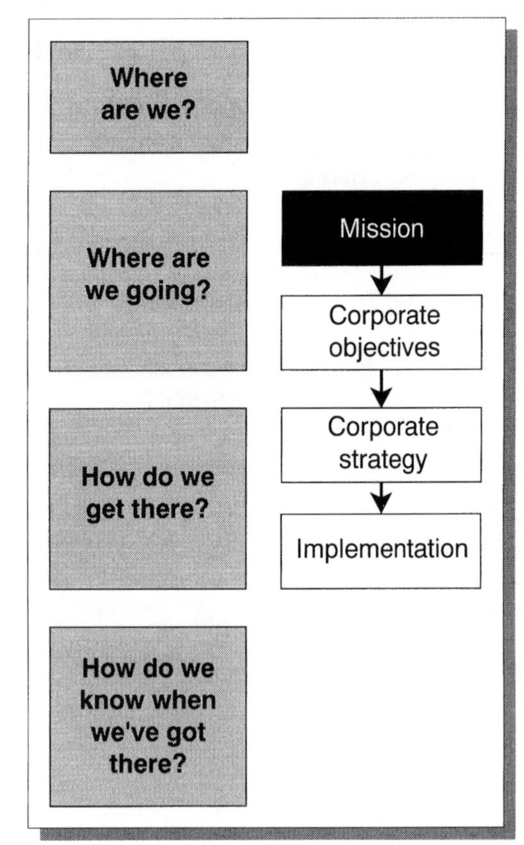

Figure 5.7 Questions asked during the corporate planning process

The same difficulties are, of course, faced by all companies. Success in developing a business strategy does not, therefore, depend upon you identifying these external forces faultlessly. You just need to do it better, or at least as well, as your competitors.

Where are we? Analysing the company's current business position is usually the most time consuming part of strategy development. The first task is to identify the company's core business function. This is the function which differentiates the company from competitors, in the eyes of customers. Searching for the core business function can be difficult and may require market research. Many companies have been forced by events to realise that the company function which customers value is totally different from what they thought customers valued. Large companies often commission on-going tracking studies (see design toolkit page 131) which periodically review customer perceptions of the company and see how they change over time. This allows them to focus more effectively their advertising of company or brand identity. Tracking studies are usually conducted by professional market research organisations and surveys hundreds of customers for each review. In principle, there is nothing to stop companies trying their own, smaller scale tracking study to discover or confirm customer opinion about the company. Once the core business function of the company is determined, the next question is broader: what are the general strengths and weaknesses of the company and what opportunities and threats does it face. This is called a SWOT analysis and it is described in the design toolkit on page 128. Finally, broader still, is the analysis of the political, economic, social and technological factors which might influence the company. Again known by its acronym, PEST, this is described in the design toolkit on page 130.

Core business function
This is whatever a company currently does that lies at the heart of its business success. The description of a core business function should be simple and concise. Ideally, it should be written in a single sentence. Identifying a company's core business function makes the mission statement much easier to write.

Where are we going? This is a question which is answered by the development of a mission statement and corporate objectives. The basis for these answers should largely be contained in the results of the 'where are we' question. The company's core business function is its main business asset. This is how customers value the company and every effort should be made to fully exploit and develop this perceived value. To develop a new core business function in the eyes of customers is likely to be a lengthy and expensive process. It is only if the core business function is a source of substantial weakness or is faced by significant threats that it should be changed. Using the results of the SWOT analysis should, in principle, be similarly straightforward. The aim is to develop strengths in order to exploit opportunities; minimise weaknesses in order to counteract threats; take action to turn weaknesses into strengths and to turn threats into opportunities. The ease with which all this can be done is often a reflection of how well and how honestly the SWOT analysis was conducted. If it is proving difficult, go back to the SWOT analysis and see if it can be improved. The results of the PEST analysis are usually the easiest to build into corporate planning. There are typically few PEST factors of direct and immediate relevance to any company's business. They are worth considering because a change in legislation or international trade agreements, for example, can have the most profound effect upon a company's business.

It is important to bear in mind that the aim in developing the mission statement and corporate objectives is to establish goals, not to decide how best to achieve these goals. This is the next task – proposing corporate strategy.

Corporate strategy

The development of corporate strategy is both the most important and the most difficult part of corporate planning. It is the corporate strategy that will dictate how changes within the company are to be pursued and it is, therefore, corporate strategy which determines how effectively the changes will be brought about. Corporate strategy is inevitably a long-term plan of action and aims to bring about change gradually and progressively. Getting the corporate strategy wrong can inflict long term damage upon the company and, to make things worse, it may not be obvious that the strategy is wrong for some considerable time. This is one of the reasons why the development of corporate strategy is preceded by such lengthy and thorough preparation, developing the mission statement and corporate objectives. The whole process of corporate planning is a part of the risk management funnel, described in Chapter 2. Risk management is characterised by a step-by-step process which focuses on progressively more specific aspects of the overall goal. At each step in the process, the best of many possible solutions is selected, thereby reducing the risk of ending up with the wrong solution. So, in corporate planning, the vision of the future for the company is projected by the company's mission. Answering the questions 'where are we' and 'where are we going' helps you to explore the many different targets which could be set in order to reach this mission and then assists in the selection of the best of these. The development of corporate strategy takes this process one stage further.

How are we going to get there? This is a classic creative thinking process and many of the techniques described in Chapter 4 can be useful in both generating the strategic options and selecting the best. A useful way of visualising the problem of corporate strategy development is the problem gap, shown in Figure 4.5. This is adapted in Figure 5.8 to relate to corporate strategy. Corporate strategy can be seen as a bridge between the company's existing business and its company mission and corporate objectives. Establishing corporate strategy is the problem gap, reaching the mission and objectives is the problem goal. The problem boundaries for establishing

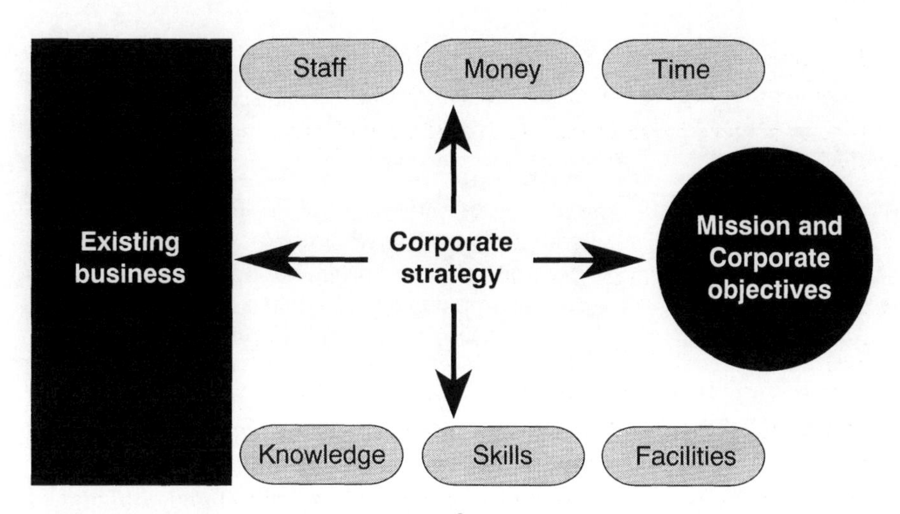

Figure 5.8
Corporate strategy as a problem gap

corporate strategy are the company factors which constrain the strategic options. The main ones are staff, money, time, knowledge, skills and facilities. Seen in this way, the problem of setting corporate strategy is one of finding ways of reaching the mission and objectives, from the position of the existing business and within the prevailing constraints.

This turns out to be a useful way to envisage the problem because the problem gap can often be quantified and potential solutions examined within this quantitative framework.. Let us return to the example of the luxury yacht manufacturer in Figure 5.6. The corporate objectives were stated as increasing revenue per unit sales by 25% and increasing profit margin to 18%. Different strategies for reaching these objectives can be analysed in terms of this problem gap (see Figure 5.10) and it soon becomes clear that only one of the three considered is satisfactory.

Another way of thinking about the development of corporate strategy is the Ansoff matrix (Figure 5.9). This is a variation on orthographic analysis (described in the design toolkit on p 88). The Ansoff matrix presents the four main ways of increasing business for a manufacturing company. With existing products, a company can either seek greater market penetration (sell more products into existing markets) or market development (sell products into new markets). Developing new products can either be to increase sales in existing markets or to diversify into new markets.

Figure 5.9
Ansoff matrix for exploring business development opportunities [7]

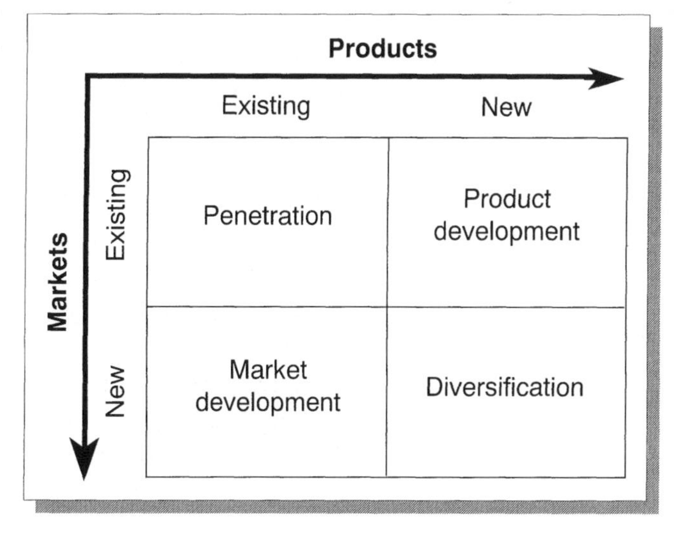

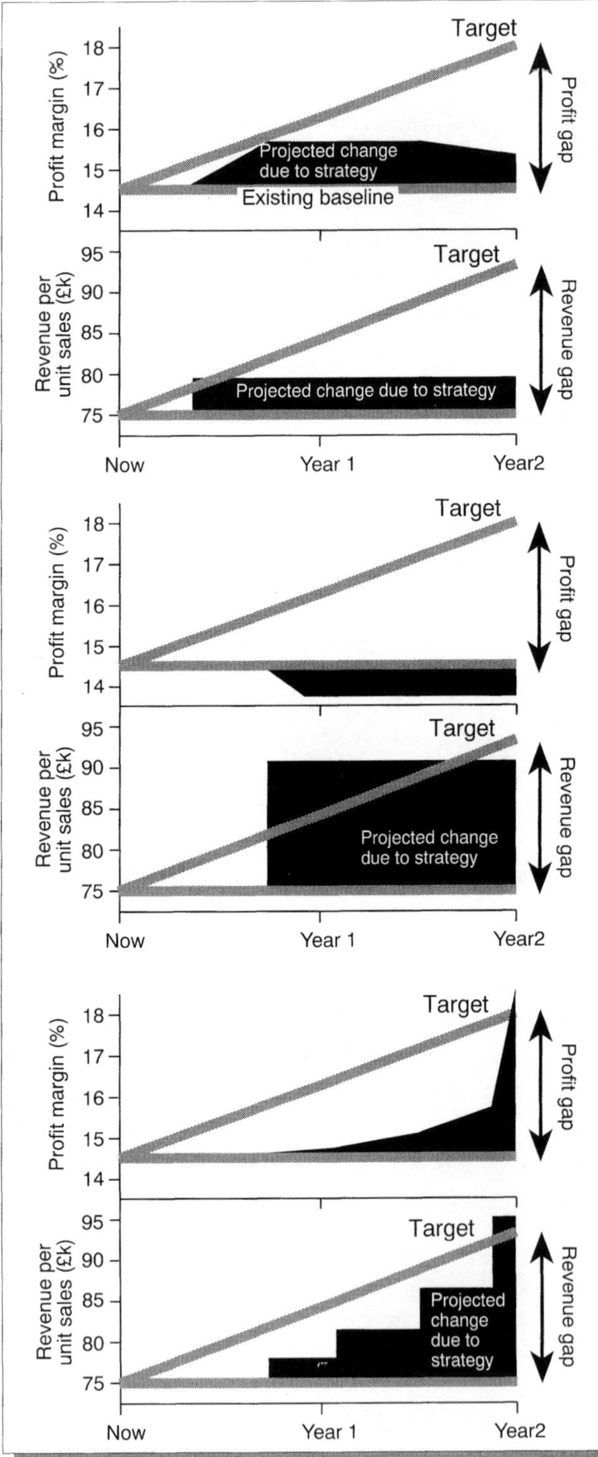

Figure 5.10
Strategy gap analysis

Strategy 1
Add high-value bought in equipment and components. This would increase both revenue per unit sales and profit, but not enough. Due to the competitive pricing that customers expect when buying equipment as part of the yacht, profit margins are squeezed. Also, equipment will always be a small percentage of the total boat price, so prospects for increased revenue are limited. It is also expected that competitors will respond by adding their own increased equipment, thereby forcing down profits after 18 months.

Strategy 2
Kill of low-priced products from existing product range, in 9 months time. This will immediately boost revenue per unit sales to almost the target level. The low priced products, however, have at present, a larger than average profit margin. Killing them off would, therefore, reduce profit margins.

Strategy 3
Gradually introduce a new platform-based range of high-value yachts. Designing to a platform-based design means having common design elements, common materials and common manufacturing processes, but leaving flexibility to customise easily to specific customer requirements around that common platform. This would require changes in the design process (CAD-based design) and introducing flexible manufacturing procedures. It will require significant investment. Upon the launch of each of these platform based designs, revenue per unit sales is expected to increase and profit margins are expected to increase cumulatively.

Implementation

Implementation takes corporate strategy one stage further and specifies who is going to do what and by when, in order to turn the strategy into action. In the language of management, implementation must be SMART [8].

- Specific – The actions required to implement strategy must be specified clearly and made the responsibility of a specific person.
- Measurable – The actions must have a clear outcome which can be measured against targets
- Achievable – Implementing strategy does not mean handing out impossible tasks. Ambitious strategies often have to be tackled in stages, beginning with a feasibility study and continuing with a series progressive steps towards full implementation
- Resourced – Implementing strategy takes time, effort and, usually money. The temptation to simply add strategic tasks to already full workloads should be resisted if the tasks are to be completed well.
- Timetabled – Achieving the company's overall corporate goals should have a time target and hence the implementation tasks must be timetabled in order to meet the overall goals.

Strategic planning of product development

We have seen how the process of corporate planning takes place and how the broadest of decision-making about the company's vision of the future is progressively and systematically translated into action. It is important, in a book on product design to see that corporate planning, even for a manufacturing company covers a great deal more than product development The outcome of product planning may be to conclude that the company should not be involved in product development. This type of decision-making must clearly happen before the company gets wrapped up in discussion and debate over what type of products it is going to develop. Where a company does, however, decide that developing new products is an important part of its mission or its corporate objectives, the

Strategic planning of product development is different from product planning

There is a risk of confusion in the use of words here. Product planning is the activity which prepares for the development of a specific product. It involves market research, competing product analysis and writing a design specification (See Chapter 6). By contrast, the strategic planning of product development is a much more general activity which decides what type of products the company should be developing and how.

remainder of the corporate planning process must take into account strategic options for product development. So, corporate planning is a more general activity than strategic planning of product development. It must be started first to see whether or not product development is needed. If it is then corporate planning and strategic planning of product development continue in parallel with the results of each influencing the other. The reason the two types of planning are described as occurring in parallel, rather than as one and the same process is that they ask different question, use different methods and lead to different decisions. Corporate planning is about the entire business of the company, not just product development. Strategic planning of product development is about what type of products to develop and how to make sure that these go towards fulfilling the corporate objectives. Strategic planning of product development is, therefore, a subset of corporate planning.

Figure 5.11
Questions asked during the strategic planning of product development

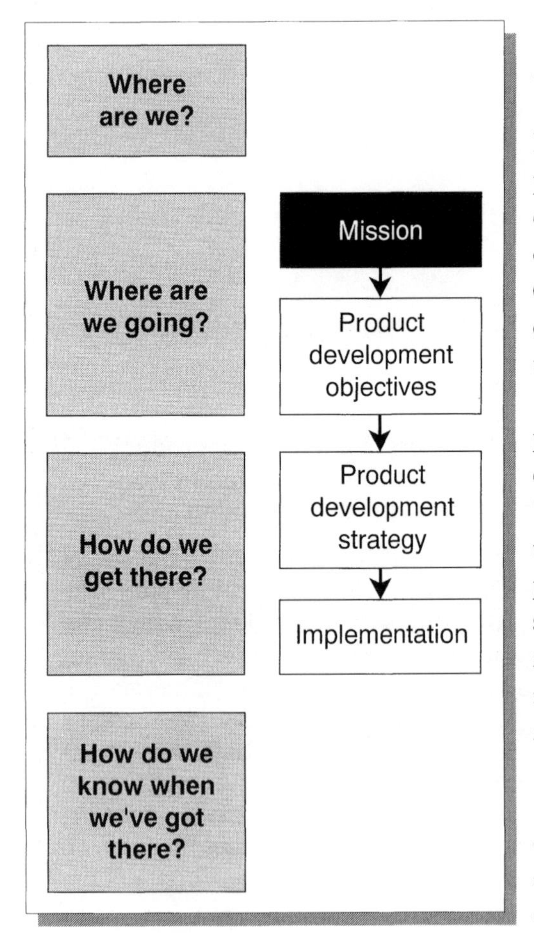

Steps towards a product development strategy

Preparing a product development strategy is a strategic planing process, just like corporate planning. It, consequently follows a similar procedure and asks similar questions (Figure 5.11). It asks them at a different level (tackling specific product development issues rather than overall business issues) and, therefore uses different methods to try to find answers.

There are three main techniques for use in developing a product development strategy. The first looks at the company's current range of products and analyses their 'business maturity' The business maturity of a product is the stage it is at in its business life cycle. The closer a product is to the mature stage in its life cycle the sooner its sales are expected to decline. Examining where the full range of current products are in their life cycle, reveals the need for new product development in the future. Product maturity analysis is described in a design toolkit on page 132.

Competitor analysis compares the business performance of the company with that of competing companies. The aim here is to discover, for short term planning, how competitors are out-performing the company and what

Specific questions	Methods	Answers	Decisions
What is the need for new products?			
Where in their product life cycle are products in the current range?	Product maturity analysis	The need to develop new products is important or urgent or both	Product development objectives
How do current products compare with competing products?	Competitor analysis		
How rapidly is the market changing?	Static/dynamic analysis		
What is the capability for developing new products?			
People? Procedures? Facilities? Money?	Product development risk audit		Product development strategy

Figure 5.12 The process for strategic planning of product development

changes to make to improve competitiveness. For longer term planning, competitor analysis tries to work out the business strategies of competing companies to be better able to anticipate future competitive threats and develop a product development strategy to counteract them. Competitor analysis is described in a design toolkit on page 134.

Finally, a product development risk audit looks at the cost of product failure and determines the amount of change to the product development capabilities of the company which would be commensurate with the magnitude of that threat. This technique is a useful way to integrate the different aspects of a product development strategy and to try out, on paper, the effectiveness of different strategies. The design toolkit describing the product development risk audit is on page 136.

Implementing product development strategy

How to implement a product development strategy is, largely, the subject of the remainder of this book. Procedural aspects of product development have been summarised in Chapter 2 and will be developed extensively in Chapters 6 to 9. The facilities required for product development are described mostly in Chapter 9, on embodiment and detail design.

This leaves the 'people' aspect of product development strategy. To say that people are the key resource in the development of new products is so obvious as to be almost facile. But despite the, almost universal, acceptance of the importance of people, the deployment of human resources for product development is often neglected or avoided. People are often seen as pre-determined in the planning of product development. Individuals hold posts which historically have had certain responsibilities. To rock this particular boat can be highly disruptive, de-motivating and a source of great inter-personal conflict. The importance of people in the product development process is, however, so great that the boat must be rocked. It need not be rocked violently and certainly it must not risk capsizing completely. Some individuals will inevitably fall overboard and others will come on-board in their place. The right people must be in the right place at the right time or all efforts to plan procedures, facilities and finance will have been in vain.

Individuals and teams for product development

We all have a vision of the ideal designer of new products. Creative and imaginative, they come up new products which transform entire markets. Paragons of visual skill, they draw, model and eventually produce products with bewitching visual form. And with an instinctive feel for the market they effortlessly produce the product customers would have dreamt of, if only they'd had the foresight to think of it . Should this wondrous creature walk through the door of your company offices, the 'people' aspect of product development would be solved in an instant. Or so you might imagine! The first and most obvious problem is that this person cannot run the company single handedly. Manufacturing, distribution, marketing and sales staff will all need to work with the new virtuoso. Exceptionally talented people have an unfortunate habit of resenting interference from others, especially if it involves imposing constraints on their creative freedom. What if the great creations cannot be manufactured? What if they do not fit into existing sales channels? What if the products are so innovative that customers fail to realise their value? At a more prosaic level, is great creative talent what is required to ensure effective quality control? Will development budgets and timetables be sacrificed in the search for perfection? And even if all of these problems can be overcome, a longer term issue is what to do when some time later the wonder-designer turns and walks out of the door again (or gets knocked down by a bus). Product development, which has soared to unknown heights, plummets to even greater depths with customer expectations of new products raised to a level no longer attainable.

For these and many other reasons, the age of the lone business saviour is over. In its place is team-working. Teams have more time than any single individual can have, they have different knowledge and different skills, having come from different backgrounds. A team decision is less prone to the idiosyncrasies of an individual decision. And whole teams get knocked down by a bus a lot less frequently than an individual.

Creating an effective product development team involves recognising the strengths and weaknesses of potential team members and putting them together in ways that complement each other. The sum of the complementary skills must be matched to the demands of the product development programme. The team-working attributes of people and the ideal composition of teams is now well understood. The general characteristics of the 'dream team' are shown on page 139 and the Belbin self-perception inventory, which allows people to judge their own team-related characteristics can be obtained in Belbin's classic book on the subject [9]. For an account of team building specifically related to new product development, see Inwood & Hammond [10].

The 'Apollo' syndrome

During his work on team performance, Meredith Belbin (on whose work most of this account of 'people' factors is based) experimentally created teams made up of especially intelligent and creative people. For his research, he labelled these teams 'Apollo', cryptically referring to their 'A' graded talents. In management exercises designed to test their team-working these teams did consistently badly, performing worse than teams made up of less talented people. The reason was that they were difficult to manage, prone to destructive debates and less able to make decisions than the other teams. This led Belbin to his pioneering work on the characteristics of people that do work well together and what goes to make up the 'dream team' (see below).

The Psion Series 3: Case study

Psion PLC is one of the few UK companies to have challenged the Japanese in the heartland of their own technological territory and won. The Psion 3 'palm-top' computer has a virtual monopoly in this market sector in the UK, a growing market share in Europe and the USA. An encouraging aspect of Psion's success is that product design was a vital weapon in their commercial battle.

The fledgling Psion Ltd. was set up in 1982 to write software for the personal computer market. By 1984 its annual turnover had reached £5 million and, in this healthy financial state, Psion began to explore new business opportunities. One of its programs, at the time, was a simple name and address book, sold on an audio cassette to be loaded into a personal computer from a tape recorder. This, they felt, could be turned into a dedicated hardware product and sold as a self-contained electronic address book which people could carry around with them in their pockets. The idea for the Psion Organiser was born. Two years later, when the first Psion Organiser was sold, it had become much more than an electronic address book. Based on an 8 bit processor and containing 2K of RAM, the Organiser was a diary, address book, calculator and programmable computer . The Organiser 1 was a modest success, selling around 30 000 units over an 18 month period. It was enough of a success for Psion to invest in the development of the Organiser 2. This was a much more sophisticated product with 32K of RAM and with several problems and inconveniences of the Organiser 1 ironed out. Business began to boom for Psion. Originally intended solely as a consumer product, an industrial market for the Organiser opened up as companies realised that, for the first time they could carry sales, stock and service data around with them in their pockets. Sales of the Organiser 2 soon rose to 20 000 unit per month. Demands emerged for different models based on the same basic hardware and software. Peripherals including bar code readers were developed. Before they had fully realised it, they had a range of products, all based on the same technological platform.

This established the corporate strategy for the future. Psion set sail on a course to be 'an architecture driven, platform based' manufacturer, producing portable computer products for both consumer and industrial markets. They sought new and radically innovative applications but they were not in the business of developing their own new technology (e.g. designing new chips).

This strategy had a profound effect on Psion at the time. The aim of developing products from a common technology base meant that they were not just looking for a new product. The 8 bit processor, upon which the Organiser 2 was based, by this time was beginning to get a bit long in the tooth. A whole new platform was needed.

The route towards the development of the new 16 bit platform was not an easy one for Psion and they found themselves becalmed or blown off course on several occasions. According to Charles Drake, Psion Director, Psion seriously underestimated the magnitude of the task they had set for themselves. This task amounted to the implementation of a complete innovation strategy. Psion aimed to develop the new 16 bit platform, and on the basis of that, develop three new products: a portable lap-top computer for the business sector, a rugged yet highly flexible computer for the industrial sector and a pocket sized computer for the consumer sector. Each product was planned with its own unique advantages over competing products and building upon the commercial and technical success of the Organiser 2, which was still selling well. Psion seemed to be travelling under full sail. Until the problems started.

The first of these was the twelve month delay by their supplier in delivering the vital 16 bit chips. Without these they could not even begin. Over the course of this delay, Psion's business prospects changed. Both Sharp and Hewlett Packard had, by now, introduced competitors to the Organiser 2 and each started to eat into Psion's market. A sense of urgency entered the product development programme. It, however, was an urgency that the development programme could do little to respond to. The task Psion had set themselves was enormous. In terms of hardware, the target was christened 'sibo', a sixteen bit architecture computer on a single board. This was leading edge technology at the time and they paid the price in development time. In terms of software, the target was multi-tasking, a system whereby all the on-board applications were open simultaneously. To change from one application to the other would be immediate, at the touch of a button. Again leading edge technology, the software development devoured over 20 man years before it was complete.

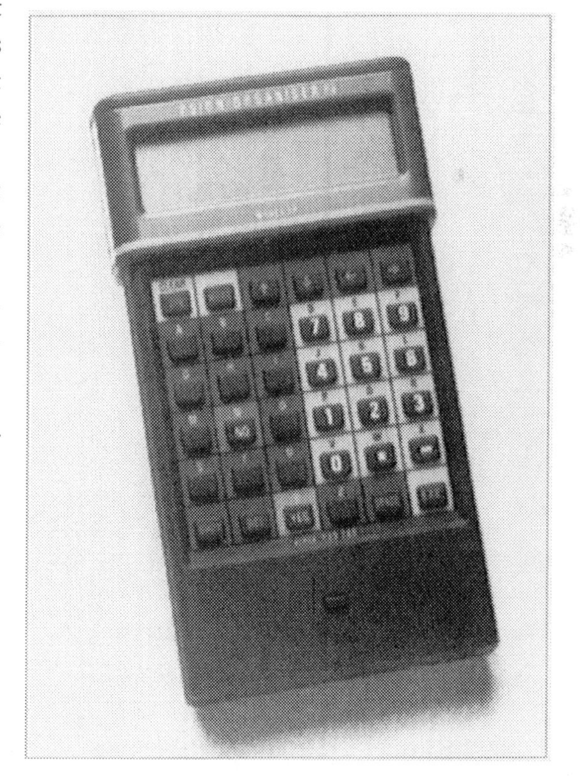

The final, but by far the most significant problem was that Psion's first fruits of these labours, the MC400 lap-top computer failed in the market. Many hours of post-mortem have tried to explain its failure. The consensus, looking back with the benefit of hindsight, is that it was a product launched before it was adequately developed. Charles Drake admitted recently that it was almost inevitable that the first product they launched using the ambitious 16 bit platform would have so many teething problems that it would fail to live up to expectations. It is probably just as well, he adds with a laugh, that it happened with a product in the highly competitive and fast moving market of lap-top computers where it might have failed anyway. But he was not laughing in 1990 when the MC400 began to falter. Psion's business plans lay in ruins. The product intended to pick up on the falling sales of the Organiser had failed to materialise. Psion's turnover almost halved between 1990 and 1991 and they reported a loss for only the second time in their history. Large scale redundancies followed.

The sense of urgency had deepened into a crisis. They needed a new product and they needed it now. At a meeting in October 1990 key parts of the specification for the Psion Series 3 were finalised. It was to be launched in 11 months time. The story of the Psion 3 continues on page 231.

Key Concepts

The need for direction

An innovative company is one that knows where it is going and how it is going to get there. This is not to deny that opportunism has an important role to play in identifying new products. But a strategy-led company will actively seek opportunities which contribute to its corporate goals. Each new product should be seen as making a significant step towards achieving the company's mission. Corporate strategy is a systematic step-wise procedure, beginning with a mission statement and then working through:

Mission statement

A company's mission is a clear and concise statement of where the company is heading and what it aims to achieve.

Corporate objectives

Corporate objectives are specific targets for change, usually at the broadest of levels (e.g. turnover growth, profit margin). Achievement of these objectives should either deliver or substantially contribute to the company's mission.

Corporate strategy

Corporate strategy explains how the company is going to achieve its corporate objectives. These can relate to all aspects of business performance (e.g. sales, marketing, manufacturing), including product development.

Product development objectives

These objectives are a subsidiary of corporate objectives but should identify what is intended to be achieved by means of developing new products.

Product development strategy

A product development strategy sets out what type of new products are sought by the company and how new product opportunities will be explored.

Strategy implementation

This plans the steps to be taken and by whom in order to reach strategic objectives

Design Toolkit

SWOT analysis

SWOT stands for strengths, weaknesses, opportunities and threats. SWOT analysis provides a simple but systematic framework for appraising the company's current business position under these headings [11]. Strengths and weaknesses are current factors and are mostly internal to the company. Opportunities and threats are anticipated future issues and are mostly external to the company. There are four stages in SWOT analysis. Firstly, brainstorming gets down on paper a long list of first thoughts, under the four headings. Many people get put off by this first stage of SWOT analysis because many issues can be categorised under all four headings. For example, the fact that customers like our latest product and are buying it faster than we ever thought possible would seem, at first glance, to be an important strength. It could, however, be seen as a weakness that this product is now contributing such a large percentage of turnover. It is an opportunity to build upon this success with future products. But it is also a threat that company growth from this product may not be sustainable. Do not worry too much about this. Write as much down as you can think of and sort it out in the next stage. Stage two clusters similar or related issues together. This stage can be made a lot easier if you have written the initial SWOT issues on Post-it notes. Stage three analyses the issues, generalises them into broader strengths, weaknesses, opportunities and threats and prioritises them. The final stage synthesises the information gathered, identifies the main issues that require changes to be made and makes decisions about how to introduce these changes. If SWOT analysis is being conducted as part of a more extensive planning process, this final stage can be deferred until any other relevant information has been gathered and analysed.

SWOT analysis is often conducted by a single senior person within a company. Although this is better than not conducting any analysis, the analysis is usually more robust and more realistic if completed by several people in the company, preferably with different job functions and different responsibilities. This can often reveal unpleasant truths about a company – which is probably why it is usually avoided.

The key to effective SWOT analysis is realism and honesty. It also needs to be broad in its coverage (try using the 'food for thought' list opposite). There is no point in concluding that you have a dynamic company, producing exciting products to the delight of customers if the person in charge of the warehouse is disorganised, incompetent and incapable of dispatching the products out of the factory gate!

SWOT analysis (Cont'd)

Design Toolkit

SWOT analysis – food for thought

Company functions
Management?
Administration?
Production?
Marketing?
Distribution?
Sales?

Core business
Markets?
Customers?
Products?
Services?
Competitors?
Suppliers?

Business performance
Market orientated?
Financially healthy?
Quality control?
Flexibility/responsiveness?
Dependence?

People
Enthusiasm?
Commitment?
Dependability?
Contentment?
Team working?
Vital to company?

Money
Costs?
Prices?
Overheads?
Profits?
Investment?

SWOT analysis – example

Strengths	Weaknesses
Existing customer base Robust distribution channels Strong brand reputation Good sales team Effective marketing	Low profit margins Heavy debt burden Dated manufacturing equipment Slow product development
Opportunities	**Threats**
Good market development prospects Product line extensions Product line monopoly in some outlets Preferred supplier status	Failure to deliver new products Product reliability problems Financial constraints on new business Staff get demoralised

Design Toolkit

PEST analysis

PEST stands for political, economic, social and technological features of the business environment which might influence or even threaten the company [12]. The acronym PEST suggests that these are all nuisance or hindrance factors to business development. New regulations to be complied with, increased taxes or import tariffs, reductions in market size due to demographic changes or the threat of new technology rendering products obsolete – these are all the stuff of executive nightmares. Often, however, PEST factors can often be of considerable business value. Deregulation is opening up many previously inaccessible markets. Innovation awards or business development loans can make unaffordable development work possible. Changing trends in consumer awareness and purchasing habits can re-vitalise stagnant market sectors. And new technology can help you to leapfrog your competitors, provided you get to it before they do!

This technique is, in truth, little more than four headings to stimulate thinking about broad aspects of the business environment that might otherwise go unnoticed.

Political – Changes in law or regulations introduced by Governments or government agencies. Can also include political changes which may stabilise or de-stabilise national markets for products.

Economic – Macro-economic issues can have a strong influence on business. These include national economic growth or recession, exchange rates, stock market fluctuations and changes in national, regional or local fiscal policy. Economic issues of more immediate significance include bank lending rates, the availability of grants, awards, subsidies or unsecured loans and trends in economic indicators such as salary levels or raw material prices.

Social – Demographic trends include changes in the population age structure, trends towards a multi-racial, multi-cultural society and increasing levels of education. Social awareness of environmental issues has had a profound effect on purchasing habits in many market sectors.

Technological – Materials, processes, control and information systems, energy sources; they are all developing at an accelerating pace.

Design Toolkit

Tracking study

A tracking study is an on-going programme of market research which monitors changes in customer perception of a company, brand or product identity. For the purposes of corporate planning, it can be useful to conduct a single study of a small number of customers to discover or confirm customers' perceptions of the company and the products it offers. Since this is market research, most of the principles described in the main market research toolkit on p 191 apply. One vitally important, though perhaps obvious, principle is that the customers must be not be aware of which company the research is focusing on. Otherwise they are likely to give a particularly glowing report of that company, to avoid causing offence. The objective in a tracking study is to obtain a customer perspective on your company, relative to competing companies. The study proceeds, as shown in the example below, by moving from general to specific questions. Usually customers are asked for their spontaneous awareness before being prompted for a more specific response.

Tracking study – example

1. Tell me the names of major chains of restaurants that you are aware of.
   (Aim – to find spontaneous awareness of target company)
2. Check off, on this list the major chains of restaurants you have heard of.
   (Aim – to find prompted awareness of target company)
3. Tell me the names of fast food restaurants that you are aware of.
   (Aim – more focused spontaneous awareness)
4. Check off, on this list the fast food restaurants you have heard of.
   (Aim – more focused prompted awareness)
5. Check off on this list the fast food restaurants you have noticed advertising over the past six months.
   (Aim – Spontaneous awareness of adverts)
6. Check off on this list the fast food restaurants your overall order of preference. (Aim – Spontaneous company preference)
7. From the fast food restaurants on this list, tell me how you rate them on:
   - i) Speed of service 1 - excellent..........5 - poor
   - ii) Friendliness of service 1 - excellent..........5 - poor
   - iii) Quality of food 1 - excellent..........5 - poor
   - iv) Hygiene 1 - excellent..........5 - poor
   - v) Value for money 1 - excellent..........5 - poor
   - vi) Exciting new products 1 - excellent..........5 - poor

Design Toolkit

Product maturity analysis [13]

Product maturity analysis reveals where each product in the company's current range is in their product life cycle and projects forward to estimate when that product is likely to start to decline in sales. New products can then be planned to replace these declining sales. This can only ever be done approximately, because market forces and competitor actions can precipitate sales decline.

The main stages of a product's business life cycle are shown in Figure 5.13. Upon introduction, the growth of product sales usually begins slowly. Once the product begins to be accepted by customers, sales growth starts to accelerate. In a retail market, this sales growth occurs when major retailers decide to stock the product. Sales growth starts to slow down when all market outlets have been fully stocked and replacement stock is ordered only as customers purchase the product. This often coincides with stiff competition between rival companies. By this stage, competitors have had a chance to catch up with a new product, or alternatively, have cut their prices to prevent losing too much market share. The period when sales growth slows down is, therefore, called the shakeout phase because some of the products previously on the market must lose ground to the new product. Provided the new product survives the shakeout, it moves into a maturity phase, during which sales remain roughly constant. They will remain there until the next shakeout period which will follow the introduction of another new product. Once dislodged from their mature position in the market the sales decline is often dramatic. Interestingly, the curve for profit sometimes follows a slightly different pattern to that of sales revenue. During the mature phase of the product life cycle, development costs have usually been recovered and advertising expenditure can drop now that the product is established. This means that profit margins can increase significantly. If, however, competing companies are in a similar position, price wars can take place, cutting profits.

The implications of the product life cycle for the strategic planning of product development are simply that a development plan should be prepared for every product in, or approaching, its maturity phase.

Figure 5.13
Product life cycle

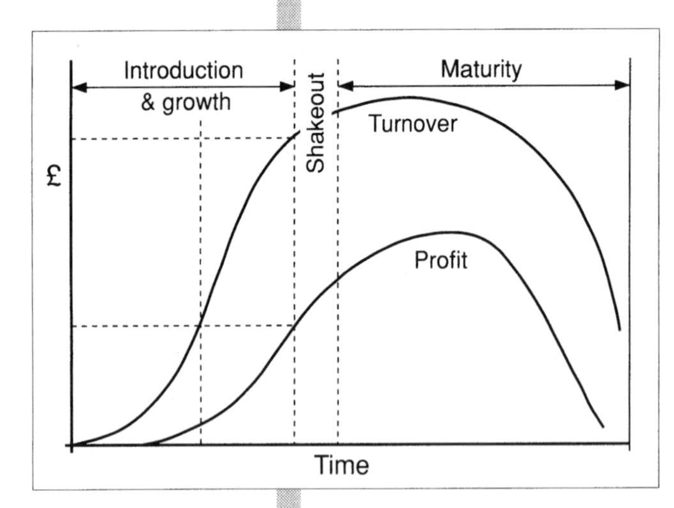

Design Toolkit

Product maturity analysis (Cont'd)

The most obvious type of development plan is to design a new product to replace the mature product. Another option, which may be preferable for a market leading product is to improve it rather than replacing it. The possible rejuvenating effect on a product's life cycle is shown in Figure 5.14. Here, a product entering the mature phase of its life cycle is returned to phases of further growth by product improvement at points 'A' and 'B'.

For a company undertaking the strategic planning of product development, the entire range of products currently on the market need to be analysed. This will produce a life cycle chart such as that shown in Figure 5.15. This company sells glass cookware, designed to be 'oven-to-table'. This type of product typically has a life cycle of several years but this is highly variable between products. The underlying manufacturing technology has been understood for decades and progress is expected to continue slowly and incrementally. Projecting product life cycles is, therefore reasonably safe.

The main conclusion from this product maturity analysis is that there is a need (which is both important to the company and urgent) to develop an improvement or replacement product for their current best-seller (product 1). Although still generating sales growth, the rate of growth is showing signs of slowing down. A shakeout may be imminent and if, for any reason, the product failed to survive that shakeout, the company's total turnover would be substantially reduced. Beyond this immediate need, there appears to be a need for intensive product development for one or two years hence. Three products, with very different lengths of life cycle and different sales value, look as though they are all in similar mid-growth phases of their life cycle. They are all being ordered by their target market outlets, although opportunities for further market penetration still exist. They are all likely to reach sales maturity in 12 to 18 months, at which time they will all need to be improved or replaced.

Figure 5.14
The effect of redesign on product performance

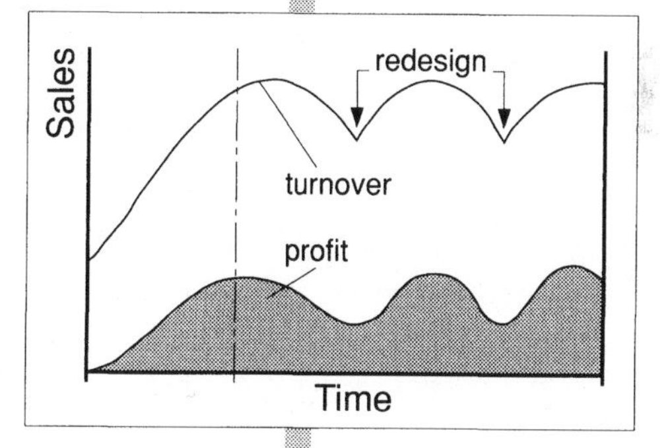

Figure 5.15
Life cycle stages for 6 products in a range

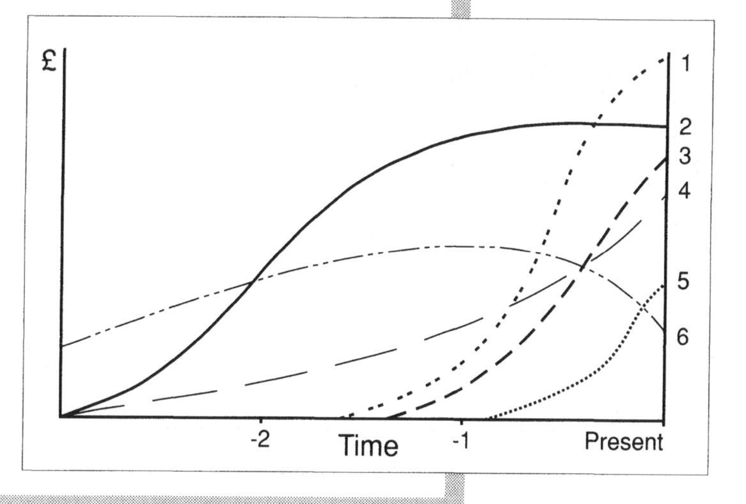

Design Toolkit

Competitor analysis

Competitor analysis examines the performance of competing companies and the product ranges they offer. It tries to establish the ways in which they have achieved business success, and where they have failed. It helps you to anticipate how they are likely to threaten your business in the future and helps you to develop more effective competitive strategies. In Chapter 6 competing product analysis will be described in detail (See page 152). This looks at individual products produced by competitors and analyses them against existing or proposed products of your own. By contrast, competitor analysis is more general and looks at the overall business performance of competitors, in relation to your own business. The key to effective competitor analysis is high quality information. Some of this is relatively easy to obtain, some requires in-depth research over a prolonged period. Publicly available sources can include company annual reports, market survey reports, sales and promotional material and the products themselves. Other sources can be sales and distribution agents, service companies and, possibly the

There are two main aims in competitor analysis:

- To learn from what competitors currently do in order to improve your own business and compete more effectively.
- To deduce their company strategy in order to predict what they are likely to do in the future and shape your own strategy accordingly.

Facts

Total turnover	Manufacturing plant capacity
Profit	Size of sales force
No. of products	Size of product development team
Average turnover/product	Marketing budget (or quantity and quality of marketing output)
Average profit/product	Major successes, Known failures
Patents	

Judgements

Core business function	Speed of past performance in new product development
Marketing mix	Quality of past performance in new product development
Relative value:price ratio	
Customer satisfaction	

Analysis conclusions

Strengths, Weaknesses, Opportunities, Threats } Relative to own company	New product development strategy
	Likely actions in next two years

Action conclusions

Immediate changes to company in order to boost relative competitiveness.	Product development strategy in response to deduced strategies of competitors

Design Toolkit

Competitor analysis (Cont'd)

most valuable source of all, former employees who now work for you!
The first step in competitor analysis is to gather together all the available facts. These should cover all the main competing companies and the major aspects of business common to your own company. Then judgements need to be made about the nature of their business and how it relates to your business. Finally, conclusions must be drawn about the changes you could make to your business to be more competitive, both now and for the future. A way of approaching these three stages is given in the 'facts', 'judgements' and 'conclusions' boxes opposite.

The marketing mix [14]

There are four ingredients to effective marketing of a product: product, promotion, price and place. These are known as the four 'P's' of product marketing.
Product: What are the product offerings from competing companies (considering both tangible and augmented aspects of the product, see below)? What range of products do they offer and how do they relate to each other in terms of common customers, common manufacturing technologies and common sales and distribution channels?
Promotion: What are the brand identities of competing companies? What features of their products do they emphasise in their promotions (quality, price, product performance). Which market sectors do competing companies target? Do they promote products differently to different market sectors?
Price: What are the pricing policies of competing companies (high value, high price or basic value, low price)? How does the price:value ratio compare with your own products? What information can you deduce about the costs of manufacture and sales in competing companies? How do their costs compare with your costs?
Place: Which sales outlets do competing companies use? How does their market penetration of these different sales outlets compare to your own market penetration (stock held or retail shelf space occupied)?

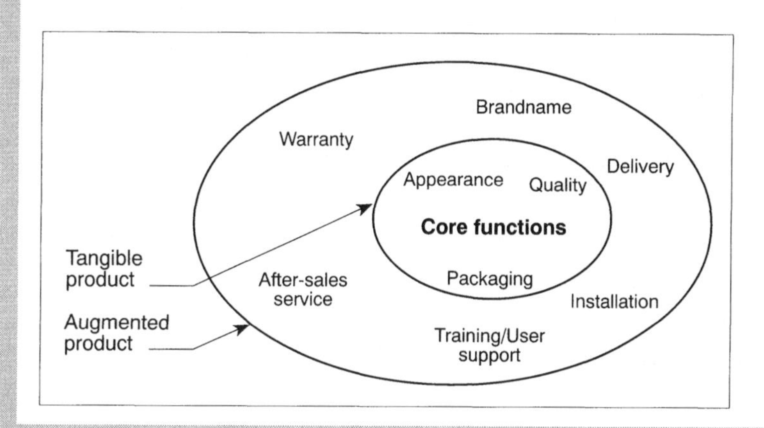

Design Toolkit

Product development risk audit

A product development risk audit is a method of exploring different product development options and assessing them in relation to the company's product development skills and historical track record. As such, it is a key part of preparing a product development strategy. In essence , this is a way of trying to match the ambitiousness of product development projects with the perceived product development capabilities of the company. Most of the measures used are estimates and many are subjective. The purpose, however, is to enable judgements to be made on which type of products to develop and what changes to try to introduce to the product development function within the company. These decisions are based on subjective judgements anyway; it is, therefore, better to have the subjective judgements based on some sort of systematic analysis than simply on the unstructured beliefs of senior managers.

A product development risk audit is conducted in two stages. Firstly, the cost of product failure is estimated, in terms of its impact upon the company's overall business. The more a product failure threatens the company's business survival, the greater the product development capability needed to take the risk of embarking upon the product's development. The second stage is to assess that product development capability.

Cost of product failure This is proportional to the product development costs and the size of the lost business opportunity? Generally, the size of the lost business opportunity will dwarf the costs of product development; otherwise why are you developing the product. The costs of product development are, however, actual expenditures and affect company cash flow. They may, therefore have a more direct and immediate effect on company business than the loss of potential future sales. Both should be assessed in terms of their threat to business survival. A broad classification into 'critical', 'significant' and 'marginal' may be sufficient. For products which are 'critical' to business survival, it is essential that all steps are taken to ensure that the product development capabilities of the company match the magnitude of the product development task. For products whose failure would pose a 'significant' threat, this is important, although less so. For those which are marginal, it may be considered that the costs of changing the product development capability of the company are not justified.

Product development capabilities These are assessed in relation to people, procedures, facilities and money. Formats for their analysis are given in the tables on the following pages.

Design Toolkit

Product development risk audit (Cont'd)

The closer design and development costs are to the company's profit margin for the projected product development period, the more critical the cost of product failure becomes. The greater the proportion of projected company turnover that is made up of projected business from the new product, the more critical the cost of product failure becomes. The failure of the new product can be judged critical from either or both development costs or lost business.

The people, procedures, facilities and financial audits should focus on either a single new product or a type of product development. The objective is to compare the skills needed from the different functions within the company with the skills

Costs of product failure

Profit margin for the projected product development period	£	Projected turnover for company if product is successful	£
Design and development costs for product(s)	£	Projected loss of business due to product failure	£

Conclusion Critical ☐ **Significant** ☐ **Marginal** ☐

People audit

Function	Creative idea generators		Technical problem solvers		Meticulous/ attention to details		Practical/ realistic	
	HAVE	NEED	HAVE	NEED	HAVE	NEED	HAVE	NEED
Market orientation								
Design and development								
Prototyping and testing								
Production engineering								
Marketing and sales								
Financial control								

Product development risk audit (Cont'd)

available in the existing staff. These can be described in terms of the level of skill required (excellence, competence, basic) and the importance of this skill to the company (essential, important, marginal).

Outcome The more critical the cost of product failure, the more important it is to take action to match the product development capabilities the company has with those that are needed. Three types of action can be taken. Internal reorganisation introduces new pro-cedures and re-assigns staff to different functions. Internal investment buys new facilities, recruits new staff or trains existing staff. External investment recruits con-sultants, forms partnerships with other companies or draws up supplier contracts.

Facilities audit

	HAVE	NEED
Prototyping		
Prototype testing		
Manufacturing and assembly		
Distribution and sales		

Procedures audit

	HAVE	NEED
Product planning		
Specifications		
Effective idea screening		
Prototype testing		
Market testing		

Financial audit

	HAVE	NEED
Design and development costs		
Investment for production		
Marketing and advertising		
Production of stock		

Design Toolkit

The design team

Meredith belbin's research has shown that the ideal team, in a business environment is one in which a mix of personalities and characters are present. Based on his questionnaire (see p 123) he identifies the following mix as the 'dream team'.

Chairman	Calm, self-confident, controlled.	Capacity for welcoming all contributions and treating them on their merits without prejudice. Strong sense of objectives.	No more than ordinary in terms of intellect or creative ability.
Company Worker	Conservative, dutiful, predictable.	Organising ability, practical common sense, hard working, self-discipline.	Lack of flexibility, unresponsiveness to unproven ideas.
Shaper	Highly strung, outgoing, dynamic.	Drive and a readiness to challenge inertia, ineffectiveness, complacency or self-deception.	Prone to provocation, irritation and impatience.
Plant	Individualistic, serious minded, unorthodox.	Genius, imagination, intellect, knowledge.	Up in the clouds, inclined to disregard practical details or protocol.
Resource Investigator	Extroverted, enthusiastic, curious, communicative.	Capacity for contacting people and exploring anything new. An ability to respond to a challenge.	Liable to lose interest once the initial fascination has passed.
Monitor/ Evaluator	Sober, unemotional, prudent.	Judgement, discretion, hard-headedness.	Lacks inspiration and the ability to motivate others.
Team Worker	Socially oriented, rather mild, sensitive.	An ability to respond to people and situations. Promotes team spirit.	Indecisive at moments of conflict.
Completer—Finisher	Painstaking, orderly, conscientious, anxious.	Capacity for follow-through, perfectionism.	Tendency to worry about small details. A reluctance to 'let go'.

Notes on Chapter 5

1. Reinertsen D.G. 1983, Whodunnit? The search for new product killers. *Electronic Business*, July 1983. The economic model used by Rhinertsen to produce these figures was later published in full in Smith P.G. and Rheinertsen D.G. 1991, *Developing Products in Half the Time*. Van Nostrand Reinhold, New York, pp28-41.
2. Design Council 1994, *UK Product Development, A Benchmarking Survey.* Gower Publishing, Hampshire, UK.
3. Adapted from Freeman C. 1987, *The Economics of Industrial Innovation.* Francis Pinter Publishers, London.
4. Schnaars S.P. 1994, *Managing Imitation Strategies.* The Free Press, New York pp 54-59.
5. Nayak P.R, Ketteringham J.M and Little A.D. 1993, *Breakthroughs!* (2nd Edition) Mercury Business Books, Didcot, Oxfordshire, UK pp 6-28 and Schnaars S.P. 1994, *Managing Imitation Strategies.* The Free Press, New York pp 168-174.
6. For a general introduction to corporate strategy see Johnson G and Scholes K 1993, *Exploring Corporate Strategy.* Prentice Hall, London.
7. Henderson S., Illidge R. and McHardy P. 1994, *Management for Engineers.* Butterworth Heinemann Ltd. Oxford, UK pp 60-62.
8. Henderson, Illidge and McHardy 1994, (see 7 above) pp 62-64.
9. Belbin M.R. 1994, *Management Teams: Why they Succeed or Fail.* Butterworth Heinemann Ltd. Oxford.
10. Inwood D. and Hammond J. 1993, *Product Development, An Integrated Approach*. Kogan Page Ltd. London pp 102-159.
11. SWOT analysis is probably the most well known and frequently used business analysis tool. For a case study of SWOT analysis and the strategic change which followed, see Henderson, Illidge and McHardy 1994 (see 7 above) pp 45-49.
12. The classic text on analysing the business environment of a company is Palmer A. and Worthington I. 1992, *The Business and Marketing Environment.* McGraw Hill, Berkshire, UK.
13. For an in-depth treatment of the evolution of both products and markets see Moore W.L. and Pessemier E.A. 1993. *Product Planning and Management: Designing and Delivering Value.* McGraw Hill Inc. New York. For a more concise summary see Henderson, Illidge and McHardy 1994 (see 7 above) pp50-56.
14. The marketing mix will be found in nearly every texbook on marketing! For a product design perspective on the topic see Urban G.L. and Hauser J.R. 1993, *Design and Marketing of New Products* (2nd Edition). Prentice Hall Inc, Englewood Cliffs, New Jersey, pp 357-378. The notion of the tangible and augmented product can be found in Henderson, Illidge and McHardy 1994 (see 7 above) p20 or Inwood and Hammond 1993 (see 10 above) p 36.

6 Product planning-opportunity specification

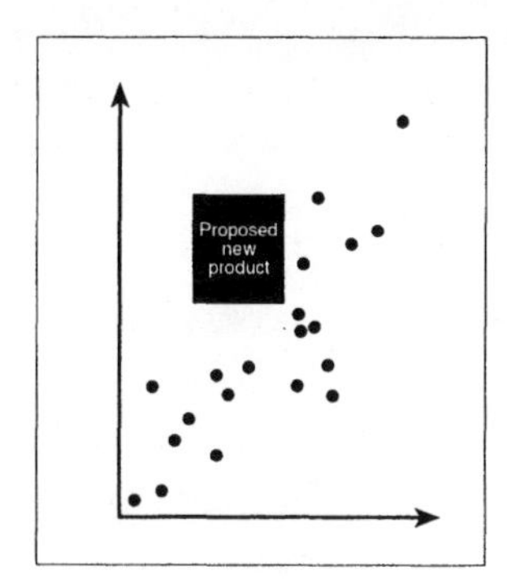

Product planning includes spotting an opportunity, conducting market research, analysing competing products, proposing a new product and drawing up both an opportunity specification and a design specification. This topic takes up more pages than any other in the book, which is appropriate because it is probably the most important. That might seem strange in a book on product design. How come the most important topic comes before the start of the actual product design process? Well, there are good reasons for this. Remember from Chapter 2 that products which are clearly and precisely specified before the start of development have three times the chance of succeeding than those that do not. Remember also that a lot of the decisions that determine what the product must do, who it must appeal to and what constraints it must meet are decided at this stage. In other words, a great many of the product design decisions are made at this stage even though the actual design process has not yet started. Remember, most important of all, that the work done at this stage is comparatively cheap. Spending a lot of time and effort getting the product planning right is time well spent, compared to trying to correct problems later.

Product planning is one of the most difficult activities in the development of new products. It can be frustrating: you feel as if you are grasping at straws trying to research and specify a product that you have not yet started to design. Most designers are itching to start bringing the new product to life by sketching and modelling. Product planning, therefore, needs self-discipline. You ***must*** take the time for thorough and carefully considered product planning if your product is going to have the best chances of success and if you are going to save yourself a lot of grief as the product develops.

The product planning process

Product planning begins with the company's product development strategy and ends with a design specification for a new product. A product development strategy, as we saw in Chapter 5 describes, in general terms, the company's approach to innovation. It proposes how that company plans to turn product innovation into business success. It describes the positioning of the company's products in the market and thereby determines the sort of new products the company seeks to develop. In short, it sets the ground rules for product innovation.

Identifying and specifying a new product opportunity from this general innovation strategy is neither simple nor straightforward. It requires the use of a whole new set of tools and techniques. It requires intellectual rigour and, as suggested above, a great deal of self discipline. The reason for this is that product planning has many possible starting points, many opportunities and constraints to take into account and many options to explore. Designers often describe this part of product development as the period of 'free fall'. You are hurtling through space with a myriad of ideas flashing past and you reach out to grasp for them, before they disappear beyond your reach. All product planning has time constraints which means that you only have so long to find the right opportunity. If, during free fall, you are able to derive a design specification systematically and thoroughly, your parachute opens – you land softly, with confidence and ready to move straight into product design. If you fail to get that design specification properly resolved, the landing will be harder. Even if you survive the fall, the product may not survive development!

Figure 6.1
The product planning process

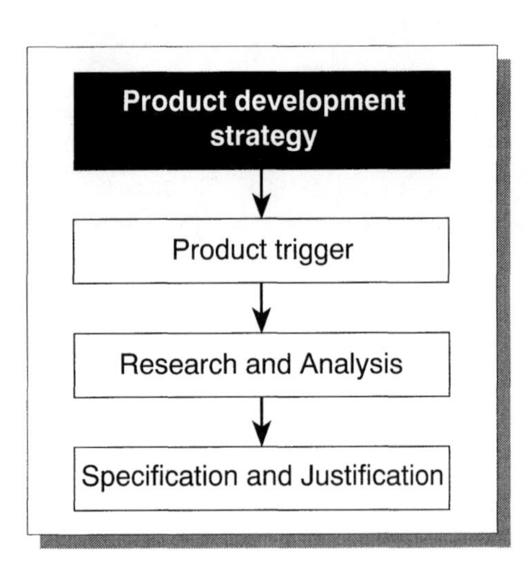

Product planning can be seen as falling into four main stages (Figure 6.1). Firstly, there is the product development strategy giving a general orientation to product planning and establishing its objectives. This can be seen as guiding the entire product planning process. Secondly, there is the stimulus to start the development of a particular product. This is called the 'product trigger'. Thirdly, the opportunities and constraints are researched and finally the proposed new product is specified and justified.

How you get from the innovation strategy to the design specification can vary considerably from product to product and company to company. Some products will be very much easier to assess and specify than others. A product which simply updates an existing product, for example, may be quite straightforward. On the other hand, a product which uses new technology, has a radically innovative design or is to sold into an unfamiliar market will be more difficult. In general, the more expensive the product is to develop the more in-depth the product planning will tend to be, in order to justify the investment. As we can see in Figure 6.2 product development projects cost from hundreds of thousands of dollars to billions of dollars. Boeing or Chrysler would, understandably require more evidence to justify their commitment of billions of dollars to develop their new product than Stanley Tools would require for their commitment of $150 000. Also, different companies vary considerably in the formality of their product planning process. Some have a variety of quality control checks relating to financial, marketing and technical aspects of the product. Others have a more informal approach and require a case to be presented, covering all aspects of product planning.

Figure 6.2
Developing a new product means very different things to different companies [1]

Company	Stanley	Hewlett-Packard	Chrysler	Boeing
Product	Jobmaster Power Screwdriver	DeskJet 500 Printer	Concorde Automobile	777 Aeroplane
No. of Unique Parts	3	35	10 000	130 000
Development Time	1 year	1.5 years	3.5 years	4.5 years
Development Team	3 people	100 people	850 people	6800 people
Development Costs	$150 000	$50 million	$1 billion	$3 billion
Sales Price	$30	$365	$19 000	$130 million
Annual Production	100 000	1.5 million	250 000	50
Sales Lifetime	4 years	3 years	6 years	30 years
Devel. Costs/Lifetime Sales	1.2%	3%	3.5%	1.5%

There is no single correct way, so long as product planning is thorough, systematic, realistic and conducive to sound decision-making in the end. For the sake of clarity, I have chosen to present a single structured approach to product planning. As with much of this book, the principles should be understood and followed as faithfully as possible. But the specific activities and the tools and methods by which these activities are carried out should be customised to suit the company and the product being developed.

Commitment - the aim of product planning

All good planning exercises should have a clear aim. The tangible outcome from product planning is getting commitment from management to begin designing the new product. The market research, opportunity spotting and design specification can all be seen as means to that end. Very useful means - they will subsequently steer and quality control the entire design process - but means, nonetheless. So, how do we get that commitment?

'Prove to me that the product is going to succeed and I will approve its design,' say the company management. 'Let me design the product and I will prove it will succeed,' replies the designer. This is the Catch 22 of all new products. The way out of it, of course, gets us back to the staged risk management funnel described in Chapter 2. Set a target specification for the new product. Take one step towards developing the product and evaluate its commercial viability against this specification. If it looks good, take the next step and re-evaluate. The problem, as far as product planning is concerned, boils down to setting the specification.

Again management and designers will have different perspectives on this. Of most interest to management is how this new product is going to differ from existing products and what business opportunities arise from this differentiation. For the designer, a more detailed specification is needed describing what features the product must have and how can they be made. Another difference is that management will want to set the business objectives and stick to them. Unless you maintain a clear vision of the business opportunity presented by the new product you may simply end up with a solution looking for a problem. Bad news for business! The designer, on the other hand, will want a degree of flexibility in the specification. Aiming for a specific technical objective always depends upon assumptions about the range of possible solutions. The design of a compact camera, for example, would normally specify the durability of the camera casing to ensure that it stands up to reasonable wear and tear over several years. This specification, however, is based on assumptions which do not include the possibility of a disposable camera, discarded after a single film. It nearly always happens, during the course of design, that new solutions will emerge. Having to stick rigidly to an original target might rule out any new, and possibly much better, solutions.

What we need, therefore, is two levels of commitment (Figure 6.3). Firstly, commitment to the business objectives for the product. These business objectives must focus on a commercial opportunity for a new product identified in the marketplace. They must specify what the product must achieve, in business terms, to exploit that opportunity. They must also present a financial justification for the proposed investment in developing the new product. The document containing these business objectives is called an 'opportunity specification'. Secondly, commitment must be made to a more flexible specification of the specific technical features of the new product. These technical features, regardless of how they may be modified during product development, must always remain sufficiently focused to ensure that the agreed business objectives are met. This technical description of product objectives is called a 'design specification'.

Separating the opportunity specification from the design specification has several advantages.

- It forces attention to be focused firmly on the business objectives presented by the new product before getting carried away by how exciting the design opportunities are.
- It can streamline decision-making. Senior management within a company should be content to approve the business objectives described in the opportunity specification. They can then delegate responsibility for technicalities of the design specification to the product development team.
- Most importantly, it achieves a nice balance between effective quality control and innovative freedom. Provided the technical specification remains faithful to the business objectives, it can be changed, if necessary, to take account of new ideas which emerge during the product design process.

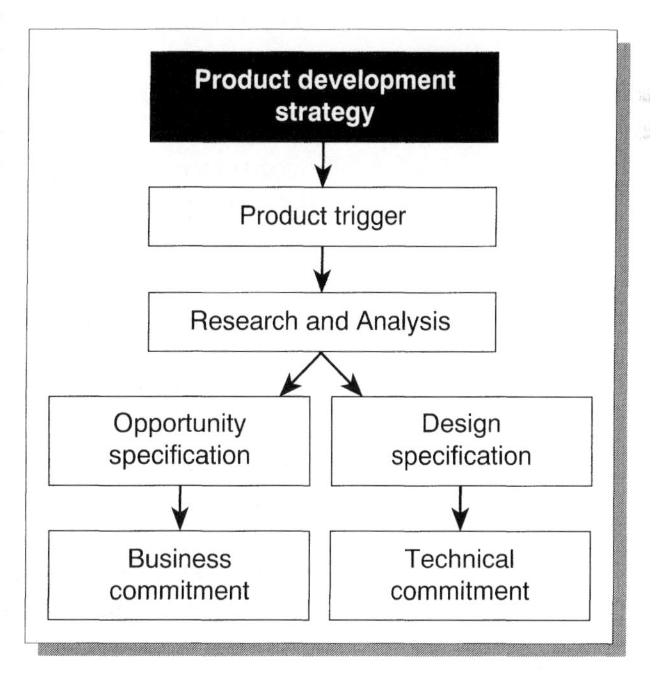

Figure 6.3
The refined product planning process

One rather obvious, but very important difference between an opportunity and design specification is the level of detail that they specify about a product. An opportunity specification need not make any commitment to the form or function of the new product. It can simply identify the opportunity that is to be exploited and the costs and benefits of that exploitation. How the new product is designed in order to exploit the opportunity can be left until later. The design specification, on the other hand, needs to provide specific design targets. It must contain sufficient detail to allow the designer to know whether or not the product is fit for its intended purpose. To be able to set effective design targets, the designer must clearly need to know what sort of product is being aimed for. Often, this is not known until concept design is complete. Which brings us to the most important reason for separating the opportunity specification from the design specification. The opportunity specification is prepared at the very start of the product development process. It makes the case to management that there is a viable business opportunity to be exploited, even though it is not yet known what design features the new product will have. Approval of this opportunity specification allows concept design to be completed with the aim of developing a set of functional and styling principles which exploit the specified opportunity. Once these design principles are established, the product can be specified in more detail. Specific design targets can sensibly be produced now that it is known what type of product is going to be developed. In other words, the design specification can be written after concept design is complete.

The product planning process is, therefore, split into two sections in this book. In the remainder of this chapter I describe the development of the opportunity specification and show how initial commitment to product development is secured. Then, in Chapter 8, following the chapter on concept design, we will go on to examine how the design specification is prepared.

What is an opportunity specification?

Let us consider, in a little more depth what we are trying to achieve with an opportunity specification.

- To secure commitment to development of the product, it must fulfil two functions: it must describe the opportunity and then justify it in business terms.

- To be seen as satisfactory, a business opportunity must present good prospects of profit for the company.
- To be profitable, a product must sell in sufficient numbers to exceed its development costs.
- To sell, a product must offer customers a clear benefit over existing products. New products which only manage to be as good as existing products, offer customers no incentive to change their established buying habits and are therefore likely to fail.
- To offer such a benefit, there must be significant product differentiation between the new product and its competitors.

This tells us a lot about what the opportunity specification must contain. At the heart of the opportunity specification is a simple and concise statement called a 'core benefit proposition' [2]. This describes what the customer will see as the benefit in buying the new product in preference to competing products. For some products the core benefit will be determined by specific market pressures. Manufacturers of small batteries for consumer goods, for example, openly compete and advertise their products on the basis of battery life. The core benefit for a new battery is, therefore, almost certain to be 'longer battery life'. For the majority of products, however, the core benefit will be less constrained. A new children's car safety seat, for example, might have the core benefit of 'easy (one handed) belt fastening' and the development effort would consequently go towards the ergonomics (and, of course, safety) of the belt fastening system. Alternatively, it might have the core benefit of 'lowest price for excellent safety specification' or it may have a multifunction core benefit, 'lowest price for excellent safety specification and widest range of optional toy accessories'. The core benefit proposition should be a simple common sense statement describing the advantage the product will have relative to others in the market. It should be the sort of statement that customers would understand, and may, indeed, form the main advertising line for subsequent marketing of the product.

The opportunity specification, as a whole, needs to contain more detail than this. It must describe the full range of factors which will determine the product's commercial success. Given that the product has the advantage specified in the core benefit proposition, how does it compare in terms of other aspects of function, in terms of price or in terms of appearance. So, returning to the child safety seat example, let us assume that we have selected the core benefit 'easy (one handed) belt fastening'. Clearly customers will only pay so much for that benefit and to make the business opportunity clear, a pricing position should be specified. A precise figure may not be

> **Percentiles**
>
> Percentiles are a way of describing a numerical position in a range. For this pricing example, imagine that there were 100 competing products on the market. To specify a price no greater than the 75th percentile of competing products means that the new product will be priced at no more than the 75th most expensive of these 100 products.

necessary at this stage; it should be sufficient to say that the new product will be priced at no more than a particular percentile (e.g. the 75th percentile) of competing product prices. Since this new safety seat is a high value product with a new feature (easy belt fastening), customers are likely to expect other high value features - perhaps luxury cushioning in high quality fabric. Any such additional but important features need to be mentioned in the opportunity specification. Any other selling advantages for the new safety seat also need to be included, even if they have nothing to do with its core benefit. This might, for example, be the first safety seat to make extensive use of recycled plastic in its manufacture.

Always bear in mind that the opportunity specification does not necessarily need to be described quantitatively; this can be left to later, when the design specification is written. It should be quite sufficient to relate to other products on the market. So, the new product will be better than competitor A, will include listed features from competitors B, C and D, and, as we have said above, it will be priced no higher than the 75th percentile of all competing products. Also, the opportunity specification does not need to be a comprehensive list of all aspects of the new product. It just needs to cover the key factors in making the product a commercial success. Ways of obtaining the relevant information and arriving at these decisions will be described later in this chapter.

Opportunity justification

So far, we have a core benefit proposition for the new product, which has then been developed to describe more fully, the opportunity presented. This should be a sufficient description of the product opportunity to allow the company management to understand what is being proposed. Of course, if the new product is to be sold in a market unfamiliar to the company, background information on the characteristics of that market will also have to be provided.

The next part of the opportunity specification is to justify the opportunity. This is mainly concerned with financial matters. But let us start with the non-financial aspects and get them out of the way first. These concern the company's ability to manufacture, distribute, market and sell the new product. This is really only of importance if any of them are likely to be a problem or require previously untried procedures. Will the product involve new manufacturing or assembly processes? Can it be quality controlled using existing methods? Does the product present unusual transport or shipping problems? Can it be sold through existing distribution and sales channels? Will current marketing methods be appropriate? If no difficulties are foreseen in these areas, then a simple statement to say so is all that will be required.

Financial justification requires four aspects of the new product to be specified.

- What are the variable costs for the product? These are the costs for manufacture, distribution and sales per unit of the product sold. They are described as variable because they vary in proportion to sales volume.
- What are the fixed costs for the product? These are the costs of designing and developing the product and of developing tools and facilities for its manufacture. They are described as fixed because the do not vary with sales volume.
- What is the target price for the product and hence what is its margin over costs?
- What is the projected life-cycle of product sales? From this, how long will it take to recover the fixed costs, before it starts to become profitable and what is the total projected profit from lifetime sales?

As you might imagine, obtaining all this information before the product has been developed is far from easy. There are, however ways of dealing with this and these, too, will be described later in the chapter.

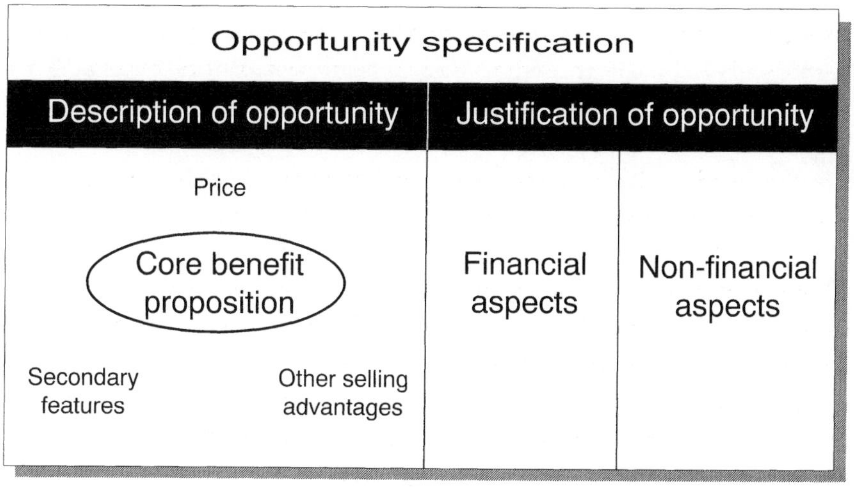

Figure 6.4 Contents of the opportunity specification

The contents of the opportunity specification can, therefore, be summarised as shown in Figure 6.4. Techniques for preparing an opportunity specification are given in the design toolkit on page 198.

Researching and analysing the opportunity

Research and analysis is generally the time-consuming part of product planning. Research is conducted to identify, evaluate and justify the opportunity. The problem with research is knowing when you have done enough. In theory, you could go on for ever, seeking that extra nugget of information to reveal a slightly better opportunity than you had before. Clearly, some sort of goal must be aimed for. The guiding principle in all product planning is that a product opportunity is satisfactory when it confirms the commercially viability of the product and demonstrates consistency with the company's product development strategy. The proposed new product must make progress towards reaching the company's product development objectives. This, in turn must be in line with the company's corporate strategy, must make progress towards corporate objectives and must be aiming towards the company's mission. We are back to a version of the risk management funnel (Figure 6.5). The research objective must be to explore product opportunities which fit the product development strategy. Such an objective is far from specific and this is the stage of product development which was described earlier as 'free-fall'. But, at least, within the framework of strategic planning, you know where you want to land and you will know, once you have landed, whether you are in the right place.

Figure 6.5
Risk management funnel from mission to product opportunity

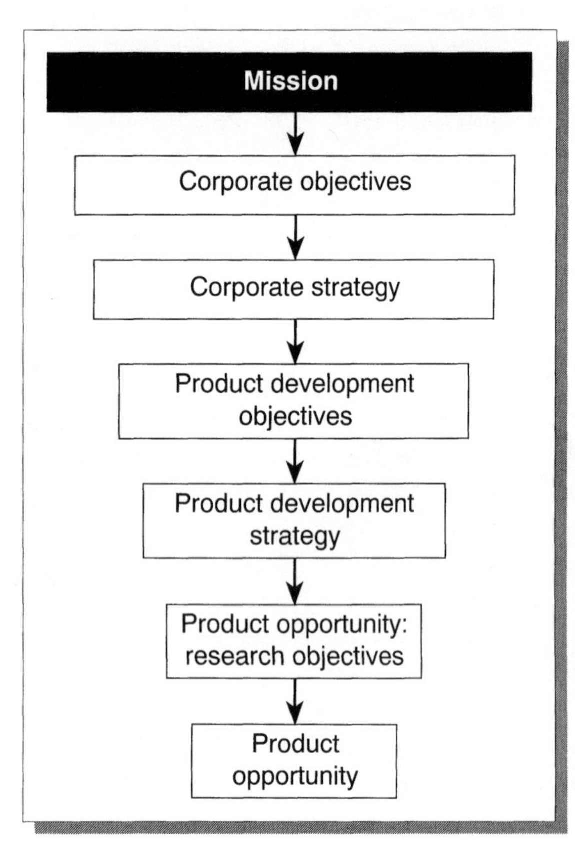

There are three main sources of information for researching a product opportunity.

- The demands and wishes of customers, discovered by 'market needs research'.
- The competition offered by existing products discovered by 'competing products analysis'.
- The technological opportunities for designing

and manufacturing the new product discovered by 'technology audits'. A significant business opportunity exists only when there is an identifiable difference between i) the demands and wishes of customers and ii) the products offered by competitors. This business opportunity can only be exploited when the available technology allows a product to be made which satisfies the previously unsatisfied market demands.

The way in which you start to research a product opportunity very much depends upon what triggered the need for the new product.

Product triggers

Product triggers fall into two main categories: market pull and technology push. Market pull refers to the demand by the market for a product or product features not currently offered by your company. This market pull may be recognised in two ways. Firstly, competing products may be edging ahead of your present products, creating a market demand for you to enhance your products in order to catch up. Secondly, there may be market needs not currently satisfied by any existing product. These two types of market pull require to be exploited by the imitative and pioneering innovation strategies described in Chapter 5. Technology push refers to the availability of a new technology creating an opportunity for product innovation. This new technology could be a new material, a new manufacturing process or a new design concept.

Surveys of the commercial success of products triggered by either market pull or technology push show that market pull products are three times as likely to succeed [3]. This is consistent with the importance of market orientation in new products which has been emphasised throughout this book. Not even the most technically awe-inspiring new product will sell if customers do not want to buy it. This does not diminish the importance of technology push as a trigger for new product development. It just means that the technological excellence of an idea is not sufficient, on its own, to make a new product successful. Thorough and detailed market research is needed to ensure that the new technology is fulfilling a real customer need.

A significant business opportunity exists only when there is an identifiable difference between i) the demands and wishes of customers and ii) the products offered by competitors. This business opportunity can only be exploited when the available technology allows a product to be made which satisfies the previously unsatisfied market demands.

The process of product planning is, therefore, quite different for products with market pull and technology push triggers. If triggered by technology push, you need to explore the commercial opportunity which it can exploit. This requires market needs research and competing products analysis but the need for research into available technologies is less important because you have already identified the technological opportunity. If you start with a business need, triggered by market pull you need to find the product to exploit it. This requires technological research and may be inspired by competing products analysis but the market needs research is less important because you already have the market opportunity. I say 'less important' because the need for such research is not dispensed with. Identifying one opportunity is very different from identifying the best opportunity. Market research may well come up with other unsatisfied needs and some of these may be better than the first one you came up with.

Short cuts in product planning, and particularly in the underlying research, are always tempting. Under severe time pressures they may be inevitable.

But there are two important things to remember:

- The better the product planning, the more chance the product has of commercial success.
- The more time spent in product planning the more time will be saved later in product development.

When you first begin product planning, the start point – the company's product development strategy – may not give rise to either a market push or a technology pull trigger. It may simply identify a general need for innovation, in order to continue to meet changing market needs, keep up with competitors and exploit emerging technologies. This means that researching the product opportunity begins with a 'clean sheet'. Market needs must be explored, competing products must be analysed and technological opportunities must be identified. For the remainder of this chapter I will assume this to be your starting point. This allows me to describe all aspects of the research which may, at some point, be needed for product planning.

Competing product analysis

In practice, it is usual to start analysing competing products before researching market needs. This can give the subsequent market needs research a much clearer focus and allow more structured and meaningful questions to be asked to potential customers.

Competing product analysis sets out with 3 general aims:

- To describe the variety of ways existing products will compete with the proposed new product.
- To identify or evaluate opportunities for innovation.
- To set targets which the new product must meet in order to compete effectively.

These aims are pursued by analysing the characteristics of products which are likely to compete with the proposed new product. This means products which customers may consider buying in place of your new product. On the basis of this analysis, and the subsequent market needs research, decisions must be reached about the product opportunity. Competing product analysis must, therefore be conducted in a way conducive to that decision-making.

Figure 6.6
What is competing product?

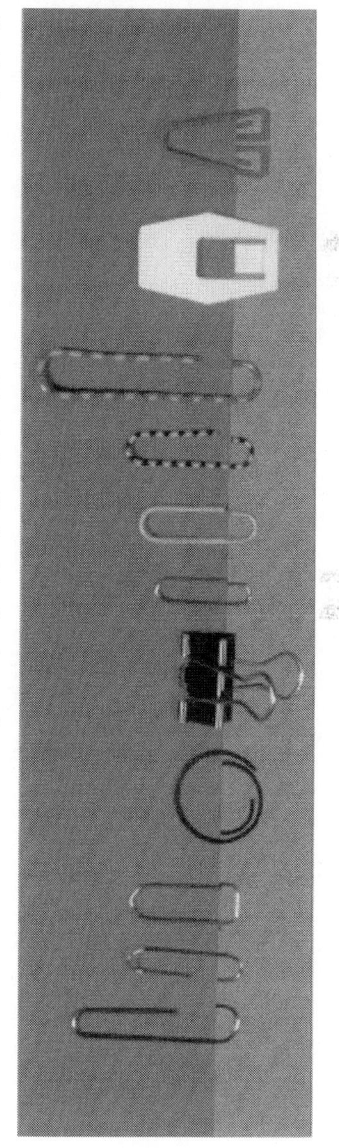

The problems which competing product analysis must overcome are:
- Deciding what constitutes a competing product.
- Establishing what characteristics of competing products to study.
- Deciding what criteria to use to set targets for the new product.

Deciding what constitutes a competing product is not be as straightforward as it might seem. A Rolls Royce, for example, may safely be concluded not to compete with a compact, economy car. But where do you draw the line? Even with something as straight-forward as a paper clip, what are the competing products. If you define a paper clip as a device for temporarily holding several loose sheets of paper together, competing products could include tags, split-pin fasteners, bull-dog clips, dress-making pins, staples and spring-clip spines (Figure 6.6). With a little more lateral thinking you might also include a clip-board, envelope, pocket file or even a filing cabinet! The initial judgement must be based on common sense. Few customers are going to walk into a shop with the intention of buying a packet of paper clips and then suddenly decide what they really want is a filing cabinet! More specifically, where to draw the line on competing products should be firmly based on market forces. The choices customers face when making a purchasing decision will obviously depend upon the range of products stocked in the target sales outlets for your proposed new product. These 'stocked' competitors must, therefore, be the focus for all competing products analysis.

Judgement is also needed in deciding what characteristics of competing products to study. What you want to know about competing products obviously depends upon how you see these products competing with your proposed new product. If the company's strategy is to produce basic, low price products, then the price of competing products and the design factors which determine their cost of manufacture will be of greatest importance. If, on the other hand, you have a high price, high value strategy, you focus attention on how the competing products offer value to the customer, in terms of both performance and appearance.

Setting targets, as described earlier, is the main aim of product planning. In preparing an opportunity specification, the type of targets you need to set are those of relevance to the business performance of the proposed new product. These generally amount to price and measures of product value (as perceived by the customer). A worked example of the decision-making for competing products analysis begins on page 169.

Progressive companies conduct competing product analysis on an on-going basis, rather than just before the development of a new product. Some even make a point of buying every new product in their market sector as soon as they are launched, thereby maintaining a 'library' of competing products. Analysing this product library can reveal trends in competing products and any significant new development in a competing product can act as the trigger for the development of a product to catch up with the competition. Other companies delegate this task to more specialist consultants.

Market needs research [4]

Understanding the needs of customers is absolutely fundamental to identifying, specifying and justifying a product opportunity.

Market needs research can be based on four main sources of information:
- In-house market intelligence
- Library research
- Qualitative market surveys
- Quantitative market surveys.

One of the main assets of an established company, operating in a familiar market, is its knowledge of that market. This must be used to the full in developing an opportunity specification. A difficulty can arise if the people involved in product development are not the people with the best knowledge of customer needs. The company sales force or the service personnel who maintain and repair products may have a much better understanding. They should know what customers want, the extent to which the company's existing products provide what they want and how this compares with the performance of competing products. This information can be distilled in a variety of ways: formal interviews or meetings, informal discussions, carefully prepared questionnaires or by asking for 'wish lists'. A wish list is, as it sounds, a description of what salesmen or service personnel would ideally like from a new product in order to satisfy their customers better. An example of how New York Life Insurance used salesman knowledge to develop a new insurance policy is given in the box above.

Lee Gammill, an executive vice-president of the New York Life Insurance Company took salesman participation to extremes in designing a new life insurance policy in 1986 [5]. His company needed a radical change in the products it was offering in order to revitalise its business. Gammill's approach was to take the company's top six sales staff, along with several specialists in different aspects of life insurance and lock them away in a room together. He told them they were not getting out of the room until they had come up with a type of policy that was noticeably different from competing products. After only two days, a new type of policy had been agreed. The salesmen thought it would have particular appeal to customers and the experts felt it would be financially viable. They were right. Within one year, sales volume was up by 80 per cent.

Company records can also provide useful pointers to customer needs. What type of products are selling best at present? Do they have any features in common? How do they differ from the products which are not selling so well? Have sales changed significantly recently? If so, why? Was it because of changes in customer preferences?

Company sources of information cannot, however, be expected to provide a complete description of customer needs. Company records will give only a patchy and incomplete picture of the changing needs of customers. Information from salesmen and service personnel will mostly be based on what customers have volunteered. Since their main interest is to sell or maintain existing products, they are unlikely to have sat down at length with customers asking what they would like in a new product. Another problem is that their information is largely historical: a particularly important suggestion from a customer three years ago may carry more weight in their minds than smaller details from the more recent past. When you research company sources you must be aware of these limitations and weigh up the information accordingly. To be sure of the value of company sources of information, and their interpretation, it is usually best to confirm them by talking

Figure 6.7 Published market research can provide a gold mine of information

The market research company MINTEL publishes regular in depth reports on products and services in many market sectors. Their 1992 report on cookware [6], for example, included a breakdown of that market into product categories such as scissors with a retail sales value of £29M and kitchen knives £19M. They also categorise products in other ways, such as by material/surface finish: non-stick cookware amounted to 68% of the market, an increase of 24% since 1988, whereas glass cookware had dropped 42% to only 4% of the market. Information of specific companies is also included: the current suppliers of non-stick cookware include Tefal (T-plus, Super T-plus, Ultra T-plus and Resistal), Du Pont (Teflon and Silverstone), Weilberger (Greblon) and Whitford Plastic (Xylan and Excalibur).

directly to customers.

The next source of information for market needs research is library research. This does not necessarily mean 'research done in a library'. Rather, it should be interpreted more liberally to mean research using any form of published information. Published reports by market research organisations can provide a gold mine of information, if they are sufficiently focused on the type of product you are considering (Figure 6.7). Sources of these are provided on page 158.

The best way, by far, to discover customer needs is to go out and talk to customers. To be of most value this should be done in a structured way using formal market research techniques. This does not necessarily mean that the research has to be lengthy or costly. With imagination, good preparation and careful execution of the research, a great deal of valuable information can be found in as little as a matter of days. How to plan and conduct market needs research is described at length and exemplified in the design toolkit on page 191. Customer surveys are of two types: qualitative or quantitative. As described more fully in the toolkit, qualitative research is exploratory and largely judgmental – you are seeking customer's judgements and opinions on their needs and how these needs are satisfied by current products. Quantitative research, on the other hand, is more specific, more precise and a more quantified estimate of how customers are likely to respond to a proposed new product. Often, both types of research will be conducted, one after the other. Qualitative research identifies the main features of customer needs and expectations and quantitative research explores these in greater depth to allow sales projections to be calculated and market positioning of the new product to be finalised.

Before leaving the subject, two related criticisms of market research, commonly made by product designers need to be mentioned. The first is that market research constrains design opportunities to the lowest common denominator of customer taste. The second is that customers can never say they want a truly innovative product that they have never even imagined before. To my mind, this is a criticism of poor or unimaginative market research rather than a criticism of market research in principle. This argument, developed in more depth on the opposite page, makes the assumption that market research is not simply conducted in order to play safe and design products to match the average expectation of the average customer. Indeed, the reason I refer to market ***needs*** research here and consider market testing as a separate issue is the belief that you must start with an understanding of the fundamental needs of customers. From this, you extrapolate what kind of product will satisfy these needs, using all the

Market research: the 'Morito factor'

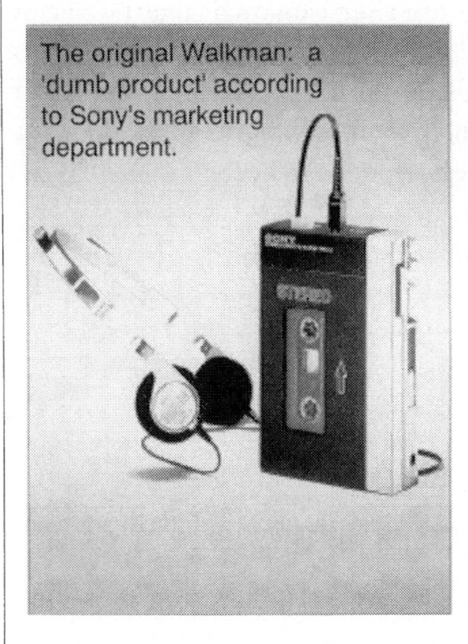

The original Walkman: a 'dumb product' according to Sony's marketing department.

In a company such as Sony whose charter is to 'Do what others have not done', the chairman, Akio Morito, believes that customers cannot say they want a product which they have never even imagined before. Morito faced exactly this problem in the now-legendary development of the Sony Walkman [7]. During the 1970's when the Walkman was first prototyped, tape recorders were products which played **and** recorded sound. Presented with a machine which not only lacked a recording facility but which also could be used by only one person, through headphones, Sony's marketing department concluded that this was 'a dumb product!' Fortunately, Morito had tried the Walkman himself and loved it. The marketing department's skepticism was over-ruled and a great success story resulted.

But where does this leave market needs research? Sony knew a lot about customer needs for tape recorders and concluded from this research that the Walkman was not viable. But what assumptions was their research based on? The need to record is explicit in the name tape recorder. To ask customers whether a tape recorder needs to record is self-contradictory. Ideally, market research should ask questions at a variety of levels from general to specific. Finding out how, when and where customers want to listen to music calls for very different research to finding out whether a tape recorder needs to record. At the time of the Walkman's development, Sony did actually have marketing data which could have been interpreted as evidence of a market need. 'Music-on-the-move' was a recognised trend at the time and the need for portability in tape recorders was only being addressed by putting carrying handles on otherwise unwieldy products. Furthermore, it seems likely that if the marketing department had explored customer's use of recording facilities, they might have found it less essential than they thought. Customers rarely record anything outside the home and, at home, most would have a second hi-fi system.

Deciding what to ask customers depends upon two factors: i) the extent to which product opportunities are constrained (e.g. by in-house manufacturing methods) and ii) the extent to which customer's perceptions of their needs may be constrained by what is currently available on the market. The more freedom you have to design your new product in radically different ways from existing products, the more you have to ask basic questions to discover the underlying needs of customers. The more constrained you are to designing a product which functions in similar ways to existing products the more specific your questions. You simply need to know what they think about what is already on offer. Akio Morito's doubts should not, therefore, be whether or not to conduct market research but rather what type of questions you should ask. As a rule, market research should begin with general questions on basic underlying needs and move on to more product-specific questions as the interview progresses. The general questions may identify opportunities for radical innovations and this may start you challenging pre-conceptions about the opportunities available. If this does not happen, the specific questions will provide detailed information about customer needs for the product you had previously envisaged.

creativity and imagination you can muster. If, as a result, you come up with a revolutionary product which you could not expect customers to say they want, then don't ask them if they want it. They are bound to say no or show complete indifference. Rather, work out which basic needs you have satisfied and how. Then use imagination and subtly to tease out information on whether you have satisfied a genuine need, in a meaningful way and at a cost they will accept. Customers can give a strong vote of confidence in a new product without ever having seen it or tried it. Market research can be a perfect example of the old 'garbage in, garbage out' theory. If you plan it badly and conduct it like a 'bull in a china shop', you deserve all the uninspired, constraining and possibly misleading conclusions that you will get. But market research carefully thought out and imaginatively applied can be the product designer's greatest asset.

Technological opportunities

Ways of identifying technological opportunities vary considerably depending upon the technology and how rapidly it is changing. It is only possible, in the space available, to briefly outline how to identify technological opportunities and leave it to individual designers or companies to explore relevant technologies in greater depth.

In principle, technological opportunities can be identified in four ways, ranging from specific to general:

- Competing products analysis
- Benchmarking
- Technology monitoring
- Technology forecasting

Competing product analysis is a good way to make sure you do not fall behind your competitors. It does require that the competing products are analysed in sufficient detail to identify technological innovation. Where competing products are monitored only by marketing specialists, new technologies may slip through unnoticed. Provided technology specialists are also involved, new technologies should be easy to identify. This, however, is only a catch-up strategy. For companies seeking technological leadership in a market, more in-depth research is required. The second level of technology opportunity spotting is to benchmark relevant technologies. Benchmarking [8], explores leading edge technologies currently used in all market sectors.

Figure 6.8
Sources of information on technology monitoring

Your greatest asset in researching new technologies is a good librarian familiar with this sort of information. Librarians at Universities which specialise in technology are usually a good bet. Several computerised databases are available for searching, at a cost, through the 'Dialog' system. The type of information accessed can include published research arranged in subject databases, current research projects (e.g. BEST covers British engineering science and technology), published patents and recently completed MSc and PhD theses (dissertation abstracts). On line searches generally have a call fee (usually a connection fee and a further fee per minute on-line) plus a charge for every record printed out off-line. A specific and well focused search will cost in the region of £100.

Most technical libraries will have, as standard, catalogues of current books in print (e.g. Whittaker's Books in Print), weekly and monthly magazines (e.g. Willing's Press Guide) and technical journals (Ulrich's Periodicals). Hidden away within these sources will be academic books and journals specialising in all leading edge technological subjects. Specialist magazines are often published by professional societies and lists of theses societies will be readily available, even if their publications are less easy to find. Several magazine publishers specialise in technology subjects. The 'What's New in ...' series published by Morgan Grampian covers a wide range of technology topics. The magazine 'Engineering' from Gillard Welch Assoc. also provides in-depth coverage of a wide range of technologies.

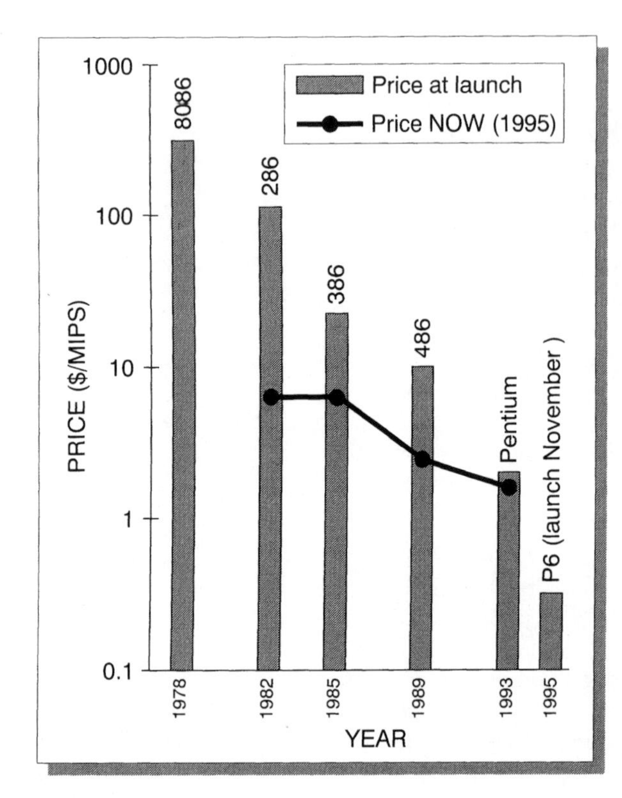

Figure 6.9
The power and price of semiconductors has been changing consistently, allowing sensible projections to be made [9]

This reveals the state-of-the-art, as it is currently used anywhere in manufacturing. The third level is technology monitoring. Most emerging technologies are extensively covered in specialist exhibitions, conferences, magazines, journals and books. Some of the most widely used information sources for finding these are given in Figure 6.8. Many Government and industry agencies go further than this and actively promote new technologies to manufacturing companies. The national or regional offices of the Government's industry department will usually be able to supply lists of their published information and a diary of technology promotion events. In-depth information about new technology is usually best sought from experts in the relevant field. Some of the most useful experts are to be found in companies supplying new materials or manufacturing equipment. It is their job to help you invest in new technologies, although obviously they will be promoting the specific types of technologies that their company sells. Nonetheless, many such companies offer excellent general guides to their type of technology and these are usually written in response to the questions their customers most often want to know. The fact that this advice and information is usually free is a particular advantage. A greater degree of independence can often be found in industry advisory agencies. These agencies, sponsored by groups of companies with a common interest in promoting a type of material or process, often supply published information and a telephone advisory service for companies.

Finding out about emerging technologies often leads to universities and research centres. With increasing commercial pressures on universities, an growing number of academics offer their services to industry. Being academics, they are generally highly specialised and may have limited commercial experience. But the nature of academic research funding usually means that they will be involved in leading-edge technologies. Guides to current research and the academics leading the research are published in many countries and a selection of these guides is given in Figure 6.8.

Finally, technology forecasting tries to anticipate future technological trends. Even in rapidly changing technologies, the trends may be predictable and can be used to set targets for future new product development and to anticipate likely changes in competing pressures. In the semiconductor industry, for example, rapid changes have been occurring for many years (Figure 6.9). Trends in these changes have been apparent for anyone who has cared to look.

For less readily available information, it may be that you are forced to rely on expert opinion rather than factually-based projections. The classic technique for seeking such opinion is the Delphi technique, described in a design toolkit on page 188. This uses a series of structured questionnaires to a number of experts to progressively home in on their consensus opinion of likely technological change.

Plasteck Ltd: from strategy to opportunity research

Figure 6.10
Plasteck's current product range

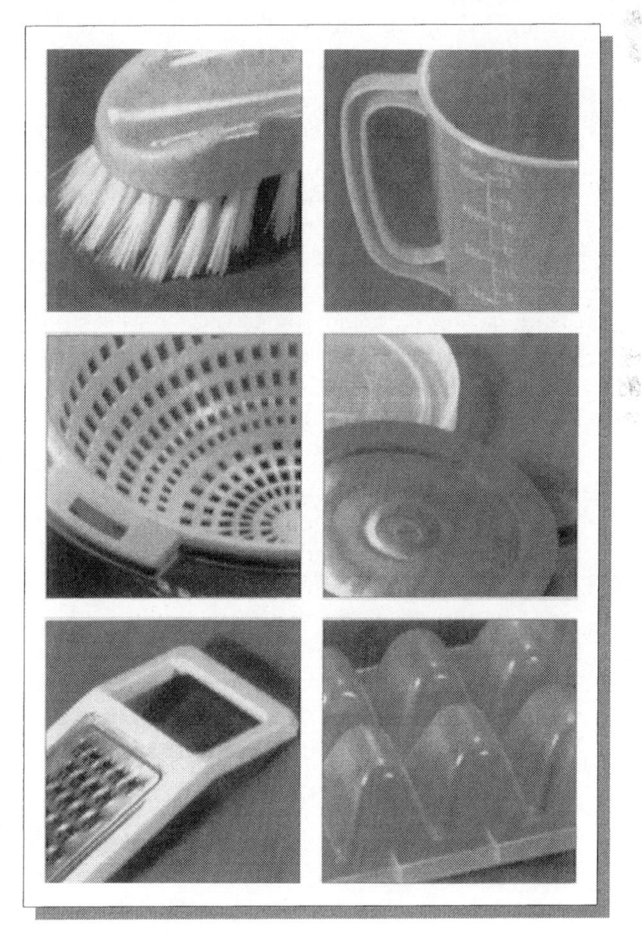

A lot of ground has been covered in this chapter so far and a great deal of information has been presented. So, it seems appropriate to have a 'time out' and work through product planning, as it has been described so far, using a worked example.

Plasteck Ltd. **[10]** is a small company (170 employees) manufacturing a range of injection moulded plastic products for the domestic household market. Most of their products are small kitchen products (Figure 6.10), known in the trade as 'domestic small goods'. They have an established distribution and sales network of hardware shops and department stores, which they supply direct. Last year they generated a profit of £900 000 on a turnover of £8 million.

During a recent financial review, Plasteck discovered that higher priced products were generating a disproportionately large share of total turnover. This encouraged the view, already voiced by some managers that Plasteck should be aiming to move up-market with their product range. At present they are at the cheaper end of the product range stocked by their main retail outlets. A move up-market would not, therefore,

Figure 6.11 Plasteck's strategy

Mission

To become the leading manufacturer of innovative small household goods, based on excellent design, excellent quality and excellent value for money.

Corporate objectives

Move up-market in order to increase turnover (by 60% over 2 years) and increase profits (by 70% over 2 years). Establish a brand identity, across the product range (over 2 years). Develop products which are appreciated by customers for their innovative design and value-for-money.

Corporate strategy

Increase investment in design and development (by 75% over 1 year). Redesign corporate image at retail level (product packaging and point of sale stands) in conjunction with representatives form retailers (implementation over 1 year). Develop new markets appropriate for new products (15% increase in retail outlets over 2 years).

Product development objectives

- Establish a brand identity which is stylish and conveys both 'innovation' and 'value for money'.
- Significantly improve product development procedures (to become more systematic, more streamlined, more innovative and better at screening new product ideas).
- Introduce at least 5 new products per year.
- Reduce product failure rate to less than 1 in 5 products launched (failure is less than one year's sales at projected sales level).
- Reduce incidences of 'over cost' and 'late to market' to zero.

Product development strategy

- Recruit two industrial designers to improve product development performance.
- Begin pro-active product planning to generate more (and better) new product proposals.
- Establish product development steering group (comprising marketing manager, development manager, production manager and finance director) and charge them jointly with reaching product development objectives.
- Conduct 6-monthly reviews, involving all line management staff, to continuously improve product development performance.

require any changes to their sales and distribution channels. They embarked upon a major strategic review, which produced, for the first time on paper, a company strategy. The main features of this strategy document are shown in Figure 6.11. The mission statement clearly indicates their intention to remain in the value-for -money side of their business. They are not, therefore, planning a radical move into the luxury end of the market. This recognises that their core business function is manufacturing injection moulded plastic products. The majority of their capital investment is in moulding equipment and key skills have been developed, getting this equipment to operate reliably and with industry leading levels of production quality. In addition, their excellent working relationships with retailers is a significant asset. The way forward seems clearly to be a move to higher value products, using their existing manufacturing facilities and retail outlets. The way to do so, they believe, is to add value by design. Their new products are going to be noticeably more innovative and unified by a brand identity although still offering excellent value for money.

They move on to developing their product development strategy. The current product range includes a total of 19 products. Product maturity analysis (see design toolkit on page 132) suggests that, of these, 4 are likely to drop in sales this year and a further 6 next year. In order to replace these, an average of 5 new products will need to be developed per year. In addition, the remaining 9 products will have to be redesigned over the next two years in order to establish the new brand identity across the entire product range. This is almost double the rate of product development achieved in either of the past two years. Clearly, some dramatic changes in new product development need to be introduced. Auditing past product development reveals a number of problem areas. From a total of 7 products launched in the past three years, three were withdrawn from sale within their first six months. Of the 7 new products, only two were launched on time (one was four months late!) and four were over their projected cost (by an average of 12 per cent). A total revision of product development procedures is decided upon. The new procedures give a target of less than a 1 in 5 failure rate, and no new products to be 'over cost' or 'late'. Two new design staff are recruited to boost product development performance. A steering group of senior managers are charged with delivering the targets set in the product development objectives. And a commitment is made to pro-active product planning.

Product planning begins with a competing product analysis. A major problem is encountered, immediately. The company's current range of 19 products covers 15 different types of product. A preliminary review reveals that over 450 products could be considered to be competing products. To analyse them all could take months. The steering group makes the decision to continue with a preliminary review to try to identify how to go about developing quality, innovative and value-for-money products. The 450 products are categorised in terms of how innovative they are (or would appear to be to customers), how good they appear in terms of value for money and how good they appear in terms of product quality. The top products in each category are then analysed in more detail to see what, if any, features they have in common and what generalisations can be made about the products judged highly. A short list of 100 products is drawn up by the design team and a sample of each product is purchased. Four non-technical staff are asked to be the judges; three females, representing the main purchasers of the products and one male. They are asked, individually, to select the twenty best products from the hundred on three separate occasions. The first time they are asked to focus on the innovativeness of the product, the second time, product quality and the third time, value for money. From the results, 22 products were selected by all four judges (6 for innovativeness, 8 for quality and 8 for value for money). Interestingly, 4 of these products were judged to be innovative, high quality ***and*** good value for money. It is decided to pursue the detailed competing products analysis on these 22 products, paying particular attention to the four all-round best products.

The competing product analysis begins by looking for features shared by the quality, innovative and value for money products. Pricing is a vital issue since increasing prices by moving up-market is a central aim of corporate strategy. It is clear that innovative and quality products are at the high end of the price range within their product category. Looking for design features in common, it is discovered that innovative products were often of a 'consolidated design'. This is where the best features of a number of different products is integrated into the design of a single product with combined benefits. Thus, in the domestic household market, some products are comfortable and easy to use, others perform their core function particularly well and others look hygienic and are easy to keep clean. The designers of innovative products appear to have analysed these benefits and built them all into a single product. The key determinant of product quality which emerged from the analysis was the product's 'finish'. They simply looked good, looked solid and looked like they had been put together well. The design team felt that a lot of effort had gone into their detail design. For plastic products, the mould cavities fitted together perfectly. There was no sign of misalignment or of flash. The joints between parts were precise and tightly toleranced. Interestingly, all of the non-technical judges commented on the 'feel' of the quality products. When questioned, they were unable to specify what it was about the feel of the products which gave the impression

Figure 6.12
Conclusions and decisions from competing product analysis

Plasteck: competing product analysis

- Innovative products are highly priced (on 4 occasions, the highest priced in their product category).
- Quality products are also highly priced, but less so than innovative products.
- According to retailers, products which had been judged to be both innovative and high quality sold well (2 were best-sellers in their product category).
- Value for money products were scattered across the price range. Some were basic, low cost products and others were high-priced, high value products.

Conclusion – Confirmation of corporate objective to move up-market into quality, innovative products (which are still perceived by customers to be good value for money).
Pricing policy decision – An innovative, high quality new product should be priced around the 90th percentile of the product price range.

- Quality products all have an excellent quality of manufactured surface finish (precise, tightly toleranced joins, textured, patterned or natural (e.g. wood) finish).
- Several innovative products consolidated the benefits offered in a number of different products.
- Products which were both innovative and high quality were judged by the design team to be particularly well styled.

Conclusions – Explore further the determinants of product quality as perceived by customers.

of quality. Studying this in greater depth afterwards, the design team thought that the weight of the product might have been important. Heavier products feel as if they are of better quality. Comparing the weight of the quality products with other products of the same category tended to support this suggestion but there was too much variation to be able to be conclusive. This was noted for further analysis at a later stage. One thing that was clear, however, is that if quality is to be judged in this way, customers must be able to feel the products at the point of sale. At present, Plasteck's products are all fully packaged, preventing this.

On the basis of this competing product analysis several parts of Plasteck's product development strategy became clarified (Figure 6.12). Developing high quality innovative products is definitely the way forward. This has the potential to move Plasteck up-market, into higher priced products with higher profit margins. Innovation would be sought by seeking to consolidate the benefits of a number of different products into a single product. Producing quality in the eyes of customers needed further research but it looked as though it could be brought about by good detail design and high manufacturing standards.

Plasteck: qualitative market research

- A predominant theme which emerged for many products was hygiene. Disliked tasks and products were often disliked because of their 'messy' operation or difficulty of cleaning afterwards.
- Customers in all groups spontaneously mentioned their dislike of three kitchen tasks related to specific products: emptying the kitchen refuse ('it's so messy'), grating food ('you catch your fingernails in the grater') and peeling potatoes ('I've just never found one that works well'. 'How come we can send a man to the moon but cannot design a potato peeler that works?').
- Presented with a sample of competing products at the end of the research, there was a general agreement with the company judges as to which products were innovative, high quality and good value for money.

Figure 6.13 Conclusions and decisions from qualitative market research

The pace of product development begins to speed up. The hunt is on for the first tangible product opportunity to emerge from the weeks of preparation. An air of excitement emerges from within the offices of the design team. The tasks are becoming clearer and they begin to run concurrently without any managerial directive to do so. Qualitative market research explores the views of small groups of customers. What kitchen tasks are most disliked? What type of kitchen products do they own? Do they find any existing products frustrating? Which have they bought recently and what do they think of them? Why did they choose these particular products to purchase? What do they look for when buying a new product? Several interesting findings turned up (Figure 6.13).

In depth discussions with company sales staff and a small number of trusted retailers revealed important case studies of product successes in the small kitchen goods market. Two, in particular, left a lasting impression in the minds of the designers. The first, a product introduced many years ago, was the Fiskars scissors. These rapidly established themselves as market leaders by having distinctive orange plastic handles moulded to fit comfortably for virtually all users. The second, was the plastic juice or milk container with the tapered base to allow it to fit into fridge door compartments. The first product of its kind to solve this particular functional problem, it became a 'dominant design' although it spawned many copy products soon after its introduction. The idea of a 'universal' product solving a recognised customer need (possibly a minor, or almost trivial need, like fitting a liquid container into the fridge door) emerges in their minds.

The design team gets together with the steering group after these research studies are complete. The time has come to decide on the type of product to be developed. The discussion immediately jumps to specific product ideas. Every team member has their own favourite and argues its case. They go round in circles for over an hour. Then the marketing manager realises that they are not making progress. He calls a halt to the rambling discussion and sends everyone back to their offices. They have 15 minutes to write down, in no more than 10 bullet points, the facts discovered so far about market needs and product opportunities. They reassemble, each grasping a single sheet of paper. Four themes quickly emerge: the, by now, well established need to develop a high-quality, innovative product; the need to exploit an identified customer dissatisfaction with kitchen products; the aim of developing a product which provides a 'universal' solution to a problem; the possibility of designing a product using a highly user-centred approach. Eventually, two of the identified customer needs are selected – the food grater and the potato peeler - and two designers are sent away to prepare an opportunity specification for each of them. The Plasteck story continues on page 177.

Selecting a product opportunity

An effective product opportunity proposes a new product which is a step closer to the company's ideal product than any product currently on the market. Selecting the right product opportunity is, therefore, a matter of analysing all the available information, synthesising several specific proposals from that information and then selecting the one that fits best to the company's goals. There are four common pitfalls (some of which are big enough to swallow companies whole!) which must be avoided.

The 'first love' syndrome Often, in the course of product planning, opportunities will be discovered. The moment of discovery is always exciting, even for the most hardened product designer. The opportunity is new, its advantages will always be seductive at first sight and the idea is ***yours***. The excitement, the sense of achievement and the feeling of possessiveness can be a compelling mixture. And many designers fall for this first idea. Like many love affairs, it may be difficult to see the faults in your 'object of desire' for quite some time. Criticism by others can provoke a jealous reaction. Once you have made an open commitment to your idea, unfaithfulness can be even more difficult. This would be admitting that you were wrong.

This 'first love' syndrome is recognised as a serious problem for companies as well as individuals [11]. Once a company has made a commitment to a product opportunity, the idea gathers momentum and often continues much further down the product development process than it should after problems have begun to emerge.

The most obvious solution is to ensure that product planning continues long enough to generate many ideas, from which, the best is selected. The problem of individuals feeling possessive about their own idea can be reduced by using the idea advocate technique for presenting ideas to an opportunity selection group. For this particular application, the technique is used to ensure that individuals do not present their own ideas. The 'inventor' briefs a colleague on their idea who then presents the opportunity without the same degree of vested interest in its success. This often results in a more independent appraisal of the idea and may even result in the inventor criticising his/her own idea during the selection process, after having seen it presented by a third party. It is important, in maintaining staff moral that the inventor is happy with the way the idea is being presented. Feelings of injustice that 'if only the idea had been presented more fairly' can lead to personal animosity and destroy a team spirit.

The 'greener grass' syndrome For a company completely familiar with a particular market, including all its difficulties and limitations, looking over the fence to the greener grass of a new and unfamiliar market can be an attractive prospect. Spectacular success stories can appear commonplace because the many failures are only known to insiders. Marketing can look easier because you only understand it superficially. Customers can appear more enthusiastic because you do not know their idiosyncrasies. And distribution may seem straightforward because you have overlooked certain key logistical difficulties. As many a sheep has discovered, the grass is rarely greener when you get to the other side of the fence, and even if it is, it rarely stays that way for long.

Entering a new market can be one of the most difficult things a company can do. Competitors, familiar with the market, can circle like school bullies to punish the newcomer. Their familiarity with the market may give them access to a variety of routes to out-manoeuvre you. Even your supposed allies, the distributors and retailers, may have tricks far too old to try on established market players. The greatest of care is, therefore, necessary before any fence-climbing is embarked upon.

The 'Concorde' syndrome Radical innovation always looks exhilarating when seen against the dull backdrop of step-by-step incrementalism. But it is an unfortunate fact of commercial reality that giant technological leaps often make small, or even backwards, steps in business development. The development of the first supersonic passenger aircraft was heralded as the way forward for the entire aircraft industry, in the 1960's. After billions of pounds in development costs and huge debt write-offs by the UK and French Governments, Concorde would have had to fly for hundreds of years in service to become a profitable investment.

What constitutes a giant technological leap for a small company may be seen as a routine incremental step by a much larger company. There can, therefore, be no universal rules. The sensible degree of technological innovation can often be established by financial cost-benefit analysis. What would it cost to make a radically innovative product and what are the likely returns on that investment. The part of this analysis which is most often underestimated is the risk of failure. Where new technology is being developed, the risks can be enormous. When, more commonly, existing technology is being applied to a new problem, the risks are smaller, more manageable and easier to estimate. The main risk then, is that the radically innovative product fails to gain market acceptance. This can often be anticipated by carefully planned and executed market research (although see the 'Morito' factor on page 157).

The 'little for lots' syndrome The crucial cost benefit ratio to consider when selecting a product opportunity is how much customer benefit is being achieved for how much cost in product development effort. Simply being able to deliver ***some*** added value for the customer is always an achievement for a new product and is something that can be very difficult to achieve. But this is never sufficient reason, on its own, for selecting a product opportunity. We saw, in Chapter 2, that the risk of commercial failure was five times less likely if the product delivered customer benefits which were both significant and important in the eyes of the customer. Products which deliver marginal benefits which the customer sees as relatively unimportant have, on average only a 20 per cent chance of achieving sales success.

A really attractive product opportunity is one which offers lots of customer benefit for little development effort. Products which offer lots of customer benefit for lots of development effort may be a viable opportunity, depending upon the company's financial strength at the time and its willingness to take business risks. Products which offer little customer benefit for lots of development effort, however, should be avoided wherever possible.

Systematic opportunity selection

Systematic selection of a product opportunity is best achieved by using an idea selection matrix. Here, a number of potential product opportunities are evaluated against the main product development targets for the company. The first step in developing an idea selection matrix is to decide on the main

Figure 6.14
Opportunity selection matrix for a child safety seat

Opportunity selection: Child's car safety seat				
			Opportunities	
Selection criteria	Weighting factor	Reference	Alternative 1	Alternative 2
Core benefit		Easy (one handed) belt fastening	Lowest price, for excellent safety	As 1 but with optional toys
Potential market size	10	0	+	+
Profit/unit sales	10	0	-	+
Product sales life	5	0	0	-
Development costs	1	0	+	-
Technical/safety risk	5	0	+	-
Market acceptance risk	10	0	+	+
Product compatibilities Manufacturing capabilities	5	0	-	-
Sales and distribution network	7	0	0	-
Design and development capabilities	3	0	+	-
Totals **Balance**		0 0	+29, -15 +14	+30, -26 +4

selection criteria. These should be taken from the company's objectives and strategy (both corporate and product development objectives and strategy). Additional and more specific criteria may be needed for a particular set of potential opportunities. Next, you need to find a reference opportunity, against which all the potential opportunities are to be compared. This could be an opportunity considered by the company some time ago and subsequently developed into a current product. In this case, the new opportunities could be assessed against risk factors and outcomes which are now known from past experience. Alternatively, one of the present opportunities (usually the one judged subjectively to be the best) could be used as the reference and all the others compared against it. The opportunities need only be categorised as 'better than', 'worse than' or 'same as' the reference opportunity, designated with a '+', '-' or '0' in the matrix. If some selection criteria are deemed more important than others, they can be weighted with a factor from 1 to 10. Opportunities can then be compared by adding up the number of '+', '-' and '0' scores (after each score has been multiplied by its selection criterion weighting number). Let us examine how a manufacturer of child safety seats could choose between the three opportunities (Figure 6.14) described earlier on page 169. Intuitively, the design team feel that the 'easy (one-handed) belt fastening' opportunity is best and this is chosen as the reference for the selection matrix. The selection criteria are agreed within the team and, after some debate, the weighting factors for each selection criterion is also agreed. Each member of the team is asked to complete the comparison of the three opportunities, using the agreed selection matrix and then they get together to compare results. Some disagreement is discovered over the specific details but, to their surprise, there is no disagreement over the conclusion. The easy fastening opportunity is not the best. The opportunity which comes out on top is the 'lowest cost, for excellent safety features'. In order to confirm this conclusion, they work together to examine each of the criteria on which this opportunity was judged better than the 'easy fastening' opportunity. Is the preference robust? Could the 'easy fastening' option be improved in some way to reverse the preference? The answer was a confident 'no'. The team moved on to explore whether the 'lowest cost for excellent safety' option could itself be improved by incorporating preferred features of the other opportunities. The profit per unit sale has little room for manoeuvre since the product's key feature is its low cost. It is clearly of vital importance that the product is not priced so low that it generates no profit at all. But this is something that will have to be explored when the opportunity is justified. The company's manufacturing capabilities were judged to be relatively poorly matched to the 'low cost for excellent safety' opportunity. This was

> ...quality control checks on product development... are primarily intended to ensure that the product exploits its identified opportunity. Discovering that the opportunity itself is wrong may only be a matter of chance...

because both equipment and processes are set up for higher value, higher priced products. This is not seen as a major difficulty. The manufacturing team have proved themselves to be highly resourceful in tackling a wide range of problems in the past and a request is made to the manufacturing manager to consider how the manufacturing process could be streamlined to minimise the costs of the proposed low-cost product.

By the systematic use of a selection matrix, different product opportunities can, therefore, be compared, ranked, developed, refined and eventually selected. The use of such rigour is vital at this stage of product development and always repays the cost of time and effort. Selecting the wrong opportunity for product development can be one of the worst mistakes a company can make. It sends product development off down the wrong track and it may not be until the product is launched that the mistake is realised. There are, of course many quality control checks on product development between identifying the product opportunity and the completion of its development. These, however, are primarily intended to ensure that the product exploits its identified opportunity. Discovering that the opportunity itself is wrong may only be a matter of chance, if, for example, later market testing contains some unexpected surprises.

Price positioning the new product

Having identified the best product opportunity, the next task is to justify it. This is always a useful check that the opportunity is indeed the best, or even financially viable, at all. Often, in the course of this justification, weaknesses in the opportunity are revealed and the product development process has to iterate round the opportunity selection procedure a second (and possibly third and fourth) time. This is a fact of life in new product development and failure to get the product right first time should never be taken as evidence that the development procedure is flawed. Indeed, detecting the weaknesses in products at this stage should indicate that the process is working exactly as it should be. Having worked through opportunity selection rigorously the first time round will mean that repeating the process will be quicker. Finding a second 'best' opportunity may, in fact, only take a fraction of the time that it took to find the first one.

Justifying a product opportunity, as we discovered earlier in the chapter, is mostly an examination of the product's financial viability. Inevitably, since the design of the product has not yet begun, this financial planning can only be preliminary and it must be refined and re-evaluated as the product

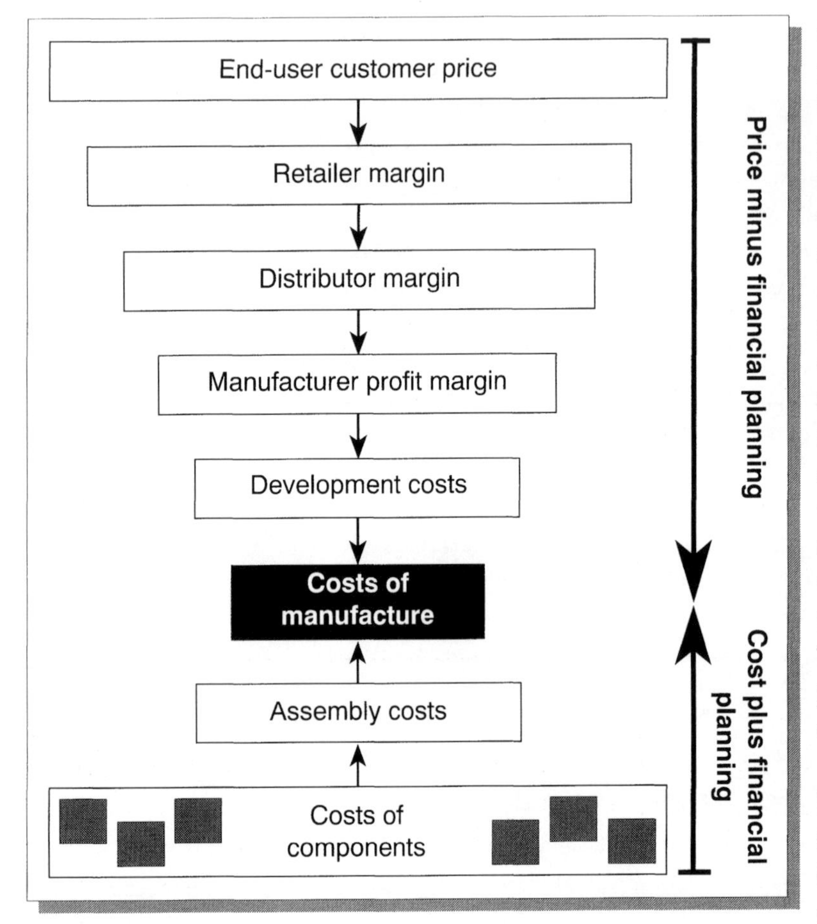

Figure 6.15
Alternative approaches to financial planning for new products

is developed. The way to tackle preliminary financial planning is by means of 'price minus' planning, a process which works backwards from the price offered to the end-user customer and subtracts retailing and distribution costs, manufacturer's profit margin and product development costs in order to arrive back at a target cost of manufacture. The alternative approach is 'cost plus' planning which builds up from the costs of manufacturing or buying-in components and then assembling them into a complete product. This is usually impossible at the product planning stage (but is discussed further in Chapter 9).

Price-minus financial planning begins with a target end-user price for the proposed new product (Figure 6.15). This should be determined by 'what the market can stand' in positioning your new product relative to competing products. The data for this must come from competing products analysis and it can often help the decision-making process to plot this data on a price-value map. A price value map for car safety seats is shown in Figure 6.16. The price axis is straightforward. It simply arranges competing products according to the price at which they are offered to end-users. For consumer products, this is their retail price. If products are offered at different prices in different outlets it may be necessary to calculate their average price. The value axis is, in principle, also simple. This arranges products according to their value as perceived by the customer. In practice, however, calculating this value can be difficult. Customers see value as being determined by a multitude of factors. They may like the way one product works but prefer the appearance of another product. To calculate values for such products a procedure similar to the selection matrix is used. A range of criteria determining product value are identified and their relative importance is estimated using a weighting

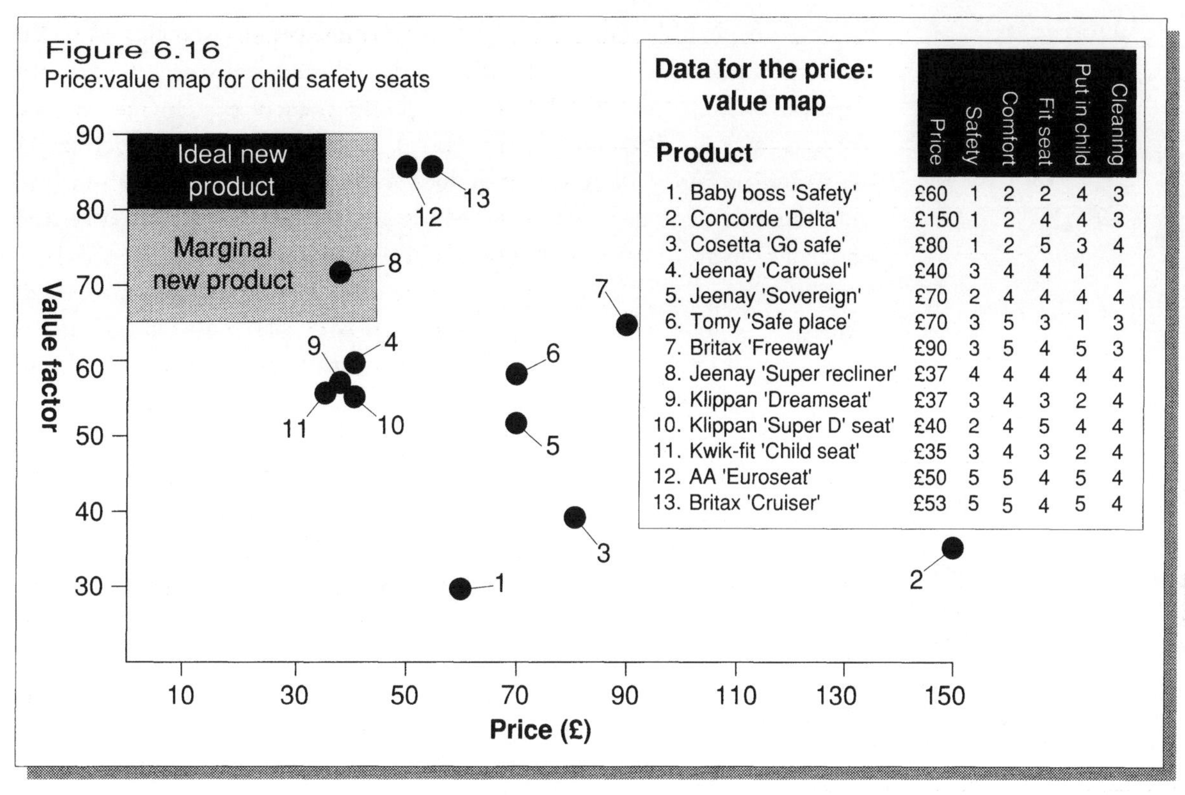

Product	Price	Safety	Comfort	Fit seat	Put in child	Cleaning
1. Baby boss 'Safety'	£60	1	2	2	4	3
2. Concorde 'Delta'	£150	1	2	4	4	3
3. Cosetta 'Go safe'	£80	1	2	5	3	4
4. Jeenay 'Carousel'	£40	3	4	4	1	4
5. Jeenay 'Sovereign'	£70	2	4	4	4	4
6. Tomy 'Safe place'	£70	3	5	3	1	3
7. Britax 'Freeway'	£90	3	5	4	5	3
8. Jeenay 'Super recliner'	£37	4	4	4	4	4
9. Klippan 'Dreamseat'	£37	3	4	3	2	4
10. Klippan 'Super D' seat'	£40	2	4	5	4	4
11. Kwik-fit 'Child seat'	£35	3	4	3	2	4
12. AA 'Euroseat'	£50	5	5	4	5	4
13. Britax 'Cruiser'	£53	5	5	4	5	4

Figure 6.16
Price:value map for child safety seats

factor. The competing products are ranked (using a 1 to 5 score rather than simply 'better than', same as' or 'worse than'). An overall value factor is then calculated as the sum of all the ranks multiplied by their weighting factor. If it is of critical importance where the new product is positioned within a range of competing products, market research will need to be conducted to ensure that calculated competing product values are an accurate reflection of customers perceptions of value. For the price:value map of car safety seats a published consumer test [12] comparing 13 different safety seats provides the data. This test compared the price, safety, comfort for the child, ease of putting the child in the seat and ease of cleaning. For purposes of this mapping exercise, safety is given a weighting factor of 10, comfort 3, fitting in car 3, putting in child 1 and cleaning 1. This gives a

Figure 6.17
Financial plan for a safety seat

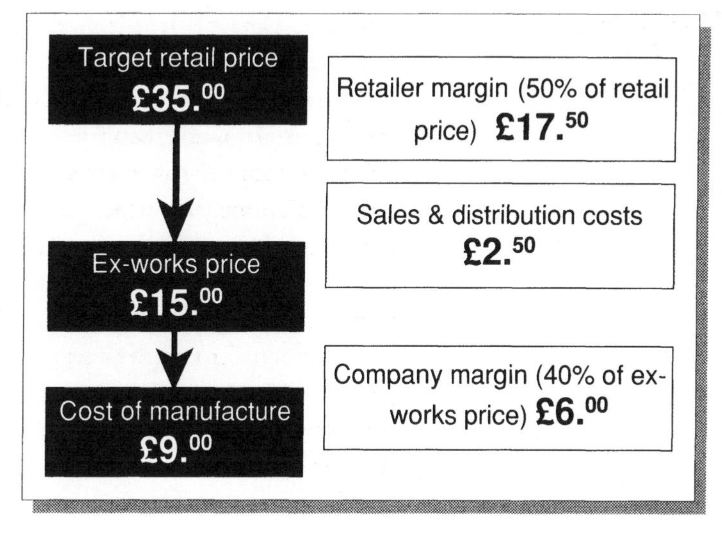

maximum possible value score of 90. The price:value map shows clearly that product value does not consistently increase (on the parameters used in this analysis) with price. The highest value products are of moderate price. The opportunity for developing a low cost, good safety seat become clear. A marginal advantage would be obtained with a seat priced at £45 or less, offering a value score of 65 or more. Ideally, the new seat should be priced at £35 or less and have a value score of 80 or more.

The final stage of justifying this opportunity is a preliminary analysis of whether or not this is achievable (Figure 6.17). Starting with £35 as the target retail price, retailers will usually demand a 50% margin. The sales and distribution of this product are estimated to cost £2.50, leaving an ex-works price of £13.50. Company policy is to impose a 40% margin on ex-works price to cover marketing, development costs and company profit margin. This leaves a direct costs of manufacture of £9.00. Analysing the best value for money product (product number 8) from the price:value mapping, it is estimated that it would cost £10.50 for the company to manufacture it, at present. To meet the target price for the new product would therefore require a reduction in costs of manufacture of at least £1.50. This is equivalent to a 14% reduction in manufacturing costs. Studying product 8 in detail there appear to be several opportunities for cost saving. Different materials could be used for certain parts, the design could be changed to reduce the number of components and the assembly could be made very much simpler. It is decided that the target price can be met and that the product opportunity is viable.

Style planning

Just because styling is the 'artistic' part of product design does not mean that it is free to follow any intuitive or creative inspiration that it chooses. Styling must be directed towards opportunities and held within constraints, in exactly the same way as the rest of the design process. Styling opportunities and constraints are of two general types. Firstly, there is a business context into which the new product must fit. A product developed for one company operating in a particular market might, for example, be completely inappropriate for another company in a different market. Secondly, there are styling issues intrinsic to the product itself. What is the product and what statement is it trying to make to customers?

These contextual and intrinsic aspects of the product together define its styling opportunities and constraints. It is the job of product planning to get agreement on these opportunities and constraints. Product planning must research and define these sufficiently to give direction to the styling process. Then they must make it clear if and when the styling objectives have been satisfied.

Contextual styling factors

There are four main categories of contextual styling factors (Figure 6.18).

Product predecessors If the proposed new product is an update of an existing product currently sold by the company then it is important to preserve the visual identity of the product. This will ensure that existing customers continue to recognise the product and purchase it again. To miss out on repeat purchases because the visual style has changed so radically from one generation of the product to the next can be a costly business mistake. The new product's predecessors must, therefore, be studied to determine which aspects of its present style need to be preserved in order to safeguard repeat purchases by customers.

Company or brand identities can give a similar degree of purchasing confidence to customers. If customers have already purchased products from a company or brand range, then visually identifying a new product as part of that range will boost customer confidence. For some brands or companies a theme of visual styling will be obvious. Certain colours or combinations of colours may uniquely identify a brand. Products in a range may share common components, such as buttons, knobs, handles or indicators. There may even be aspects of the overall form of the products which identify them as part of a range. Being of the same physical proportions, having the same radius of curvature or a distinctive set of angles may be what makes a product range distinctive. Establishing what determines a company or brand identity is, therefore an important part of product planning.

Figure 6.18
Contextual styling factors

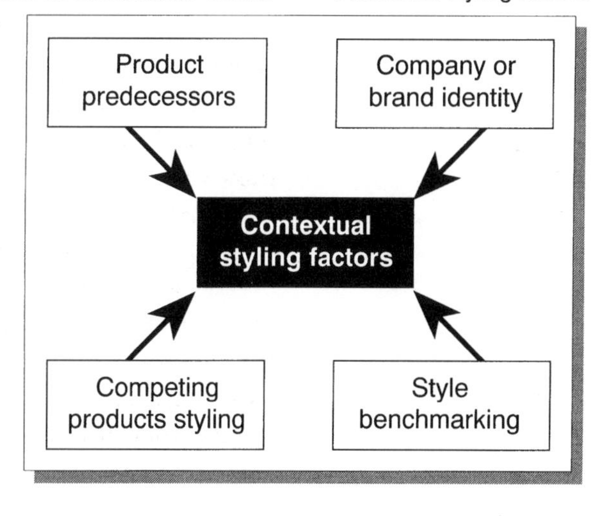

The style of competing products on the market can provide a great deal of food for thought in setting styling objectives. What is the general standard of product styling in the market for the new product? Is styling a strong feature of existing products or are well styled products uncommon? What are the prevailing styling themes? What styling messages do the existing products convey? Are these messages about the function of the product (product semantics) or about the lifestyle and values of the customers (product symbolics)? Collecting together images of competing products will focus attention on their styling and will help you decide which features enhance their attractiveness and which detract from it.

Style benchmarking Studying products from different markets can help you establish style benchmarks which you will aspire to match with the style of your new product. Are there any style icons which have exactly the look you want to achieve with your new product? These styling attributes could, and indeed should, relate to all aspects of design, from the most general to the most specific. Which has the most pleasant overall form? What gives the best semantic and symbolic messages? Which materials, colours, surface textures and design detailing look best? Images of 'ideal' or benchmark products from a wide range of markets provide styling objectives for the design of the new product.

Intrinsic styling factors

In describing the appearance of a product we usually say what it 'looks like'. To say that a product looks like something means that it conjures certain images in our mind. These images could be very literal and physical - it looks like an ironing board with feet straps, or they could be abstract - it looks very loud and youthful. This is different from saying what the product is; it is a snow-board manufactured by O'Niell. The imagery conjured by the appearance of a product is that product's symbolism. Effective styling controls symbolism. This controls what observers think that the product 'looks like' and this, to some extent, controls what they think about the product

Products tend to have two types of symbolic value. Firstly, a product can symbolise things about itself. A product can look like it is sturdy, robust and durable or fragile, delicate and perishable, just by its visual appearance. Secondly, a product can symbolise things about the person who owns it. Clothes, cars, watches, pens, briefcases, mobile phones - these have all become symbols of status in the Western world.

In practice, designers use the term 'product symbolism' to describe the human values aspect of the product. To describe a product's symbolism, therefore, is to describe the personal and social values embodied in the product's appearance. The way a product symbolises its own function is known as 'product semantics' (literally product meaning). Ideally, product semantics should also make the product look like it works better than any other product. So, product semantics does not just make a product say 'this is what I do'. It also says 'this is how I do it better than my competitors'. The outcome of the planning stage of product semantics should be a list of 'this is what I do' and 'this is what I do best' statements. The way in which these semantic statements are subsequently designed into products is covered in Chapter 7.

Styling specification

The aim of product planning, in general, is to collect background information on the proposed product, define objectives for that product's development and then evaluate its commercial viability.

In relation to product styling, this means:
- researching the styling context
- exploring product semantics and product symbolism
- collecting together a documented record of these planning activities and their conclusions and using these as the styling specification for the remainder of product development (Figure 6.19).

Given this specification as a styling objective for the product, the prospects for achieving the objective, within cost and other relevant constraints can be evaluated. Styling objectives then become evaluation criteria for concept, embodiment and detail design selection. If achieving certain styling objectives is vital to the success of the product then the product development should be terminated as soon as it becomes clear that these objectives are not going to be met. This is no different from the functional aspects of the design specification and it is a key part of good product development practice.

Plasteck Ltd: opportunity specification

We left the Plasteck saga on page 161 with two opportunity specifications being prepared. Let us now follow the development of one of these opportunities, the potato peeler. As a first step in developing the opportunity

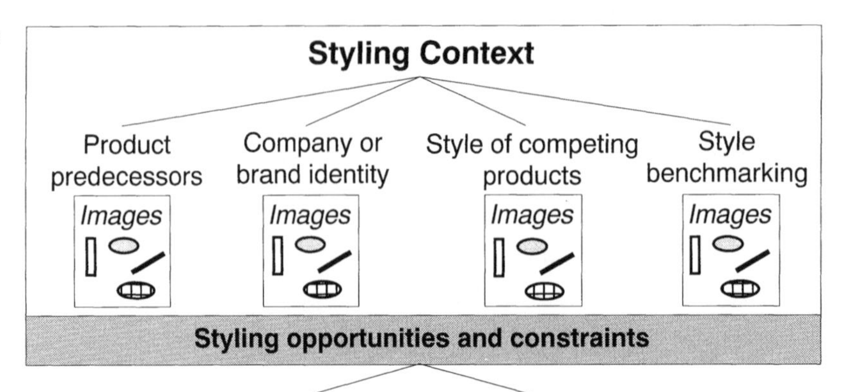

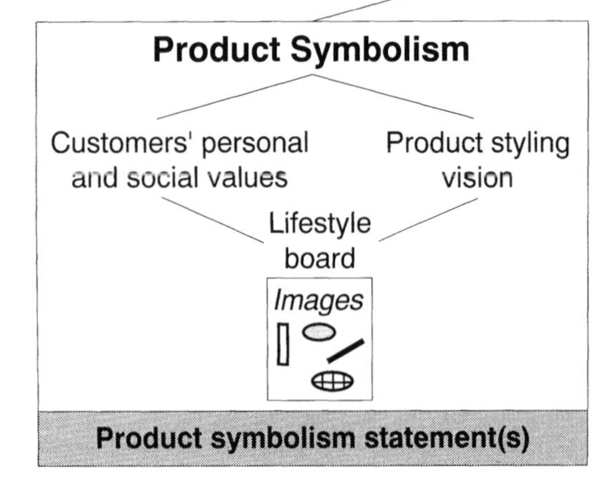

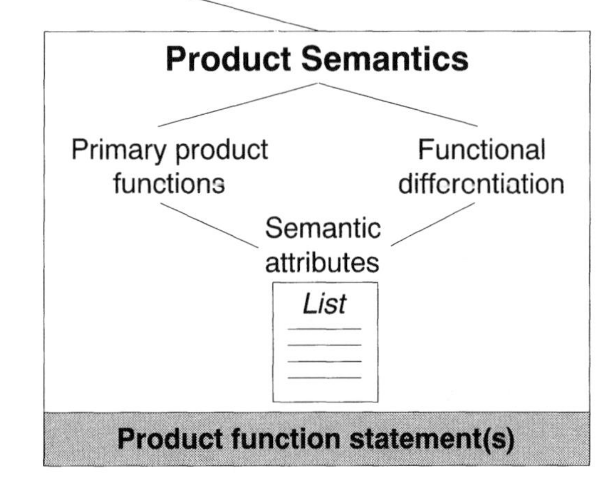

Figure 6.19 Styling in product planning

specification, the designer reviews the conclusions from the strategic planning and opportunity research so far. The relevant aspects (Figure 6.20) are written out as a reminder of the overall goals.

Before getting stuck into further research, the designer tries to establish what he is trying to achieve with the opportunity specification. A core benefit proposition is obviously needed and this will require identifying ways of adding value to existing products or reducing their cost. This in turn will need a better understanding of competing products than is currently available as well as further market research. They do not want concept proposals presented as part of the opportunity specification. They are simply looking for a way to decide between the two business opportunities presented by the potato peeler and the food grater. It is recognised that justifying the opportunity is going to be difficult, at this early stage.

Strategic planning conclusions
Mission... innovative small goods.... based on excellent design, quality and value-for-money.
Objective... brand identity... stylish, and conveys innovation, quality and value-for-money.
Product development performance... no product failures... all products on time and to cost targets.
Innovative, high quality product priced at 90th percentile of competing products.

General opportunity research conclusions
Universal product derived from consolidated design.
Excellent quality of finish on products... good surface finish and texture (good 'feel'), product weight (feels solid) and excellent detail design (no flash or moulding misalignment, tightly toleranced mating parts).

Specific peeler conclusions so far
Hygiene is important. Peelers are not currently perceived to work well.

Figure 6.20
Plasteck: conclusions so far

Competing potato peeler analysis

For the earlier, more general competing products analysis, only 4 out of the 100 products collected were potato peelers. This range now needs to be extended and a shopping expedition plus a call for help to friends and relatives produces a horrifying 43 further peelers (Figure 6.21)! Although all are distinct products (rather than simply the same peeler with a different coloured handle) many share the same basic design features. Six distinct design 'families' are identified. Of the differences between the families, most of the variation in design can be accounted for by two variables, i) blade type and ii) the relative configuration between blade and handle. The six design families with their different categories of blade type and blade to handle configuration are shown in Figure 6.22.

The peeler collection provides a wealth of information for analysis. The categorisation into design families offers a way of structuring the analysis. The problem, however, is that we cannot be sure that the difference between a good and bad peeler is all, or even mostly, due to the differences between design families. Since all design families are still on the market, we must assume that each offers some distinct advantage. Good and bad peelers may, therefore, differ solely on their design details – length of blade, radius of curvature of the handle or the sharpness of the gouge, perhaps. There are simply too many peelers to be able to complete an exhaustive analysis of such details without knowing what to look for. Market research is needed to provide a focus for any further competing product analysis.

Potato peeler market research

After informal talks with 9 company staff who claim to use potato peelers frequently, certain predominant customer needs begin to emerge. The peeler must work well. When asked what they meant by this, the features most commonly mentioned were: quick to use, sharp blade, and comfortable to hold. The peeler must also be hygienic: it should have a clean appearance, have no dirt traps and rinse clean after use. When asked whether or not they considered that potato peelers work well, they confirmed the reports from the earlier market research that they do not. When asked to be more specific, most people referred to failures of the needs they had just described: blunt blades, awkward or cumbersome to use and difficult to clean. A few new issues were raised: some blades are easily bent or fall out of their sockets, the gouge for removing the eyes of the potato is difficult to use and some peelers are no use for large potatoes because the cutting edge of the blade is too short. One perceptive interviewee also commented that dark coloured peelers were difficult to find among the potato peelings. She suggested that this was probably done intentionally so that peelers were accidentally thrown out with the peelings more often, thereby increasing peeler sales!

During the course of these discussions, a plan develops for testing the competing peelers. Having identified the main customer needs, a number of peeler design features could be identified which appear subjectively to satisfy these needs to different extents. By then carefully selecting peelers which vary along these design features and having them tested by volunteers, the predicted relationship between design and satisfaction of needs could be tested. (This research plan is based on 'quality function deployment', a technique introduced in Chapter 8).

The plan is finalised. A set of customer needs is listed and a related set of design features identified which are thought to determine how well these customer needs are satisfied (Figure 6.23). This research is not intended to be a comprehensive test of customer needs or peeler design features. It simply seeks to explore the key customer needs which are not, at present, sufficiently well understood. 12 peelers are selected to give a wide range of the four different design features. Some are swivel and others fixed blade. Some have the blade in line with the handle, others perpendicular to the handle. Some

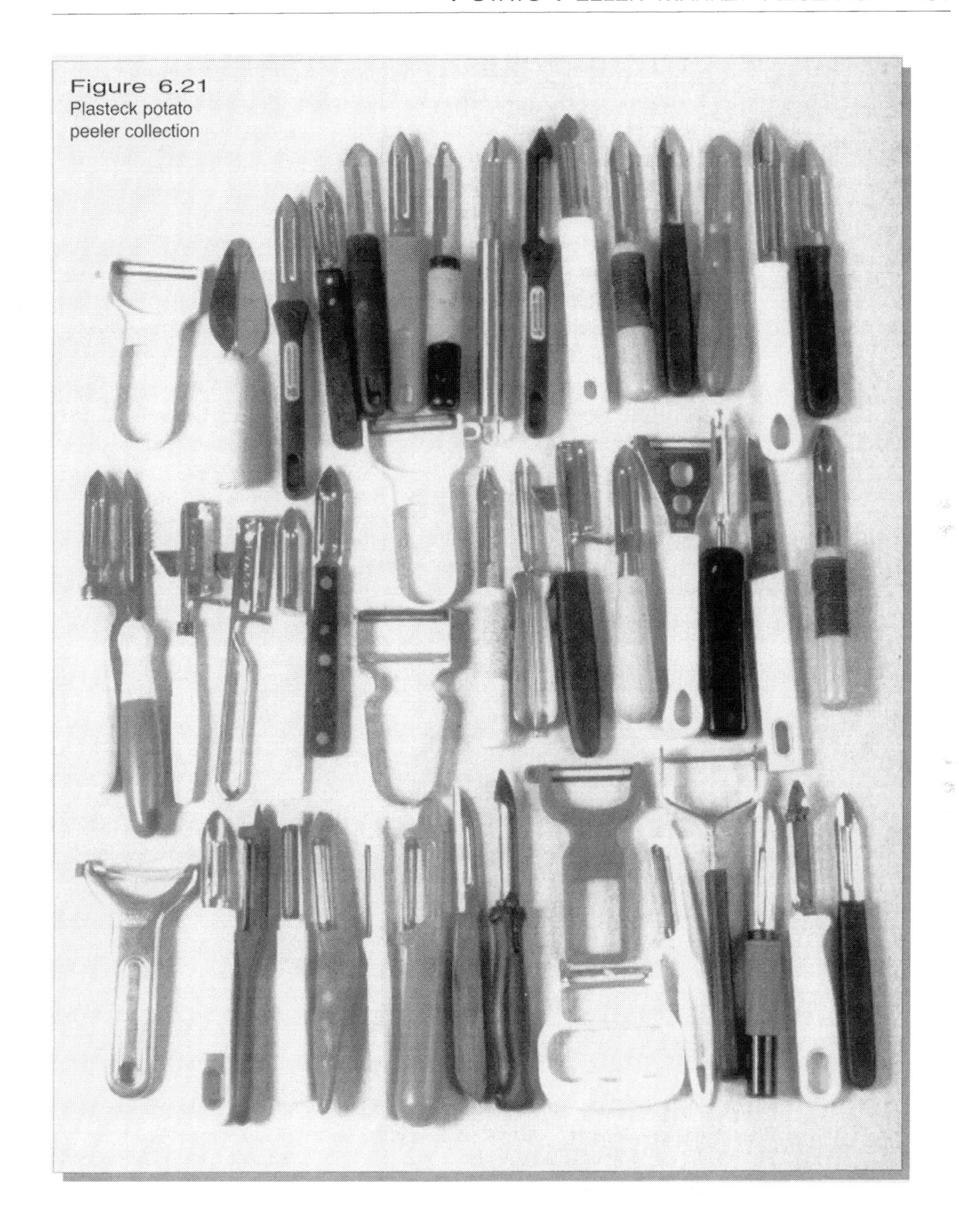

Figure 6.21
Plasteck potato peeler collection

have sharp blades, others blunt. Some have large handles, others small. A group of 12 staff (not including the 9 used earlier) are each asked to test every one of the 16 peelers and rate them on a 1 to 5 scale for each of the 3 customer requirements. They are told nothing about how the design features are expected to relate to customer needs. Before starting the tests, they are questioned about their initial impressions of each peeler. This was to compare their pre-test and after-test impressions to see if they change.

The study was set up with the intention of obtaining numerical values for customer preferences for different peelers. These numerical values were then going to be used to quantitatively determine the importance of the different design features. Unfortunately, customer perceptions of peelers proved more complex than was allowed for in the design of the test. The numbers obtained are of little value. But the discoveries made in the course of the data collection are of immense value. By the end of it, some real breakthroughs have been

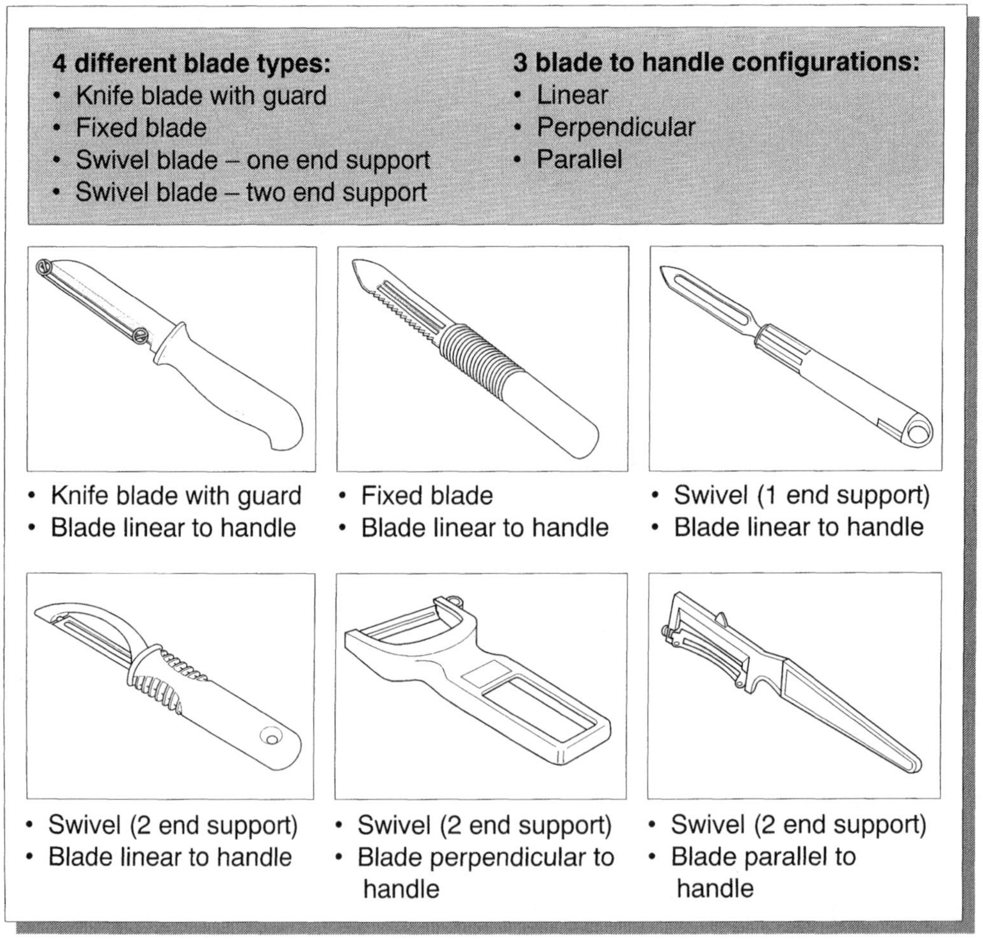

Figure 6.22
Potato peeler design 'families'

Design features / Customer requirements	Type of blade	Handle to blade configuration	Sharpness of blade	Size of handle
Quick to use	Swivel better than fixed?	Unknown	Sharper blade is faster?	Unknown
Easy to cut	Unknown	Unknown	Sharper blade is easier?	Not relevant
Comfortable to hold	Not relevant	Linear design is best?	Not relevant	Unknown

Figure 6.23
What is the relationship between certain key customer requirements and design features? A plan for peeler research

made. The first breakthrough is the discovery that 'customers' reported striking differences between swivel peelers and fixed blade peelers, irrespective of the handle to blade configuration. Most of them were accustomed to only one blade type and admitted they had never tried the two types side-by-side. Most of the swivel peelers made nice easy cuts but the slices were thin and narrow. The fixed blade peelers made deeper, wider cuts. So, answering which was quicker to use was difficult. The swivel blade peelers were quicker for making each slice but more slices were required to peel the potato. On balance, they felt that the fixed peelers were quicker. One way in which swivel peelers were, however, judged to have a clear advantage was on ease of cutting around the shape of the potato. With regard to handle to blade configuration, there were, again, quite fundamental differences. With a linear configuration, most users reported that they had more control of the slicing action. A few volunteered the fact that they used their thumb to guide the blade while slicing. With the blade perpendicular to the handle this was not possible. The linear configuration was voted most comfortable to use by the majority of users.

Although the trial certainly did not work exactly as planned, it achieved what was most important: it allowed an opportunity to be identified. That opportunity was to develop a peeler which combined the advantages of both swivel and fixed blade designs. This would peel quickly by having a fixed type of blade, peel easily by having that blade swivel and peel comfortably by having a linear configuration of blade to handle. At present, the designer has no idea what design of peeler would achieve this but that does not matter.

That was the job of concept and embodiment design. It looked like it should be achievable by means of in-depth analysis of the existing designs and a generous measure of creativity.

Potato peeler opportunity specification

Now the opportunity specification has to be prepared. This document has to convince firstly the steering committee and then the managing director that a new peeler is a viable business opportunity. To do so, it has to find the right balance between technical and commercial information. The technical opportunity seems clear. A consolidated peeler design featuring the advantages of both fixed and swivel blades would bridge a market previously divided into those customers who prefer one type of peeler or the other. The obvious risk of failing to please either group could be managed by market research at a later stage. The company's expertise in injection moulding should be able to cope with the manufacture of the handle and an existing subcontractor already produces stamped metal parts and hence could supply the blades.

The commercial opportunity was much more difficult. Phone calls to retailers of Plasteck's existing product range produced market information on peeler sales in general (Table 1 of Figure 6.24). From this it is possible to make estimates of upper and lower sales figures. In addition an upper and lower sales growth trajectory is plotted (Table 4 of Figure 6.24), based on the sales growth of other similar products launched by Plasteck in the last 3 years. The price positioning of the product is based on a price:value map derived from the competing product analysis and market research conducted so far (Graph 1 of Figure 6.24). Knowledge of the cost structure of Plasteck's sales and distribution channels allows a price-based financial plan to be drawn up and a target cost of manufacture to be estimated (Table 2 of Figure 6.24). The viability of this cost of manufacture is assessed through discussions with Plasteck's own production engineers and the metal-stamping subcontractor. Development costs for the new peeler are estimated from the designer's own experience of developing similar products (Table 3 of Figure 6.24). With all the financial data assembled, a financial model is developed on a simple spreadsheet and critical values for sales, costs, and development time are calculated (Table 4 of Figure 6.24). The opportunity specification is shown in full in the following three pages (Figure 6.24). It concludes that the opportunity is viable, although market research will be necessary to verify the sales projections.

Opportunity specification

Customer needs

- Customers generally show strong purchasing fidelity to either fixed blade or swivel blade designs
- Customers are aware of problems with both designs

Core benefit proposition

- To develop a new peeler design which features the advantages of both fixed and swivel blade designs but has the disadvantages of neither (a consolidated design).

Marketing aspects of opportunity

- Bridges the previously segmented peeler market
- Peelers are universal kitchen products (every one of 40 company staff questioned owns a potato peeler)

Sales aspects of opportunity

- Potato peelers are stocked by an estimated 95% of Plasteck's existing retail outlets (estimated by sales manager)

Manufacturing aspects of opportunity

- Handle injection moulded in plastic by Plasteck, metal blade stamped and ground by Stamp-press (existing component supplier)

Major marketing risk

- New peeler fails to satisfy either group of customers (fixed blade or swivel blade purchasers)

Management of this risk

- In-house market research on visual rendering of proposed new product, confirmed by customer market research on prototype of new product

Figure 6.24 Plasteck's opportunity specification for a new potato peeler (continues on the subsequent two pages)

Opportunity justification

Table 1: Sales data and sales projections for new peeler

Retail outlet and (No.of outlets)	Current: Total peeler sales	Current: Best peeler sales	Current: Worst peeler sales	Projected: Best sales estimate	Projected: Worst sales estimate	Projected: Annual sales best	Projected: Annual sales worst
Hardware shops(495)	6	2	0.1	3	1	62955	12648
Superstores (85)	53	24	9	30	7	125504	10615
Department store(91)	37	13	1.8	16	6	71370	12965
Mail order cat. (5)	146	58	13	73	23	17684	1904
Others (191)	5	2	0.2	2	1	21400	4079
Total						**298912**	**42210**

These figures are estimated from weekly sales data provided by a sample of each of the different types of retail outlet in Plasteck's current list of stockists (5 hardware shops, 2 superstores, 2 department stores 1 mail order catalogue and 2 other categories of outlet). Best sales estimate = 1.25 x current best sales. Worst = 0.5 x (current best - current worst)

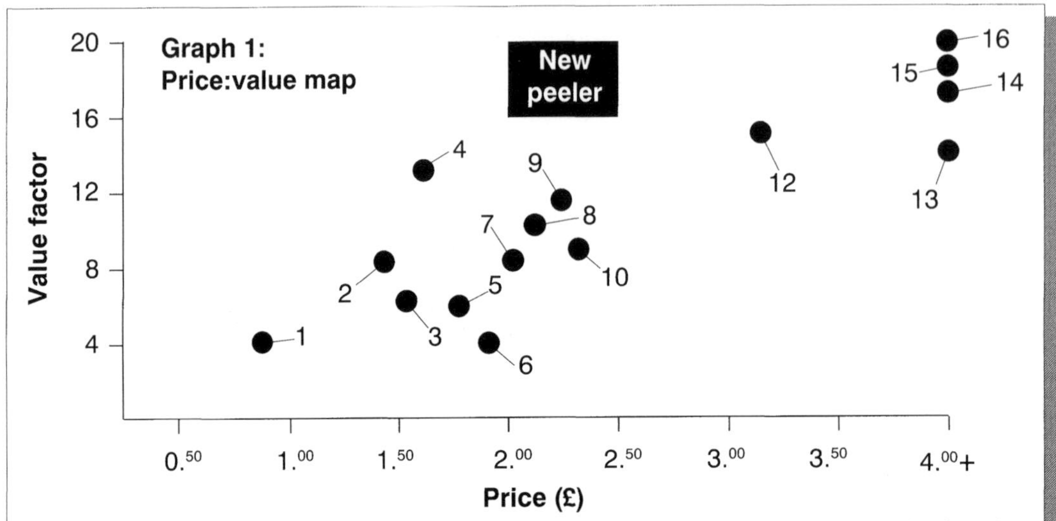

Average retail price versus value, calculated as the sum of value scores (1 to 5) for i) perceived quality of handle, ii) perceived quality of blade, iii) perceived function of handle and iv) perceived function of blade. All judged by 12 in-house 'customers' before testing the peelers.

Table 2: Price-based financial plan

Target retail price	From price:value map	£2.20
Retailer's margin	50% of retail price	£1.10
Wholesale price	Calculated from above	£0.88
Wholesaler's margin	20% of wholesale price	£0.22
Distribution cost	£4/carton of 1200 units	£0.03
Ex-works price	Calculated from above	£0.85
Margin over m'facturing cost	60% of ex-works price	£0.32
Target manufacturing cost	Calculated from above	£0.53
Sales and marketing costs	50% of margin over m'fact cost	£0.16
Profit margin	Remaining 50% of margin	£0.16

Table 3: Development costs

Development stage	Time Man days	Cost @ £190/day	Non-labour costs
Concept design	10	£1900	£300
Design specification	5	£950	£0
Embodiment design	10	£1900	£1000
Detail design	15	£2850	£1000
Production engineering	15	£2850	£30000
Totals	**55**	**£10450**	**£32300**
Grand total			**£42750**

Table 4: Projected sales performance of new peeler

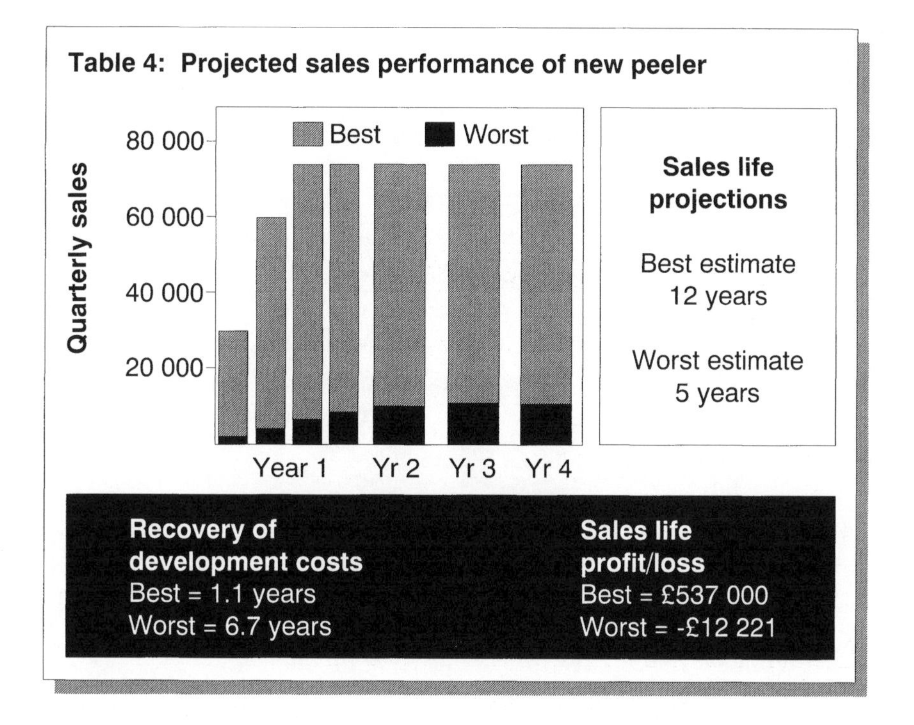

Recovery of development costs
Best = 1.1 years
Worst = 6.7 years

Sales life profit/loss
Best = £537 000
Worst = -£12 221

Conclusions

1. It is proposed that the development of an innovative design of potato peeler is commercially viable for Plasteck.

2. Best estimates for sales and sales growth suggest a break-even point after 1.1 years and a profit of £49 000 per year thereafter.

3. Design efforts should be directed towards the development of a consolidated design which combines the advantages of fixed blade and swivel blade peelers whilst avoiding their respective disadvantages.

4. A total of 55 man days of design effort is estimated to be required plus a tooling investment of £30 000, giving a total development cost of £42 750. Of this, £6 500 would be required to develop a working prototype for final market testing. The further £36 700 investment for detail design, production engineering and tooling would be committed only if the market testing suggested a level of sales which is likely to recover development costs in less than 2 years.

Key Concepts

1. Product planning occurs in two stages:

i) preparing the opportunity specifications at the start of product development and then

ii) preparing the design specification following concept design.

2. Opportunity specification

The opportunity specification seeks commitment from company management to continue with product development. It summarises the commercial opportunity presented by the proposed new product and then gives a commercial justification of that opportunity.

3. Core benefit proposition

The essence of a commercial opportunity is described in a 'core benefit proposition', a statement which identifies the way in which the new product will be differentiated from competing products.

4. Opportunity justification

The opportunity is justified by proposing a price position for the new product based on the relative price and value of competing products. From this price, a target manufacturing cost can be estimated and the feasibility of being able to manufacture to that cost can be assessed. Then sales growth can be estimated from which the time to recoup development costs can be calculated and sales lifetime profits estimated.

5. Preparing the opportunity specification

An opportunity specification is researched and prepared on the basis of competing product analysis, market needs research and technology auditing. At the end of all of this, you should end up with a product opportunity which fulfils an important and valued customer need, with a product which is differentiated from competing products and which is technologically feasible to manufacture.

Delphi technique

The Delphi technique was developed by the RAND corporation in the 1950's and the first application was used to obtain an expert consensus on the severity of the Soviet nuclear threat to the USA. The Delphi technique elicits ideas from participants by means of a series of highly structured and progressively more focused questionnaires. Often the selected participants are leading authorities on a subject and hence are likely to be widely scattered and probably difficult (or expensive) to bring together. The Delphi technique was, therefore, developed to be conducted by post, with the participants giving their answers in writing. There is, however, no reason why the Delphi technique should not be used face to face with participants giving their responses verbally.

Having identified a problem, a questionnaire is compiled to obtain an initially broad range of ideas from the selected participants. The responses are then collated centrally, summarised in the form of a second questionnaire which is then sent back to the participants. This second questionnaire is intended to clarify and expand on issues, identify areas of agreement and disagreement and make a first attempt at establishing priorities. Often the participants will be asked to vote on specific proposals derived from the first questionnaire. The results of the second questionnaire are then summarised and sent back in the form of a third questionnaire which aims to establish a consensus view on the issues raised and the best potential solution. Again participants may be asked to express their views by voting on specific propositions. The results of the third questionnaire are taken to represent the distilled ideas and opinions of the participants. These ideas and opinions are then taken as a basis for decision-making on the best possible solution to the stated problem. The Delphi technique is an excellent way of picking the brains of a range of people who could not otherwise be brought together for a face to face discussion. It can be a useful and highly disciplined approach to problem solving for any group of people. It does, however, require a considerable period of time (usually no less than 2 months) and the dedicated and skilled efforts of a coordinating team. The value of the material collected by the Delphi technique is generally only as good or as bad as the expertise of the people compiling the questionnaires and interpreting and summarising the answers.

Delphi technique example An electronics company is considering investing in new technology to be able to develop products with neural computing systems. They are seeking information and advice about this new

Design Toolkit

Delphi technique (Cont'd)

Round 1 questions

1. How do you perceive the future of neural computing technology? Do you have any data on the projected growth for this technology sector?
2. Are there any particular applications or market sectors for which you see neural computing playing a particularly significant role in the next five years?
3. Can you provide, or refer me to, any case studies of neural computing applications which you consider to be state-of-the art?
4. Who are the market leaders in the provision of neural computing development systems?
5. What do you perceive to be the main obstacles for a company such as mine starting to develop neural computing applications?
6. Do you think it is a good business investment to invest in neural computing at this time? Please outline the main reasons for your advice.

Round 2 questions

1. It was suggested that there is likely to be substantial growth in the market for neural computing in the area of automatic signal processing for both domestic and industrial applications. Do you agree? Yes/No
   Have you any data or personal estimates of the likely size of this market in i) the UK, ii) Europe, iii) USA or iv) Worldwide?
2. The main obstacle for my company developing neural computing systems was seen to be acquiring the required level of staff knowledge and skills. For staff with a good current knowledge of electronics and computing would you estimate that the necessary training (full time) would take: less than 2 weeks, less than 2 months, less than 6 months or more than 6 months?

Round 3 questions

1. I have focused in on the development of automatic signal processing products as the main justification for investing in neural computing technology. Who, in your view are my main competitors and how would you judge their strengths and weaknesses?
2. I have begun to explore training-providers for my staff in the area of neural computing. Please asses by giving a score out of ten the importance of training in the following areas:
   i) Development systems hardware ii) Development systems software
   iii) Applications systems hardware iv) Applications systems software
   v) Signal pre-processing systems

and rapidly changing technology and decide to conduct a Delphi survey of leading authorities.

Round 1

A letter explains the company's background, its plans for the future and its interest in neural computing. It asks for their help, explaining the Delphi procedure and offers a fee for their assistance. The first questionnaire is enclosed and includes the questions opposite

Round 2

The main results from the first round are summarised and the second questionnaire is sent.

Round 3

The results of round two are summarised and the third and final questionnaire is sent. This is intended to focus in on final decision-making answers.

Market needs research

Design Toolkit

Market needs research is a set of methods for finding out what customers are looking for in a particular type of product. It seeks to determine whether consumers perceive a need which is not satisfied by current products on the market. It requires access to consumers (or other market specialists with an in-depth knowledge of consumers' perceptions and buying behaviour in the relevant market sector). It is essential to have a structured and well thought out approach to the questions to be asked and the analysis of consumer reactions. It provides evidence of market need (or its absence) from which you can critically evaluate the viability of a new product proposal. At this stage, relatively little resource will have been committed to the new product. It is, therefore, of vital importance that if a product is likely to fail due to insufficient market demand, it should be identified at this stage and its development killed off. Market needs research methods can be split into in-house, library, qualitative and quantitative categories. In-house and library research are described on p 159 in this chapter. This toolkit examines techniques for qualitative and quantitative research.

Planning market needs research Market needs research aims to inform a decision which cannot otherwise be made. Using this criterion rigorously should avoid market needs research being done unnecessarily or without a clear focus. That focus can be broad (e.g. what do customers want that they cannot obtain from exiting products) but it must exist. The planning of market needs research establishes that focus by determining what product planning decisions are required. Firstly, the research requirements are established, partly based on existing beliefs about market need but particularly focused on the critical areas of uncertainty upon which the success of the new product depends. Next, the research methodology is established, describing the category(s) of research to be used (e.g. qualitative or quantitative) and what questions are asked. Then sampling is decided, identifying the target market sector (age, socio-economic status, geographical distribution etc.) and the sample size. Measurement methods determine, how the questions are to be asked (e.g. face-to-face or telephone) and data analysis lays down how the collected data will be processed. The final planning stage, is to decide how the results are to be interpreted and translated into decisions (e.g. Go/No Go in new product development).

Figure 6.25 Market needs research [4]

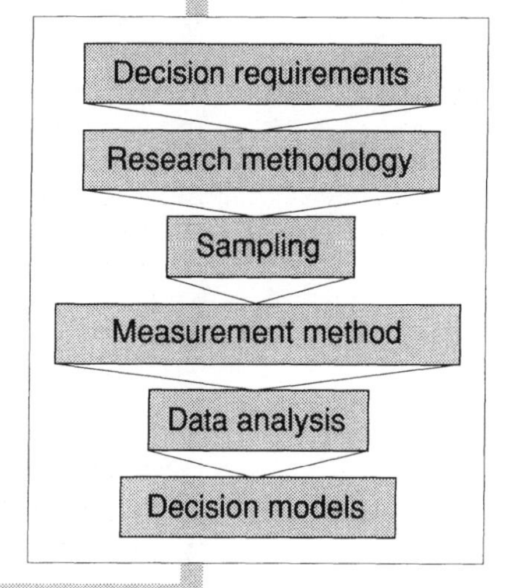

Design Toolkit

Market needs research (Cont'd)

Research objectives The most important starting point is the decision about what market information is needed. This must be based on the perceived opportunities and threats which will determine the success or failure of the proposed new product. Verifying the 'core benefit' (see p 146) should always be fundamental in these considerations. The key issues which emerge should be written down and discussed to try to establish a consensus of opinion.

The issues requiring research will be:
i) those upon which the success of the product most critically depends and
ii) those over which there is greatest uncertainty.

A summary of the identified market research requirements is a useful quality control document at this stage, ensuring clarity and consensus over the research objectives.
Since the purpose of market research is to allow business decisions to be made the research objectives must be described in ways which will inform, support or refute these business decisions.

Research methods Having identified the market research requirements, the next stage is to decide what type of research will be best suited to satisfying them. Qualitative research can give coverage of a wide breadth of issues and can study in-depth consumer's perceptions of products in a market. It is, however, limited to small sample sizes and consequently provides only a 'snapshot' of likely market response. Quantitative research asks a small number of specific questions to a large number of people. It seeks quantitative estimates of market response; what proportion of the market perceive a particular market need? what percentage would pay extra for added features and how much extra? For large product development projects, both types of market research will be used. Qualitative research will explore the perceptions and needs of customers and focus in on specific questions to be answered by quantitative research.

People issues Both qualitative and quantitative market need research is about interviewing people. Selecting these people is a key part of the market research process and is worth spending time and effort to get right. In general terms, subjects must represent the range of consumers which you intend to target with the new product. If the target market is not well

Market needs research (Cont'd)

understood, the selection of subjects must initially cover a wide range of market segments and focus progressively narrower as the research proceeds. Many companies use in-house employees for preliminary market research. Others maintain contact lists of known users of other products in the company's range. The assumption underlying this approach is that existing brand users are more likely to try out new products. Professional market research companies are another alternative and these usually retain the services of a number of consumers who can be called upon to gather together groups of friends and neighbours for market research studies. Finally, subjects can be selected by cold-calling, either face to face or by telephone (the type of market research most of us have experienced as consumers). The advantages and disadvantages of these subject selection methods is shown below (Figure 6.29).

Figure 6.26
Different subjects for market needs research

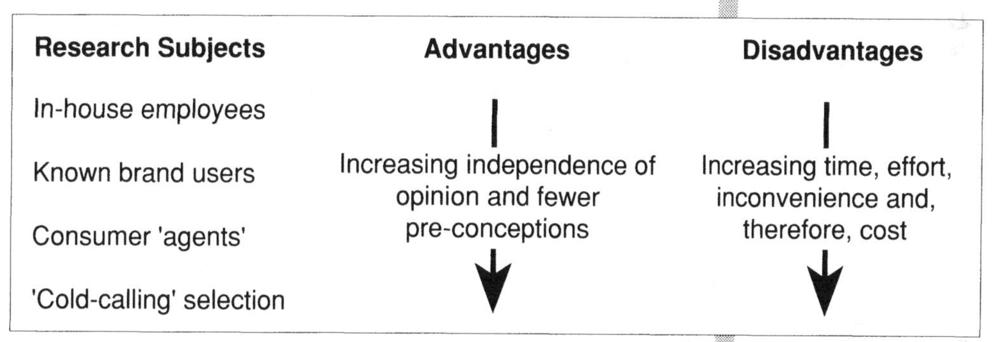

Research Subjects
In-house employees
Known brand users
Consumer 'agents'
'Cold-calling' selection
Advantages
Increasing independence of opinion and fewer pre-conceptions
Disadvantages
Increasing time, effort, inconvenience and, therefore, cost

Asking questions and getting answers Both qualitative and quantitative research is also about getting the required information by asking carefully considered questions. Market research is never simply about collecting evidence to support pre-conceived ideas. It must be open minded and give the subjects equal opportunity to answer in different ways. Questions are often best phrased in leading ways, and offer a range of alternative answers. This helps to structure the data analysis and interpret the results. The questions must not, however, be weighted in favour of any of these alternatives. One common way of mistakenly weighting questions is to give subjects the opportunity to simply agree with the questioner. It is an unfortunate aspect of human nature, from a market research viewpoint at least, that people tend to give the answers they think you want to hear. To ask 'Do you prefer product X over product Y?' tempts the subject to simply agree. Asking 'Which product do you prefer, product X or product Y?' forces the subject to make an unweighted decision. Making questions

Design Toolkit

Market needs research (Cont'd)

un-weighted is not just a question of wording. Intonation in the way the question is asked can be just as damaging. Consider the effect intonation can have in the simplest of questions. '***Why*** did you do that?' 'Why ***did*** you do that?' 'Why did ***you*** do that?' 'Why did you ***do*** that?' 'Why did you do ***that***?' Another key mistake which can weight the subject's responses is to tell them at the start of the interview the purpose of the research. It is normal, therefore, to give only a very broad indication of the aims of the research at the start - 'I would like to ask you some questions to find out your views on mobile telephones' - and leave the more detailed explanations for the end of the interview. A method of asking questions which is used widely in market research is called attitude scaling. This asks subjects to express their attitudes on a scale provided to them either verbally or on paper. A 'Likert' scale rates the subject's degree of agreement with a proposition on a five or seven point scale.

I believe that mobile telephones are an essential tool in running a modern business'

	Strongly Agree	Agree	Neither agree nor disagree	Disagree	Strongly disagree
Do you	________	________	________	________	________

Another method is semantic differentiation, where subjects position their opinions between two extreme positions.

'I believe the memory functions on my mobile phone are .....'

Over-complicated and difficult to use | ___|___|___|___|___|___|___| | Well designed and perfectly suited to my needs

A further method is to ask subjects to allocate points in proportion to their preferences

'Allocate 100 points between these two brands of mobile phone to reflect your preference'

Nokia __________ points Sony __________ points

Qualitative research This aims to give an in-depth understanding of the market perception and market needs of a small number of consumers (Figure 6.30). The research is conducted on either a one-to-one basis or in

Market needs research (Cont'd)

Design Toolkit

small groups called 'focus groups' comprised of about 5 people. The research is often done in someone's home to give a relaxed atmosphere conducive to open discussion. Empirical research by Griffin and Hauser [13] has shown that the percentage of total customer needs identified through qualitative market research increases in predictable ways as the number of subjects increases and as the number of research analysts increases (Figure 6.31).

Feature	Description
Informative and explanatory	Give subjects the opportunity to say what they really feel. Find out their reasons. Encourage feelings, perceptions and beliefs, not just the results of their direct experience. Try to get positive and negative opinions. Seek out surprises and opinions which conflict with company pre-conceptions.
Exploratory and loosely bounded	Don't stick too firmly to the script. Pursue all in interesting lines of discussion. In focus groups, let subjects feed off each others views.
No final answers	Don't attempt to force final answers from the interpretation of qualitative research. Instead, be informed, guided and focused by the discussions.
Dependent upon moderator skills	The moderator plays a key role in effective qualitative research. He/she will draw the line between free expression of views and uncontrolled rambling. The moderator must have an in-depth understanding of the aims of the research.

Figure 6.27
Key features of qualitative research

Figure 6.28
Dimensions of qualitative research

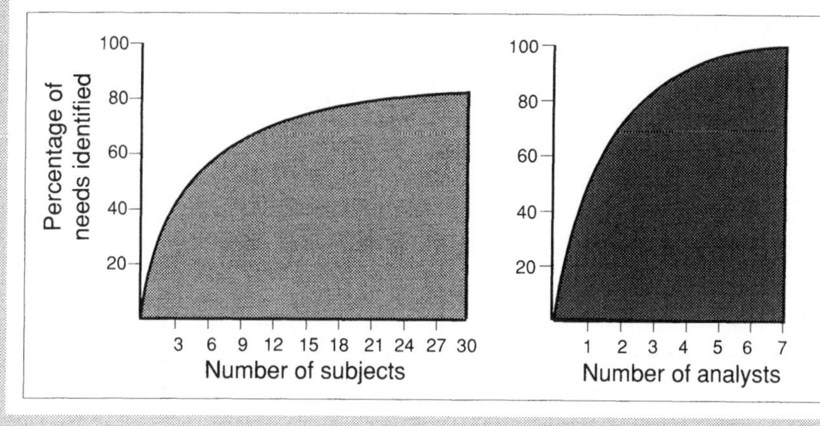

These data suggest that between 20 to 30 subjects should be surveyed and the results should be studied by at least 2 analysts to get most value from qualitative market research. The effectiveness was no different for one-to-one interviews or focus groups.

Design Toolkit

Market needs research (Cont'd)

Quantitative research This aims to give definitive answers to specific questions by surveying a statistically representative sample of consumers from a target market group. The research is conducted either face-to-face or by telephone and sticks rigidly to a carefully devised questionnaire. A typical quantitative research study will interview a minimum of 100 people. Quantitative market research is a highly complex subject and can be used to develop sophisticated models of customer behaviour. It should not be tackled lightly and a more in-depth understanding than can be provided in this book should be sought before embarking upon any research. Further reading is cited in the chapter notes [4].

Figure 6.29
Key features of quantitative research

Feature	Description
Definitive and specific	Usually based on previous qualitative research, quantitative studies focus on a small number of key issues. These issues are tackled by asking a fixed series of questions by means of a questionnaire. The use of the questionnaire must be identical for all subjects.
Numerical and statistically oriented	Quantitative research is usually used to make business projections (e.g. on market penetration and sales volumes) and hence the results must give a representative view of the market. This involves statistical design of the survey to facilitate statistical analysis of the results. Sample size can be precisely calculated given an estimate of the variance of results and the desired accuracy of the business projections.
Informs and facilitates quantitative business decisions	The questions must be designed to give clear, unequivocal, quantitative answers, of direct relevance to business decisions. Preliminary testing of the questions should ensure that answers given are meaningful and to the required accuracy and precision.

Opportunity specification

Design Toolkit

An opportunity specification is a concise written document describing the market need for a proposed new product and the business opportunity presented by that product. It is used, firstly, by management as a basis for deciding whether or not to make an initial commitment to the new product and subsequently, as a summary document to keep the product development team focused on the opportunity they are setting out to exploit. An effective opportunity specification covers two aspects of the proposed new product:

- It describes the business opportunity presented by the proposed new product, focusing on the product's core benefit proposition.
- It provides a justification of that business opportunity in terms of projected sales, margin over manufacturing costs, development costs and projected break-even point (at which development costs will be recovered).

Stages in the preparation of an opportunity specification

The first stage in preparing an opportunity specification is to identify the new product's core benefit proposition. This proposition (Figure 6.25) is identified through the use of competing product analysis (see p 152),

Figure 6.30
What is a core benefit proposition?

A core benefit proposition is a simple and concise proposition of the main benefit enjoyed by the customer from buying the new product instead of any competing product.

Competing product analysis
Analysis of current products on the market aimed at identifying gaps in the market where customer needs are not fully met

Core benefit proposition
- A simple and concise statement,
- describing a significant and important market need,
- not currently satisfied by existing products and
- which can be satisfied by available technology

Market needs research
Research into customer requirements aimed at discovering needs which are currently unsatisfied

Technology audit
An assessment of technological opportunities for satisfying customer needs in novel ways

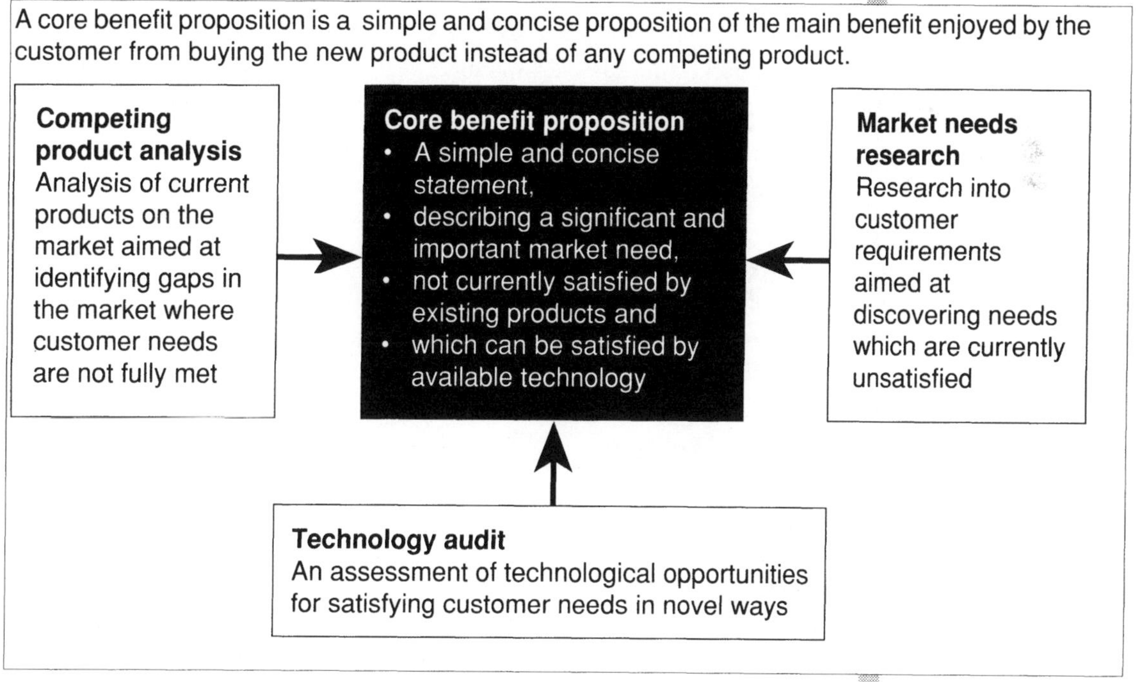

Design Toolkit

Opportunity specification (Cont'd)

market needs research (see design toolkit on page 191) and technology auditing (see page 158 and toolkit on p 136). The three main types of product opportunity (increased value product, reduced price product, market gap product) are shown in Figure 6.26.

Examples of core benefit propositions include [2]:

- American Express traveller's cheques – Prestigious; accepted everywhere; prompt replacement and complete protection if lost.
- Hewlett Packard Laserjet – Quietly prints documents with excellent print quality on several media; easy to use and maintain, reliable and flexible.
- Silkience self-adjusting shampoo – A shampoo that provides the appropriate amount of cleaning treatment (automatically) for different parts of your hair; cleans roots without drying the ends of your hair.

The next stage is to develop other aspects of the opportunity description. This will include:

- Other, secondary benefits of the proposed new product over other products on the market,
- Other features required to maintain the product's proposed position in the market, relative to other products,
- Key marketing sales and manufacturing aspects of the new product,
- Any identified risks in the development of the new product.

The next stage is to justify the opportunity. The first and most important step is to 'price position' the product. This means finding a price for the new product which is competitive with other products and reflects the value of the new product. Pricing is one of the most important decisions to be made about the new product. Price has been described as a 'dangerously explosive variable' and economics research has shown that the price of a product has a 20 times greater effect on sales than advertising, for example [14]. The way to price position a new product is to plot a price:value map

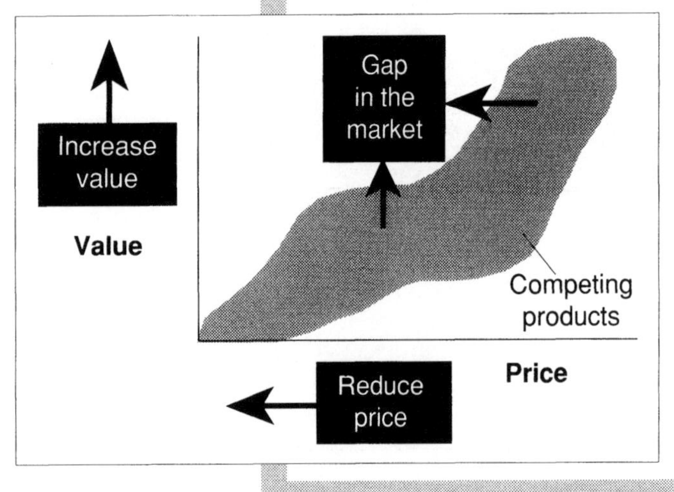

Figure 6.31 Three general types of product opportunity: Increase value, Reduce price or Identify gap in the market

Opportunity specification (Cont'd)

of competing products as shown in Figure 6.26 (another example is on p 173).

Once price is established, a target cost of manufacture can be calculated from 'price minus' financial planning. This takes the retail price of the product (or price to end user if it is not a product sold through retail outlets) and subtracts the margins and costs for each stage of the supply chain for that product (Figure 6.27). From this, the target cost of manufacture can be assessed to see if it is possible to deliver the intended core benefit proposition (and other secondary benefits) for that cost. Provided that it is, the final stage of opportunity justification is to complete a financial model of development costs and the time projected to recoup these development costs from margin on sales. This model should also give a projection of the expected sales life of the new product, its annual level of profit after development costs have been recouped and its lifetime profit projection. This financial modelling is exemplified for the potato peeler example on pages 185 to 187.

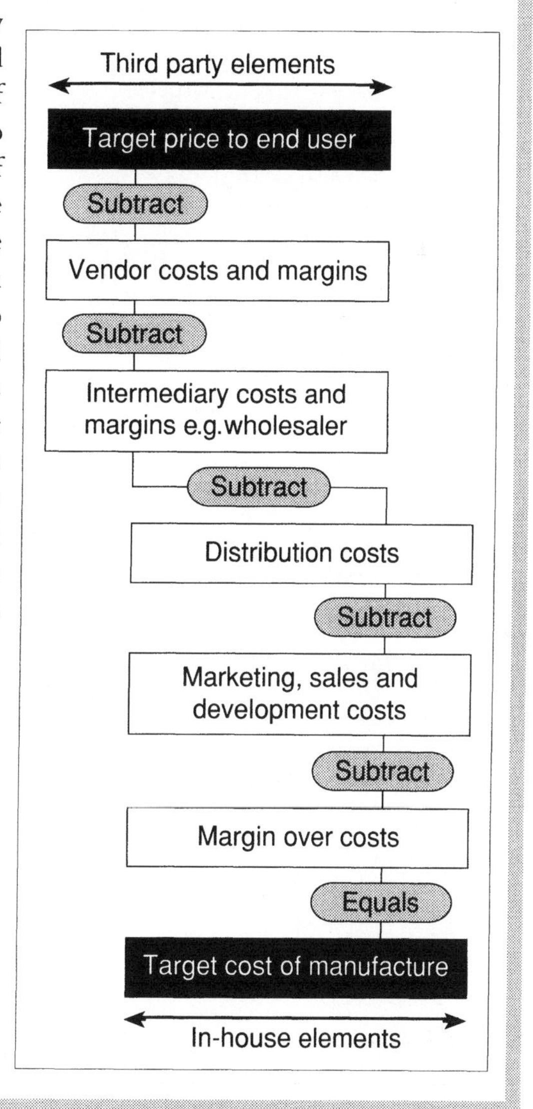

Figure 6.32
Price-minus financial planning

Notes on Chapter 6

1. Ulrich K.T. and Eppinger S.D. 1995, *Product Design and Development.* McGraw Hill Inc. New York, p 6.
2. Urban G.L. and Hauser J.R. 1993. *Design and Marketing of New Products (2nd Edition)*. Prentice Hall International Inc. Englewood Cliffs, New Jersey, p164.
3. Urban and Hauser 1993 (see 2 above), p 30.
4. The classic text on market research is Churchill G.A.Jr. 1992, *Basic Marketing Research* (2nd Edition). The Dryden Press, USA. For more product-oriented accounts see Urban and Hauser (see 2 above) or Moore W.L. and Pessemier E.A.1993, *Product Planning and Management: Designing and Delivering Value.* McGraw Hill Inc. New York.
5. Waterman R.H. Jr. 1990. *Adhocracy: The Power to Change*. Whittle Direct Books, Knoxville, Tennessee.
6. Mintel Market Intelligence, *Cookware*. July 1992, p 8-9, Mintel International Group, London.
7. The story of the development of the Walkman is told by Morito himself in Morito A. 1991. Selling to the World: the Sony Walkman Story. In Henry J. and Walker D. (eds) *Managing Innovation.* Sage Publications Ltd. London pp 187-191. It is also described in Nayak P.R., Ketteringham J.M. and Little A.D. 1993. *Breakthroughs!* (2nd Edition). Mercury Business Books, Didcot, Oxfordshire, UK pp 94-111.
8. A good account of benchmarking for new product development is given in Zangwill W.I. 1993, *Lightning Strategies for Innovation.* Lexington Books, New York, pp 60-73.
9. The data on semiconductor prices was provided by the editorial department of *Electronics Today International*.
10. Plasteck Ltd is a fictional company but its products, market and product development objectives are based on those of the manufacturing companies which DRC has been working closely with over the past 4 years. We have disguised the facts to maintain commercial confidentiality.
11. Proctor T. 1993, Product Innovation: the pitfalls of entrapment. *Creativity and Innovation Management* 2: 260 - 265.
12. Data from Which? report '*Sitting Safely*', August 1993, pp 34-38. For more information on Which? reports call 0800 252100 (UK only).
13. Griffin A. and Hauser J.R. 1993, The Voice of the Customer. *Marketing Science*, Vol 12.
14. Diamantopoulos A. and Mathews B. 1995, *Making Pricing Decisions: A Study of Managerial Practice.* Chapman & Hall, London, p7.

7 Concept design

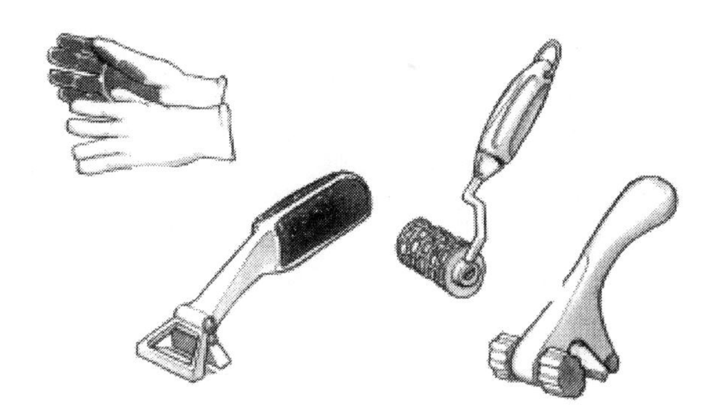

Concept design aims to produce design principles for the new product. These should be sufficient to satisfy customer requirements and differentiate the product from others on the market. Specifically, concept design should show how the new product will deliver its core benefit proposition. Prerequisites for effective concept design are, therefore, a defined core benefit proposition and a good understanding of both customer needs and competing products. Armed with this information, concept design sets about producing a set of functional principles for how the product will work and a set of styling principles for how it will look.

The concept design process

There are two simple secrets to successful concept design. Firstly, generate lots and lots of concepts and secondly select the best. Concept design is the stage of product development which usually demands the greatest creativity. It is at this stage that inventions are invented. Truly innovative designs rarely jump 'out of the blue' by chance. Remember the messages from Chapter 4: creativity is 99% perspiration and 1% inspiration and hence good preparation is vital for solving problems. Preparation has, so far been quite extensive during product planning but even more is needed for concept design. For every product that succeeds there are likely to be several sketch pads full of rejected concepts testifying to the creative labours involved. The

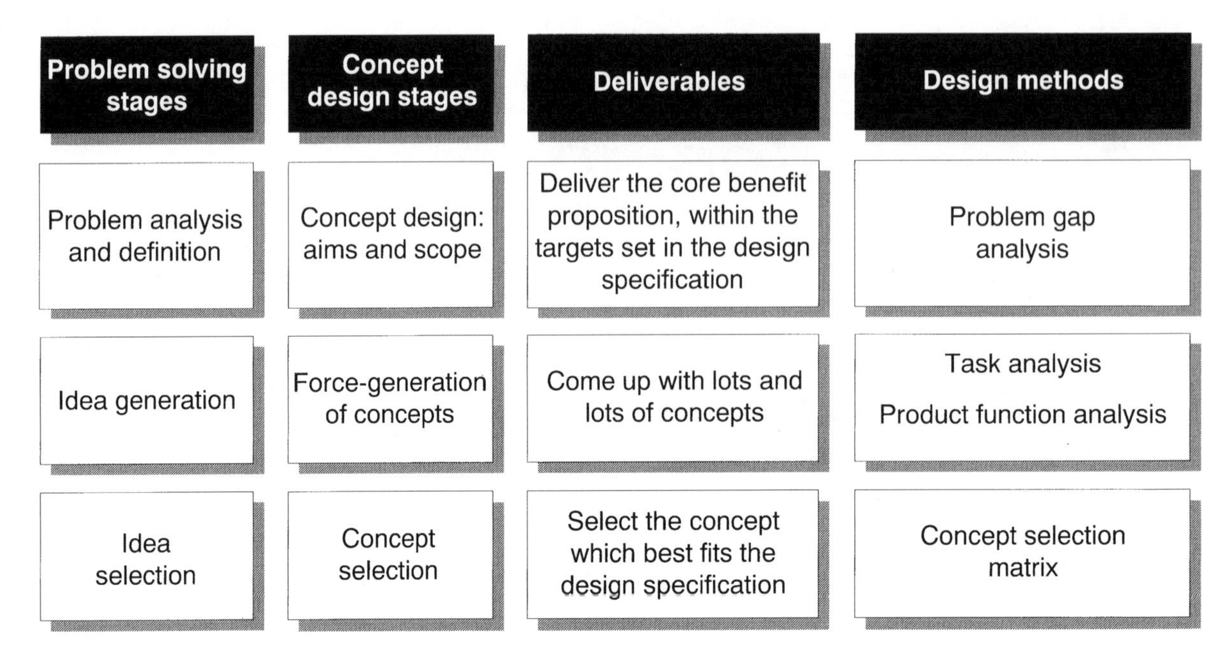

Problem solving stages
Concept design stages
Deliverables
Design methods
Problem analysis and definition
Concept design: aims and scope
Deliver the core benefit proposition, within the targets set in the design specification
Problem gap analysis
Idea generation
Force-generation of concepts
Come up with lots and lots of concepts
Task analysis
Product function analysis
Idea selection
Concept selection
Select the concept which best fits the design specification
Concept selection matrix

Figure 7.1 The concept design process

route to creative success follows a path similar to the general principles of creative thinking. Figure 7.1 shows the stages of problem solving and, alongside, the corresponding stages of concept design, the deliverables from each stage and the design methods available.

Establishing the aims and scope of concept design

The aims and scope for concept design differ greatly for different products. Mostly this is due to the conceptual freedom being constrained to different extents by different product opportunities. If, for example, the identified opportunity is to rapidly produce a reduced-cost version of an existing product, coming up with a whole new set of working principles for the product may be counter-productive. If, on the other hand, all products currently on the market fail to satisfy an identified customer need, then a fundamental re-think of design principles may be essential.

Establishing the aims and scope of the concept design requirement can be tackled by plotting the problem gap (a technique introduced on page 72 and used again on page 117). If product planning has been done effectively, it should have collected together all of the information needed to guide, goal-orient and constrain concept design. So, why do it again in a problem gap format? During product planning attention should have been focused primarily on customer needs and secondarily on the feasibility of manufacturing the product. This should, by default, have determined the goals and scope of concept design, to some extent. But now, when the process of concept design is about to begin, you need to re-examine the implications of product planning for concept design, crystallise them in your mind and check that they are sensible, meaningful and useful. So, problem gap analysis acts as a distilling and focusing procedure. It also acts as a check on the completeness of the product planning conclusions.

Remember that concept design aims to develop a set of working principles concerning the overall form and function of the product. Problem gap analysis must constrain itself to aspects of the design and opportunity specifications relevant to these aims. It is often easier to conduct problem gap analysis in reverse for the present purposes: in other words begin with the goal and work back through the constraints to the current business position in the company. The goal of concept design, as we have said earlier, should be obvious. It should be to produce a set of functional and styling principles which deliver the core benefit proposition, as stated in the opportunity specification. You need to check that it does provide a useful objective for concept design. Imagine a few concepts which you intuitively feel would be perfect solutions – are they adequately described by the core benefit proposition? Imagine a few solutions which fit perfectly the core benefit proposition – do they all intuitively feel well matched to the company at the present time and to your understanding of customer needs. If they do not , the core benefit proposition may be poorly worded or, worse, completely mis-conceived. Obviously, concept design should not commence until everyone involved is satisfied that the core benefit proposition is an appropriate and well defined goal.

The next stage of problem gap analysis (in the present 'working backwards' approach) is to explore the problem boundaries. What constraints are appropriate to be imposed on the range of potential concepts which could be generated. These are the design constraints imposed on how the core benefit of the new product is to be delivered in a commercially realistic way. These constraints are established to try to ensure that product development is not over-ambitious, in relation to the existing business position of the company. A typical constraint might require that the new product can be manufactured using the company's existing facilities or those of its established suppliers. Another might be that the new product is suitable for sale by existing sales outlets, or alternatively, that it specifically gives access to new potential sales outlets. Yet another could be that the new product incorporates a specific new component or technology. All of these should have been referred to in the opportunity specification. You should, therefore go through the opportunity specification, picking out all the requirements which constrain concept design. Remember that concept design is a fundamentally creative process and only goes as far as proposing functional and styling principles for the entire product (as opposed to specific components of the product, which occurs in embodiment design). So, take care when specifying these product constraints. It is all too easy for a company to be conservative in its product development ambitions and thereby prevent more radical solutions from even being considered. It is important to try to keep as many options open as possible for concept generation and then impose rigorous concept selection procedures once the ideas have been generated. At the same time there is no point in generating concepts which you know, at the outset, are never going to be accepted. This is not only a waste of time, it can also be highly demoralising for the designer. (The Psion case study on p 231 is a good example of how concept design can be constrained by the product opportunity).

In the first instance, therefore, for concept generation it is often useful to relax the constraints in the design specification to some extent, to allow greater creative freedom, as depicted in Figure 7.2. This means thinking along the lines of 'I eventually want to end up with X, but for the purpose of concept generation I am prepared to consider Y'. The principle, here, is

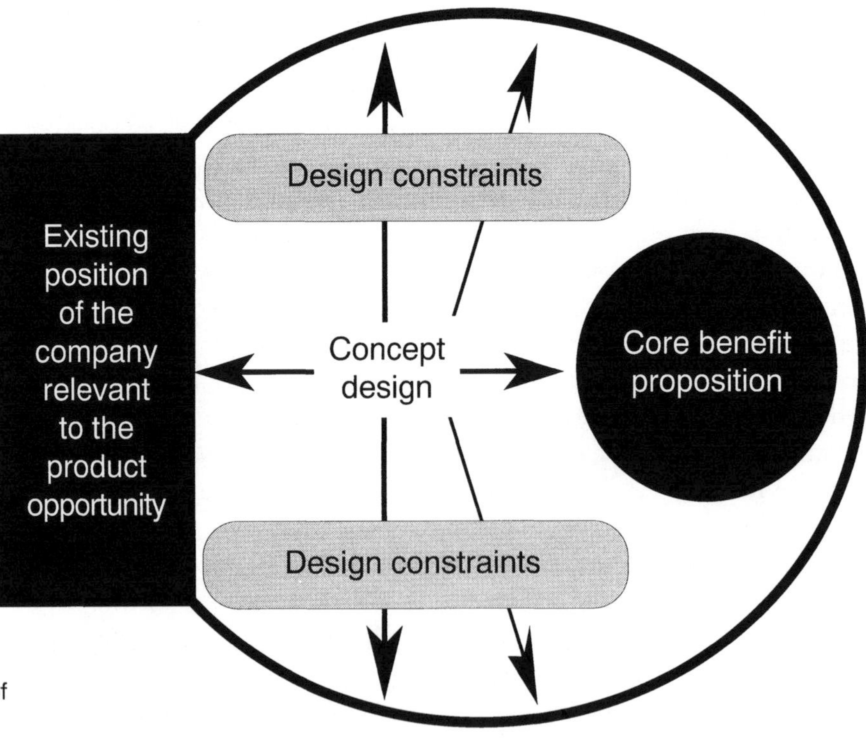

Figure 7.2
The problem gap as a way of constraining concept design

that allowing this extra creative freedom may allow you to come up with a wider range of concepts which can subsequently be more tightly constrained during concept selection.

Once you have identified realistic design constraints think through them, in the same way as you thought through the core benefit proposition. Are these constraints consistent with your intuitive feel for customer needs and current company objectives. Is anything too constraining, or perhaps not constraining enough. Can you think of any potential concepts which would be wrongly excluded by the constraints imposed. Can you think of any unsuitable concepts which the constraints would currently allow.

Having worked backwards through problem gap analysis, from the problem goal, through the problem boundaries, you should end up at the problem start point. For the present purposes, this is the current business position of the company. Clearly, the aims and scope of concept design need to be consistent with the company's mission, its corporate objectives, corporate strategy, product development objectives and product development strategy. Of course, it was the aim during product planning to specify both the opportunity and the product design to be consistent with these. Now, at the start of concept design, you have the chance to confirm this. Do the concepts that you have been imagining so far fit with the company's overall aims and strategies. Can you think of any concepts that would be perfect for the company which have somehow been missed by the product planning process? Do any concepts which meet the goals and constraints you have just established for concept design clash with the company goals? Again, certain aspects of the opportunity specification or concept design aims and scope may need to be reconsidered.

This problem gap analysis may all seem rather repetitive and tedious, when you are itching to get on with concept design. But nearly every experienced designer will have had the bitter experience of generating wonderful new concepts for a product, only to find that they have been heading in the wrong direction and have to start again. The first time that bitter experience happens, you will wonder why on earth you did not take the hour or two to go through problem gap analysis and save yourself the grief!

Force-generation of concepts

Good concept design requires the use of intuition, imagination and logic to come up with creative solutions to the, now well-defined, problem. The main difficulty in concept design is freeing your mind sufficiently to come up with original concepts. As discussed in Chapter 4 on creative thinking, this involves overcoming the creative blocks which arise as a consequence of conventional modes of thinking. Three concept generation methods are described in this chapter: task analysis (see below on this page), product function analysis (see p 210 and design toolkit pp 236-238), and life cycle analysis (see p 213 and design toolkit pp 239-240). Together these methods can be thought of as ways of 'force-generating' concepts. By means of structured techniques, they force you to:

- Reduce the concept design problem to its core elements (concept abstraction)
- Use structured thinking methods to analyse different aspects of the concept design problem and generate a great many possible solutions based on this analysis.

(Another force generation technique is described in Chapter 9 on embodiment design. Product feature permutation is primarily concerned with the arrangement of product parts although it may also be useful for concept design see p 275)

Simply using imagination and intuition may allow you to generate a handful of new concepts. Using these techniques increases the number of concepts generated to many tens or even hundreds of concepts. To maintain consistency, concept design techniques will be illustrated using Plasteck's potato peeler problem.

Task analysis

Most products are designed to be used, in some way, by people. When examined in detail, the product-user interface for even the simplest of products is often complex and rarely well understood. Consequently, this aspect of product design often provides a rich source of inspiration on concept design. Task analysis explores the interaction between the product and the person who uses it by observation and analysis and then uses the results to generate new product concepts. It gives the designer first hand experience of how customers actually use products. Through this, it stimulates concept generation to improve the user interface and paves the way for the subsequent application of ergonomic or anthropometric design methods. Task analysis covers two very important but highly specialised aspect of product development: ergonomics and anthropometry. The word ergonomics is derived from the Greek word 'ergon', meaning work and so is literally the study of work. Early ergonomic studies did, indeed, study people in their working environments but now ergonomics is used much more widely and loosely to refer to the interaction between people and built artefacts. Ergonomics is a research topic in its own right and covers aspects of anatomy, physiology and psychology, as well as being applied to design. A comprehensive account of current knowledge in ergonomics would fill far more pages than this book has in total and is, therefore neither possible, nor would it be particularly useful. By far the best way to tackle ergonomics, for most product designers is on a 'need to know' basis. If you become involved in a design project which relates to a specific type of interaction with a product, then you start researching current knowledge of the ergonomics of that task (sources of information on ergonomics is cited in the chapter notes [1]). For most design work, however, a sufficient insight into the person-product interaction can be gained by observing people performing the relevant task and, from this, deriving a first hand understanding of the issues involved.

'If a piece of industrial equipment was designed to fit 90% of American men, it would fit roughly 90% of Germans, 80% of Frenchmen, 65% of Italians, 45% of Japanese, 25% of Thais and 10% of Vietnamese'
P Ashby [2]

Anthropometry is the measurement of people. When designing products for people to use, it makes inescapable sense to use measurements of people as a basis for product dimensions. A great many published sources provide data on the dimensions of different parts of people. Everything from the standing height of an adult to the diameter of a child's little finger can be found in most good design libraries. The problem with anthropometry is not, generally, finding the data. It is knowing how to use it. Again a comprehensive account of the subject is not possible in these pages but a summary is cited in the chapter notes [1].

Task analysis is a simple, almost common sense approach to studying both the ergonomic and anthropometric aspects of products. It is common sense because it sets out simply to observe people in their typical use of products and follows this observation up by questioning the user about how they perceive the product to work. Specific issues arising from task analysis can often be tackled in greater depth by manipulating some aspects of the

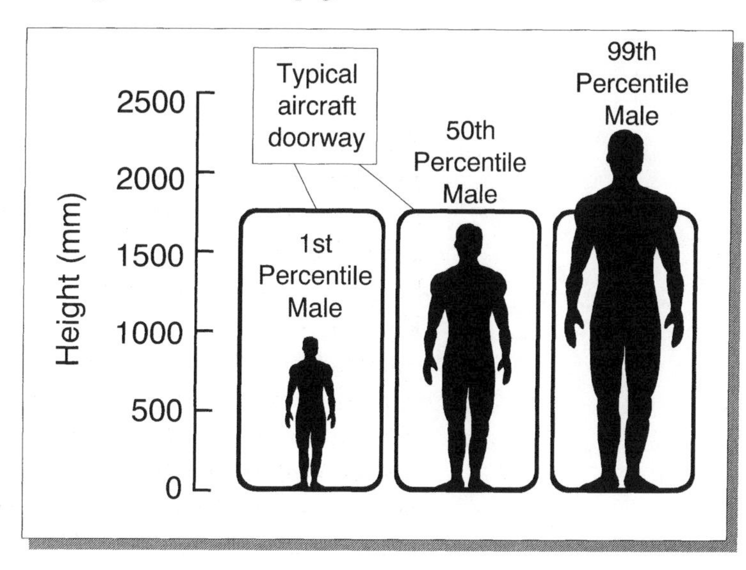

Figure 7.3
Anthropometry data is widely available [1]

product's function and asking people to use the modified product. Despite being a common sense approach, it is something that designers rarely take the trouble to do. As a result, products suffer problems in the way they interface with users far more often than they should.

Plasteck – peeler task analysis

Plasteck applies task analysis by observing a number of people peeling potatoes with different peelers and using photography to characterise different types of peeling actions. It soon becomes clear that the same product can be used in different ways by different people. One of the most obvious differences is that some people peel by slicing the potato towards their bodies, whereas others slice away from their bodies (Figure 7.4). This simple discovery reveals a new purpose for double sided peeler blades.

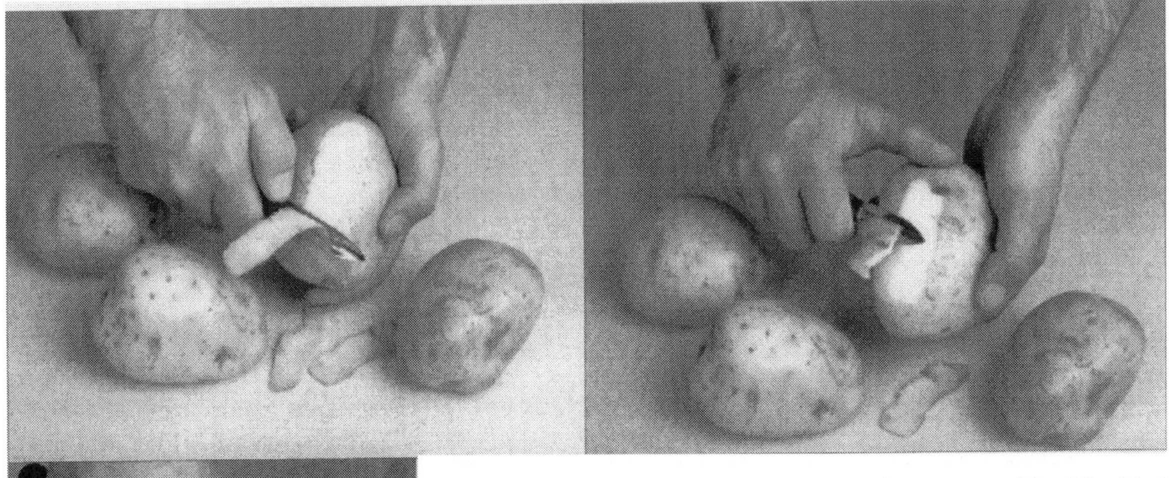

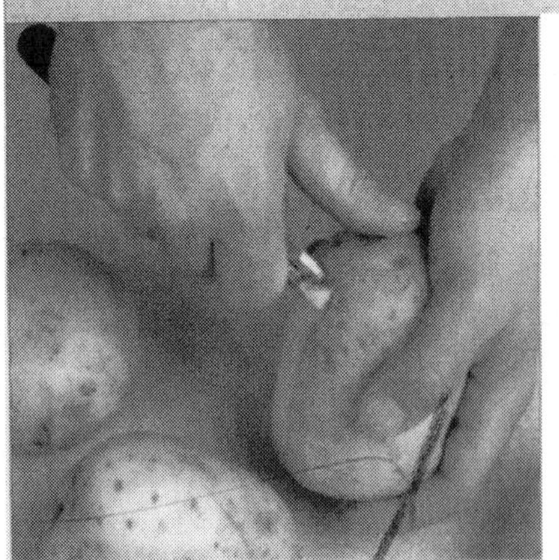

Figure 7.4 'Push' and 'pull' potato peeling and design problem in use of the gouge

Previously they had thought that the sole purpose of double sided blades was to accommodate left and right handed people. The possibility of producing left and right handed peelers is now dismissed in favour of having a double sided blade which accommodates left and right handed users ***and*** 'push and pull slicers'. Looking at how people used the gouge revealed what was considered to be a common but quite significant failure in peeler design. Many users were seen to move their grip off the handle and down over the blade in order to remove the eyes from potatoes (Figure 7.4). When asked about this, they said that the gouge was too far away from the handle to give the necessary control. Some

Figure 7.5
'Dry' and 'bowl' potato peeling

also commented on the fact that keeping their grip on the handle would have increased the risk of the gouge slipping off the potato and into their other hand.

Looking at the overall task of potato peeling more broadly, revealed two distinct 'strategies' (Figure 7.5). Some users peel potatoes dry and rinse them after peeling ('dry' peelers), whereas others fill a bowl with water and repeatedly immerse the potatoes while peeling ('bowl' peelers). Both strategies were claimed to have disadvantages. Dry peelers said that repeatedly turning on and off the tap was a nuisance and that often, with badly soiled potatoes, they missed bits of peel and after rinsing had to finish little bits of peeling and rinse a second time. Bowl peelers, on the other hand, said that ending up with a bowl of dirty water filled with potato peelings was messy. The bowl then had to be emptied into a sink to drain and then dripping and dirty peelings had to be lifted out of the sink and discarded. Disposing of dry peelings would be preferable, they thought. Some mentioned problems of hygiene in immersing peeled potatoes in dirty water, others mentioned problems of putting too many potatoes into the water and having to put wet potatoes back in the vegetable rack. Yet others mentioned the hassle of finding the last potatoes among all the peelings. Generally, however, each group were convinced that their method was best, on balance.

Two new concepts emerged from the task analysis. Firstly, the location of the gouge was wrong. It had to be moved to allow users the control and safety they sought without requiring them to grip the blade. Secondly, a wash-and-peel concept was devised in which the peeler has a hose connection to the tap and water jets over the blade. This might offer a way of overcoming the problems expressed by both 'dry' and 'bowl' peelers. Concept sketches for these two ideas are shown in Figure 7.6.

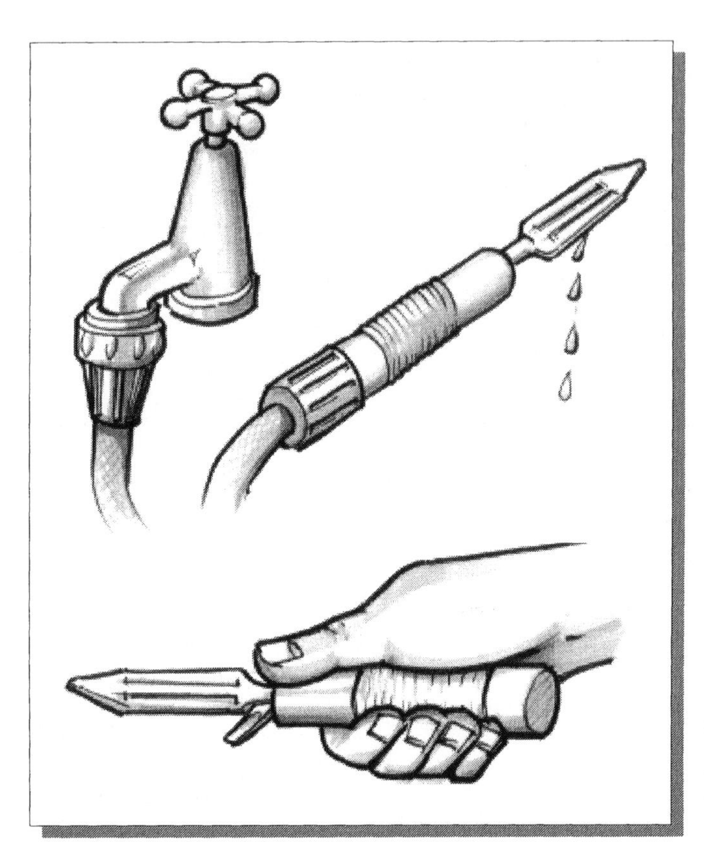

Figure 7.6
New peeler concepts derived from task analysis

Product function analysis

Task analysis is primarily a descriptive technique and, as such is a valuable first step in concept design. Not only does it give rise to interesting new concepts it also familiarises the designer with the use of the product by the customer. Now, however, we move on to techniques which are more analytical in their approach and they delve deeper into the reasons for products being designed the way they are.

Product function analysis is a powerful technique, which can be used on its own for concept design, or used as the first step in two other design methods, value analysis (see p 214) and failure modes and effects analysis (design toolkit on p 294). For the purposes of the present chapter, it is sufficient to describe how it is used on its own. As described in the design toolkit on page 236, product function analysis is a fundamentally customer-oriented technique. Throughout, it presents the functions of the product as perceived by the customer and as ranked in importance by the customer. For products with complex, or not properly understood customer functions, it will have to be based on formal market research. For a product, such as a potato peeler, the design team's knowledge of how customers use the product derived from task analysis is likely to be sufficient. The primary function of a potato peeler is to 'prepare potatoes for cooking'. It does this by removing the skin and eyes from the potato, the skin being removed by a series of depth-limited cuts and the eyes by gouging. These functions are achieved by holding the handle of the peeler and either slicing to cut the skin or inserting the peeler point and levering or twisting to remove the eyes. These simple statements describe the functions of the peeler, from its primary function, to prepare potatoes for cooking, right through to the specific actions by which this primary function is achieved. As shown in Figure 7.7 they can be represented in a product function diagram or function tree. This function tree is read by moving down the tree and linking each level to the one below using the word 'how?' How are potatoes prepared for cooking? By removing the skin and eyes from the potato. How is the skin removed? By cutting the skin, depth-limiting the cut and following the contours of the potato surface. How is this done? By holding the peeler handle and slicing with a depth limiting blade, whilst swivelling the blade. Alternatively the tree can be read by moving up the tree and linking each level to the one above using the word 'why?' Why do you insert the peeler point and twist or gouge? To gouge a hole in the potato. Why do you gouge a hole in the potato? To remove the eyes from the potato. Why do you remove the eyes from the potato? To prepare it for cooking.

Once the product function analysis is complete, new concepts can be generated by exploring how each function could be achieved differently from a conventional potato peeler. In general, the more you try to find alternative ways of achieving the higher order functions (i.e. those higher up the function tree) the more you are challenging the basic assumptions upon which potato peeler design is based. So, for example seeking alternative ways of preparing potatoes for cooking is likely to reveal solutions which dispense with the need for a potato peeler entirely. You could simply wash them and cook them in their skins. Finding alternative ways of achieving the lower order functions, on the other hand, will reveal more minor design modifications to peeler design. The alternative to having a point for gouging the eyes out of potatoes might be, for example, a sharp edged tube which removes a small core out of the potato. Product function analysis can, therefore, reveal radical, blue-skies innovations by focusing on high-order functions or small incremental innovations by focusing on low order functions. This is a particular strength of concept generation using product function analysis. For design problems with few constraints you should begin with the primary function to seek radical innovations and then move down through the function tree looking for progressively more incremental innovations.

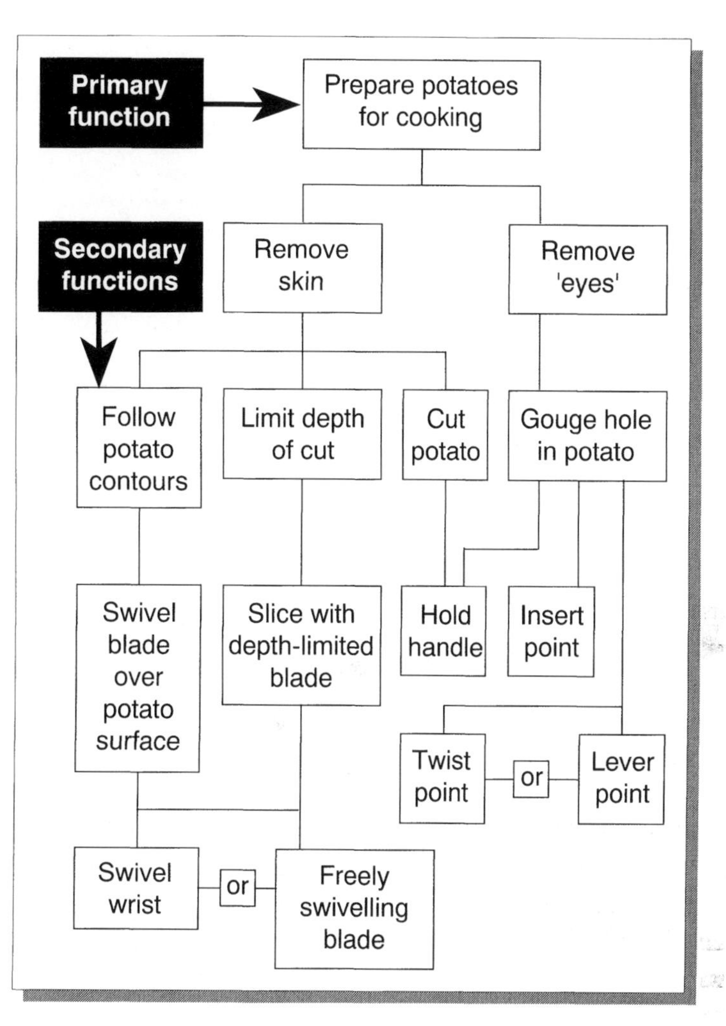

Figure 7.7
Product function analysis for a potato peeler

Analogies (described in design toolkit on pages 91-92) are particularly useful for generating functional alternatives. Figure 7.8 shows an analogy table for several of the potato peeler functions and 7 of the new concepts generated from it.

Figure 7.8
The use of analogies in concept generation

Peeler function	Functional analogies	Principle of operation	New peeler concepts
Prepare for cooking	Meat tenderiser	Reduce 'toughness' in chewing texture	Potato peel tenderiser
Remove skin	Sand down old paint	Abrasion	Potato sanding block
	Metal milling machine	Blade counter-rotates to direction of movement	Hand-held power-peeler
	Single sheet feeder for printer	Shears top sheet off pile of paper	Grip potato and impose shear force to remove skin
Limit depth of cut	Shave face or plane wood	Blade protrudes from guide surface	Potato razor
	Plough field	Surface roller limits depth	Cutting blade with surface roller
	Cheese grater	Lots of little blades	Potato grating block
Swivel blade	Door handle	Handle swivels	Fixed blade swivel handle

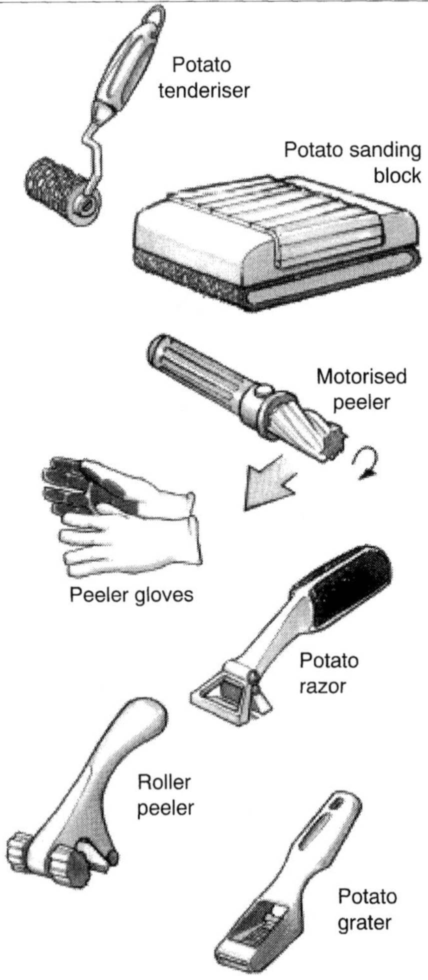

Described later in embodiment design

Life cycle analysis

Another analytical technique which can be used to force generate new concepts is life cycle analysis (see design toolkit on p 239). This technique is used most widely by designers interested in improving the environmental-friendliness of new products, but in principle it is applicable to design for all purposes. By mapping out the life cycle of a product from the time it enters the factory as raw materials, to the time it is discarded after use by the customer, the designer is forced to think about how well the product is designed for each of these life cycle stages. Figure 7.9 shows a life cycle analysis chart for a plastic plant pot. Starting with the plastic being delivered to the factory as a raw material. It works through the manufacture of tooling, the manufacture of the product, its packaging, storage and subsequent transport. At that point, the distribution chain divides and the pots are

Figure 7.9
Life cycle analysis of a plant pot [3]

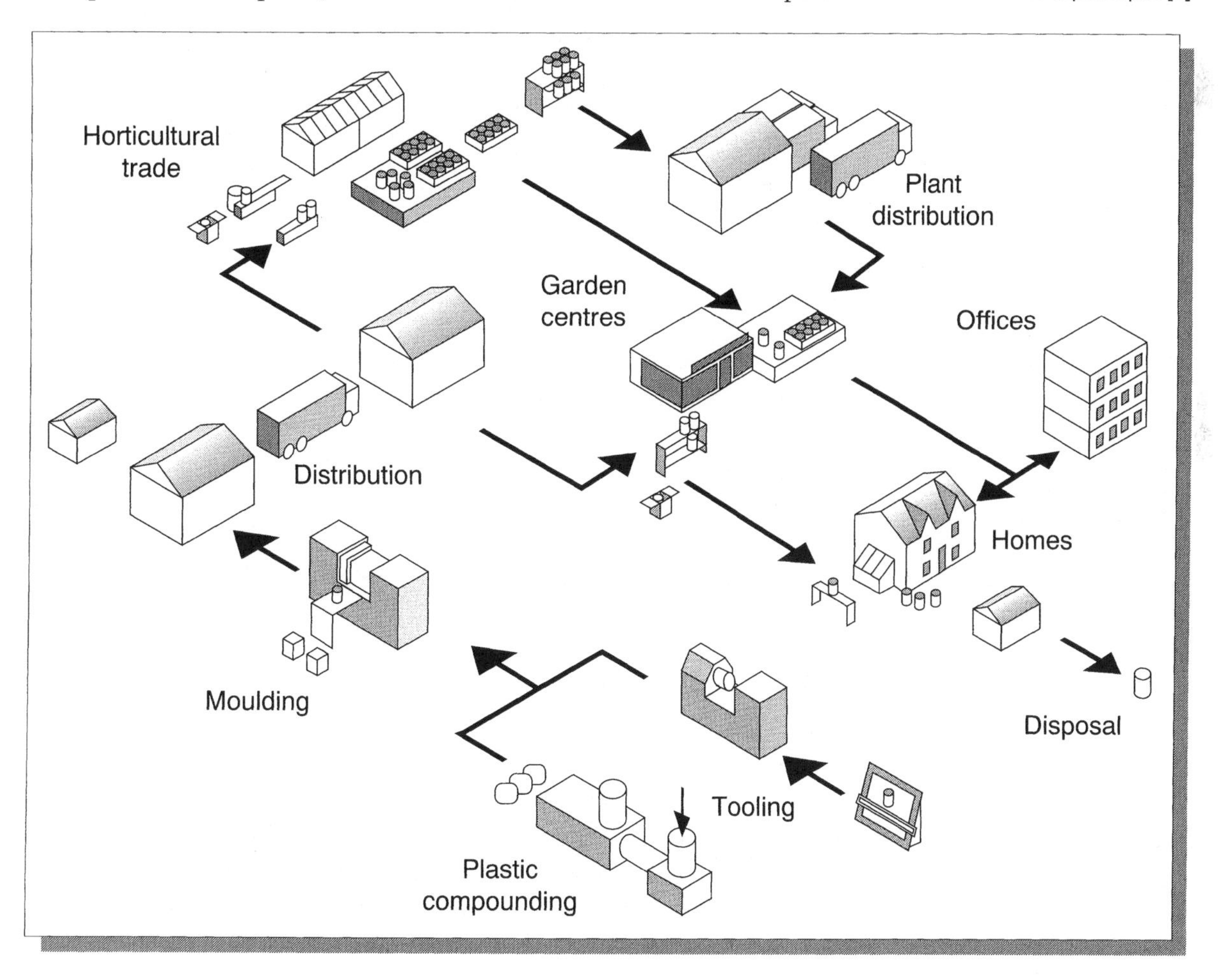

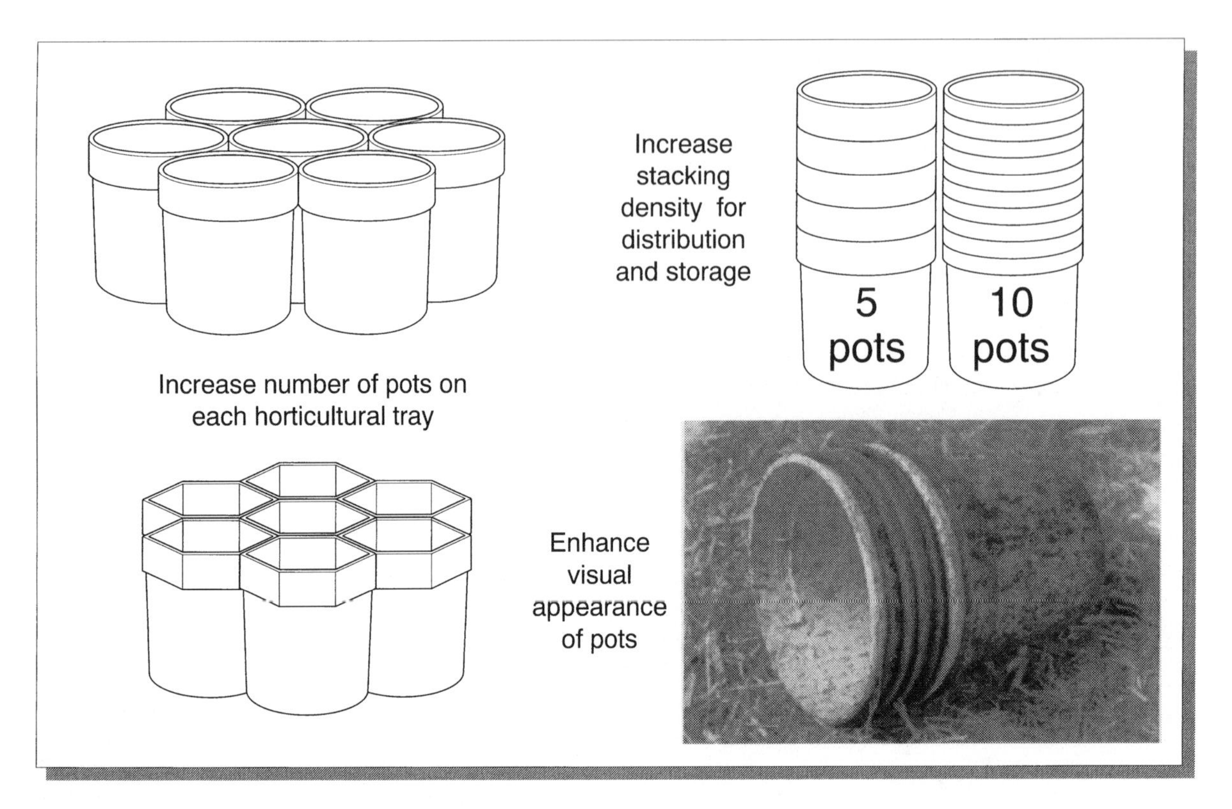

Figure 7.10 Concepts generated from life cycle analysis

either distributed to retail outlets for sale as empty pots or supplied to the horticultural industry to be potted up with plants and then sold through garden centres to the public or distributed direct into the horticultural trade. From there they will either go to private homes, to be kept in the house or a glasshouse or alternatively supplied to offices. At the end of the plant's life, the pot will either be re-used or discarded as refuse. The opportunities for innovation which arise out of this analysis (Figure 7.10) include increasing their stacking density to reduce transport costs, improving the way pots fit together on horticultural trays or enhancing their visual appearance.

Value analysis

The traditional approach to analysing the cost of products is to examine the cost of materials, manufacturing process and labour for each of the product's components. The problem with this approach is that it tells you nothing about the value of each of the components. As a result, the component with the highest individual cost may become the main target for cost reduction, despite the fact that it might make by far the most important contribution to that product's function. Value analysis overcomes that problem by including measures of value to weigh up against the measures of cost. The principle underlying value analysis is that design features which are expensive to produce should make a large contribution to customer value. Conversely, features which have little value should add little to cost. An application of value analysis to the design of a survival suit for the offshore oil industry (conducted by Multifabs Survival Ltd. [4]) is given in Figures 7.11 and 7.12.

Value analysis begins by conducting a product function analysis to obtain parameters on which customer value can be determined. Survival suits are used by oil workers during transit out to the offshore oil rigs in the North Sea. Their primary function, as their name suggests is to facilitate survival in the event of immersion in the sea if the transit helicopter ditches en-route to the offshore rigs. Secondary functions include facilitating use during normal (i.e. non-emergency) transit conditions, preserving life in emergency situations and facilitating rescue. Subsidiary functions are shown in the function tree (Figure 7.11) and for the purpose of value analysis, it is decided to examine the customer value for the survival suit using the subsidiary functions which are shaded. The next step in value analysis is to attribute relative importance to the different survival suit functions. With some products this can be done through market research. For products such as a

Figure 7.11 Product function analysis of a survival suit, as a first step in value analysis

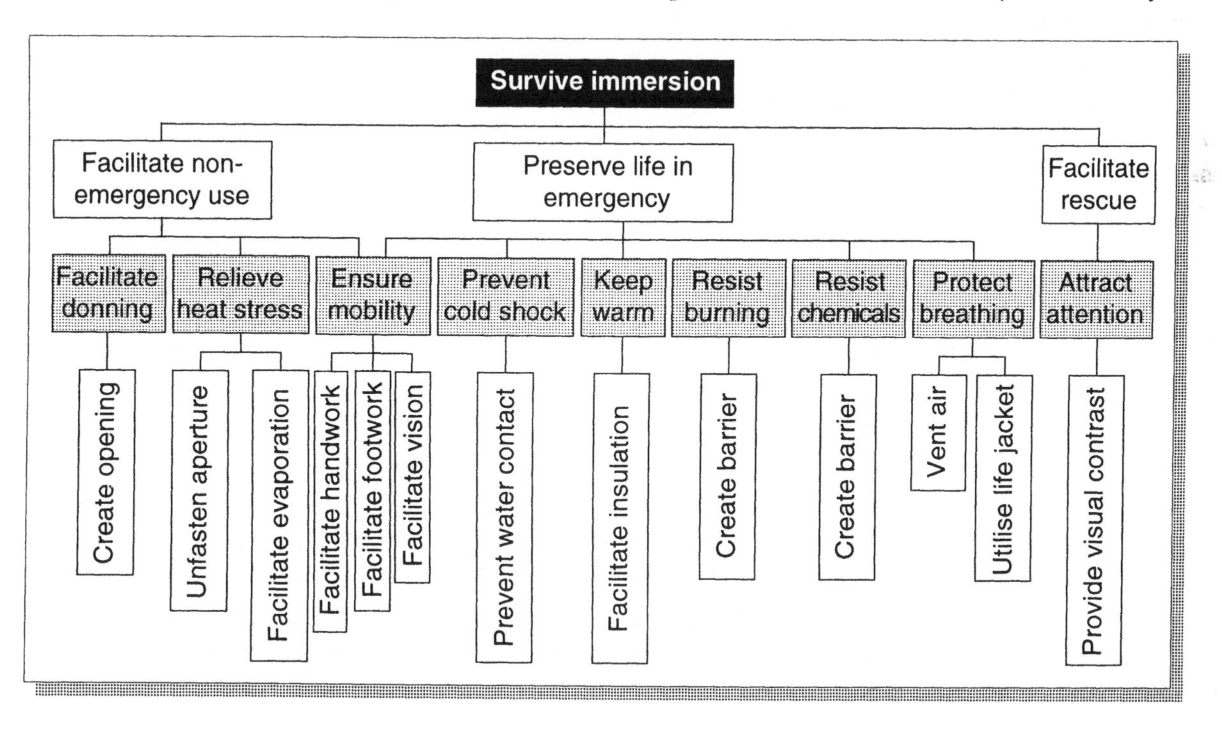

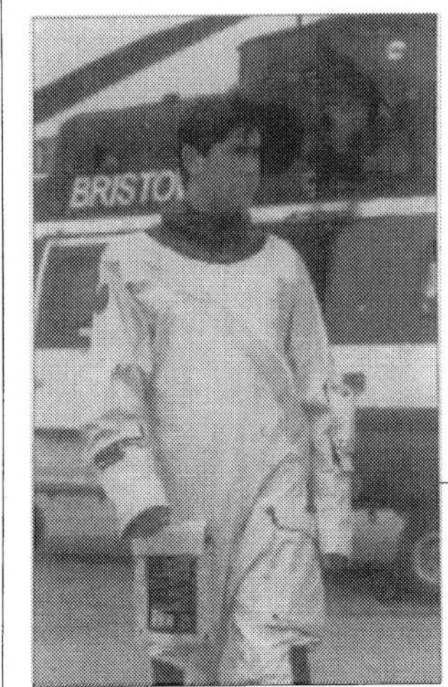

Figure 7.12 Value analysis of Multifabs (Survival) Ltd. offshore survival suit [4]

	Facilitate donning	Relieve heat stress	Ensure mobility	Prevent cold shock	Keep warm	Resist burning	Resist chemicals	Protect breathing	Attract attention	Miscellaneous
Hood	£0.45		£0.27	£5.61	£5.61	£0.53	£0.53		£0.67	
Feet units			£0.60	£8.86	£2.80					
Main body	£10.01	£20.03	£13.35	£33.38	£13.35	£9.35	£6.68		£6.29	
Wrist seals	£0.97			£2.00	£1.67			£1.00		
Valves			£0.31					£2.34		
Main zip	£33.58	£3.34			£3.34					
Gloves	£1.19		£1.34	£5.34	£5.34				£0.33	£0.53
Mitt pocket	£6.81					£0.67	£0.67		£0.67	
Mitt pocket	£6.68					£0.60	£0.60		£0.60	
Leg pocket	£5.34					£0.53	£0.53			£1.32
Strobe						£0.53	£0.53			£6.70
Washing label										£0.73
Gaitors			£2.00				£1.00	£10.09		
Becket							£0.80		£0.84	£7.38
Back/leg fastening			£9.35					£9.95		
Front zip flap						£5.62	£0.67		£0.67	
Neck seal	£1.00			£3.34	£3.34			£1.00		
Logo print										£0.73
Barcode label										£0.37
Registration label										£0.37
Testing										£2.96
(Lifejacket)								£191.35		
Function cost	£66.03	£23.36	£27.21	£58.53	£35.44	£17.84	£12.02	£215.32	£10.07	£21.12
% of total cost	14%	5%	6%	12%	7%	4%	2%	44%	2%	4%
% of function	8%	6%	13%	13%	14%	12%	4%	15%	9%	5%
	0.59	1.25	2.33	1.08	1.92	3.28	1.62	0.34	4.35	1.15

survival suit this is virtually impossible. Never having experienced an offshore emergency, most users of survival suits are likely to attribute greater importance to the suit's non-emergency performance (e.g. making it less uncomfortably hot whilst in the helicopter) at the expense of its primary function. The importance of functions, therefore, had to be attributed on the basis of what is known about survival factors in offshore emergencies and other specialist knowledge of the offshore survival suit market.

The other side of the value analysis equation is the cost of the product. To be effective, value analysis must be based on a thorough understanding of cost factors. It must, at the very least, be possible to break costs (for materials, labour and manufacturing processes) down to individual product components. This effectively means that value analysis must initially be conducted on an existing product. To explore the value implications of new designs the changes in cost and value can be estimated, altered on the value analysis of the existing product and their effects calculated.

Once both cost and the importance of functions have been quantified, the calculation of product value can commence. This is done by allocating the cost of different components to different product functions in a value analysis matrix (Figure 7.12). So, the main body of the suit, which costs £112.42 is attributed to 8 different suit functions. If the suit body did not need to be a bright colour to attract the attention of rescuers, its cost would be £6.29 less. £6.29 of the total £112.42 cost of the main body is, therefore, attributed to the function 'attract attention'. The overall value of each function is then calculated by dividing the relative importance of that function by its relative cost. This gives a range of value ratios, with large numbers indicating high value and low numbers indicating low values. Design efforts at improving the value of the survival suit are then directed towards enhancing the performance or reducing the cost of the low value functions. As a result of the survival suit value analysis, design efforts were concentrated on redesigning a new life jacket to better protect breathing but at a reduced cost, trying to facilitating donning of the suit with less cost than the current zip and preventing cold shock with reduced cost for the main body of the suit.

Concept styling

So far we have been concentrating on the functional features of the new product and now we need to move on to consider product styling before returning to the selection of the best overall concept. Our aim, within concept design is to develop styling principles for the new product. This means that the overall form of the product must take shape, although we do not have to concern ourselves, at this stage with how the individual product components will be designed to make up that form. This comes later during embodiment design. In the last chapter we saw how styling objectives could be established and how these are derived from any product predecessors, company or brand identity and product semantics and symbolics. Of these, product predecessors and company or brand identity are quite straightforward. Let us start, therefore by exploring in more detail what we mean by product semantics and symbolics.

Product semantics

Products which move fast should look sleek and streamlined. Products which are durable and hard-wearing should look robust and rugged. Products which are fun should look bright and happy, whereas products used for serious work should look sombre and efficient. This is the essence of product semantics. Within concept design it is important to establish a visual form which makes the product look as fit for its intended purpose as possible. Figure 7.13 illustrates some of the semantic statements made by the new Aston Martin DB7 car. By contrast, the Range Rover is a car with a very different purpose and hence with a very different semantic expression. Whilst on the subject of cars, it is interesting, but probably far from coincidental, that all the major marques of German cars have a ring of steel as a key feature of their logo (Figure 7.14). A sign of integrity, strength, and engineering quality, the ring of steel perfectly represents the functional qualities that the German auto industry promotes.

In a very different market sector, hair care toiletries have their own semantic messages conveyed in the styling of both their products and its packaging (Figure 7.15). Here the styling messages relate to the purity and simplicity of some products and the technical quality of the formulation of others. In such a lucrative but highly competitive marketplace in which all products perform their baisc function (cleaning hair) perfectly adequately, product styling plays a key role in product differentiation and market segmentation.

Product symbolism

Product symbolism is often misconstrued as simply pandering to the vanity of prospective product purchasers. But it is much more than that. We all have a self-image based upon the personal and social values we hold. And it

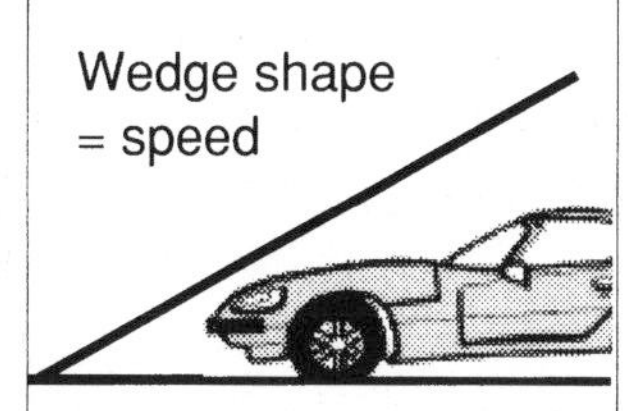

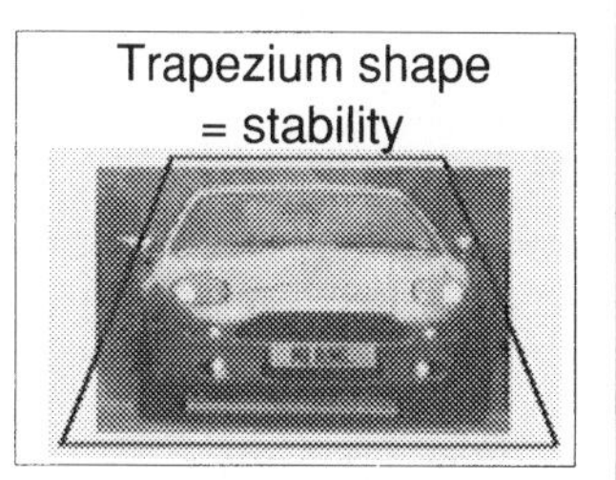

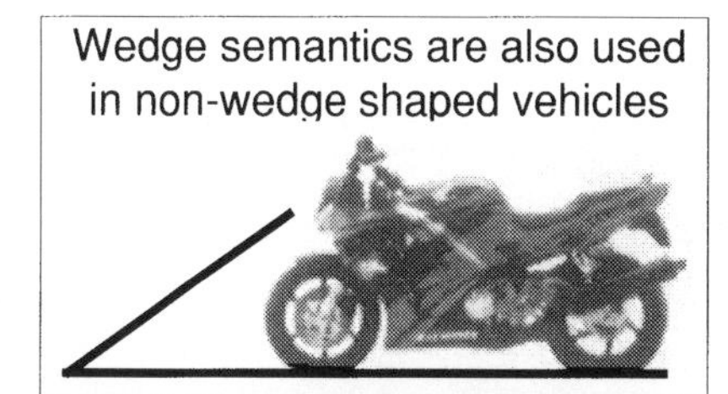

Wedge semantics are also used
in non-wedge shaped vehicles

Figure 7.13
Semantic expressions in the styling of the Aston Martin DB7 and the Range Rover [5]

Figure 7.14
All German cars make a strong semantic statement with their 'ring of steel' logos [6]

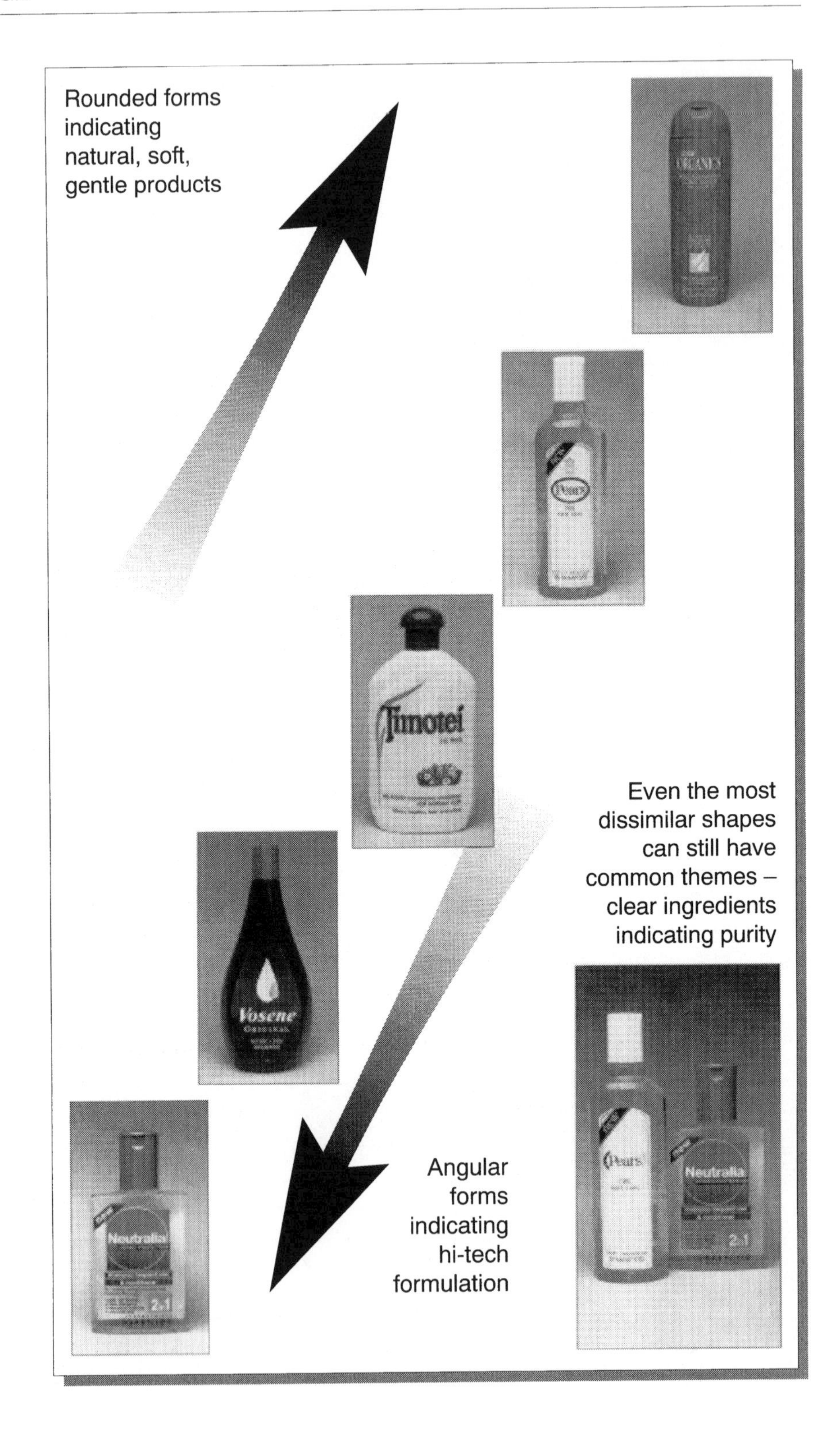
Rounded forms indicating natural, soft, gentle products
Timotei
Vosene
Even the most dissimilar shapes can still have common themes – clear ingredients indicating purity
Neutralia
2:1
Angular forms indicating hi-tech formulation
Pears

Figure 7.15 Semantic expressions in hair care products

is an inescapable aspect of human nature that we tend to surround ourselves with artefacts which reflect our self-image. The houses we live in, the cars we drive and even the dogs we own; they are all pieces in the jigsaw which together make up the visual image we project of ourselves. There is little doubt that the majority of products we purchase are bought mostly for their functional value rather than purely their symbolic value. Given the choice of two products of similar function and the symbolic value of the product can play a vital role in determining which is purchased. And a significant part of the judgement of what looks best will be determined by how well the product fits into the symbolic expectations of the customer. A stereo system built-in to a mahogany cabinet will not appeal to the son any more than a brightly coloured stereo with 100W speakers will appeal to the father.

In the car industry, the most obvious symbolism relates to the affluence of the car's owner. As we move from Lada to Lexus, the increase in price is accompanied by an increase in the status symbolised by car ownership. Part of this symbolism is simply to do with wealth. Driving an expensive car is a rather explicit statement that you are wealthy enough to afford such a car. But car styling symbolises many more subtle values than this. Figure 7.16 shows the interior of the MG with its quintessentially 'English' leather and mahogany furnishings. This symbolises the aristocratic values associated with lunch in the St James's Club, an afternoon at the House of Lords and then back home to ***one's*** Palladian mansion in Surrey. In the cosmetics industry, the symbolic values of products range from the simple, environmentally-conscious utility of the Body Shop to the decadent, exotic mystery of Monsoon (Figure 7.17).

Lifestyle, mood and theme boards

We now know what we are trying to achieve with product styling, but how do we get there? Since styling is predominantly an exercise in visual thinking, it makes sense for styling design techniques to be predominantly visual in their approach. As with all product development, we need to start off with broad objectives and narrow them down (through a risk management funnel) to specific, tangible and manufacturable forms. In principle, this proceeds through three main stages:

- Firstly, images of the target customers are collected together on what is known as a 'lifestyle board'. These images should convey something of the customer's personal and social values as well as representing the sort of life they lead. Lifestyle boards are usually idealistic in their portrayal of customers. They show happy smiling people enthralled with the joys of their lives. The hardship, drudgery and stress which occupies real lives is rarely represented because people will hardly want that reflected in the styling of a new product. A common mistake at this stage of the styling process is to assume that the new product will have only one type of customer. Few mass produced products could survive with such a narrow customer base. The key to exploring product symbolism, therefore, is to explore the range of customers that the product should appeal to and try to discover personal and social values common to them all.

- Secondly, derived from the lifestyle board, a 'mood board' tries to identify a single expression of values for the product, which will appeal to customers with the identified lifestyles. The mood of a product is the sentiment, feeling or emotion which the product engenders when first seen. It might appear mellow and relaxed (image: a fire burning quietly in the hearth) or energetic and effervescent (image: Olympic 100m final). It could be frivolous and fun (image: a fairground ride) or intense and business-like (image: a board meeting). It may be soft and comforting (image: a koala bear) or rugged and durable (image: a steam engine). A good mood board captures these feeling in images but without referring to specific product features which might limit the range of styling options considered. Images of products which are similar in either form or function to the proposed new product should, therefore, be avoided. Mood boards developed by designers often suffer from an excess of artistic licence. A mood board has an important communication role to play. It gives all members of the design team a common styling objective and allows that objective to be communicated beyond the design team to management and even clients or customers. If it is abstract to the point of obscurity, it fails to communicate and consequently fails as a mood board.

Figure 7.16
The 'English' interior of the Rover MG RV8 [5]

- Finally, derived from the mood board, a 'visual theme' board collects together

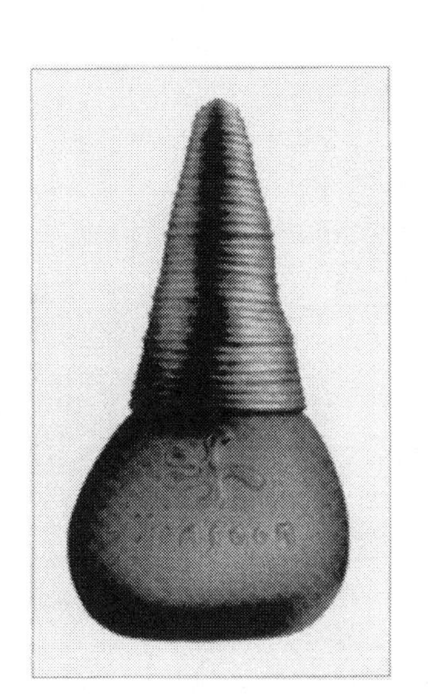

Figure 7.17
The simple utility of Body shop to the decadent, exotic mystery of Monsoon [7]

images of products which manage to convey the target mood. These products can be from diverse market sectors and have widely different functions. A visual theme board allows the design team to explore styling features which have, in the past, succeeded in their current task. These styling features should provide inspiration and a rich source of visual forms to adapt, refine or elaborate during the development of the new product's style.

Let us now examine this style development process in action, using the design of a mobile phone as an example. Mobile phones at present have a very characteristic style which reflects the professional, business lifestyles of typical users. With the recent push towards market development by the mobile phone companies, a much wider range of people are becoming users: people who do not fit into the stereotype of a mobile phone user. It is, therefore timely to explore how the styling of mobile phones could be changed to match more closely the semantic and symbolic expectations of the diversified customer base. Figure 7.18 is a lifestyle board for the new wider range of mobile phone users. By way of contrast, Figure 7.19 shows images of the lifestyle of the stereotype mobile phone user. Two moods are derived from this lifestyle board. The first (Figure 7.20) is called 'adventure' and

Figure 7.18 Lifestyles of the 'new' mobile phone user

Figure 7.19 Lifestyles of the 'old' mobile phone user

shows a rugged, outdoor, active mood. The second (Figure 7.21), called 'fun' shows a lively, playful mood. From these mood boards, two visual theme boards are developed, showing images of products which convey the 'adventure' and 'fun' moods (Figures 7.22 and 7.23 respectively). A visual theme can be better represented by a collection of products rather than simply images of these products on a board. This allows the physical feel of the product to be used in the styling design process as well as its visual appearance. Clearly, this can cause practical problems –carrying actual products to meetings is less easy than carrying a board – and it can become expensive.

Having worked through customer lifestyle, product mood and product visual theme, the styling objectives for the new product have become focused. They have narrowed from images of target users to examples of product style that these users would value. The time has now come to start generating styling concepts for the new product. As with the design of product function, concept styling should aim to produce many ideas and select the best. In particular, it is important to explore many styling themes for the new product provided they are consistent with the identified moods for the product.

Generating style concepts begins by extracting the main visual features from the visual theme board. This should never be a simple matter of plagiarism – styling should always be much more imaginative than that. In the 'adventure' theme, for example, ruggedness is conveyed by bold, chunky features such as the corrugated or castellated profile on the surface of the product. Fun is conveyed in the other theme by splashes of bright contrasting colours (admittedly not shown to their best in black and white images!). Figures 7.24 and 7.25 show the concept sketches developed for the two mobile phone moods.

The most important 'take home' message from this exercise is that products can be given very different and quite distinctive styles by following the structured use of lifestyle, mood and visual theme boards. These styles are carefully and deliberately derived from an understanding of market needs and they are focused on specific interpretations of customer value. These can all be tested by market research before they are finalised. The concepts preferred by customers can then be selected by means of concept selection techniques to make sure that they match functional requirements as well as the company's strategic objectives for product development.

Figure 7.20 'Adventure' mood board

Figure 7.21 'Fun' mood board

Figure 7.22 Visual theme board for 'adventure' mood

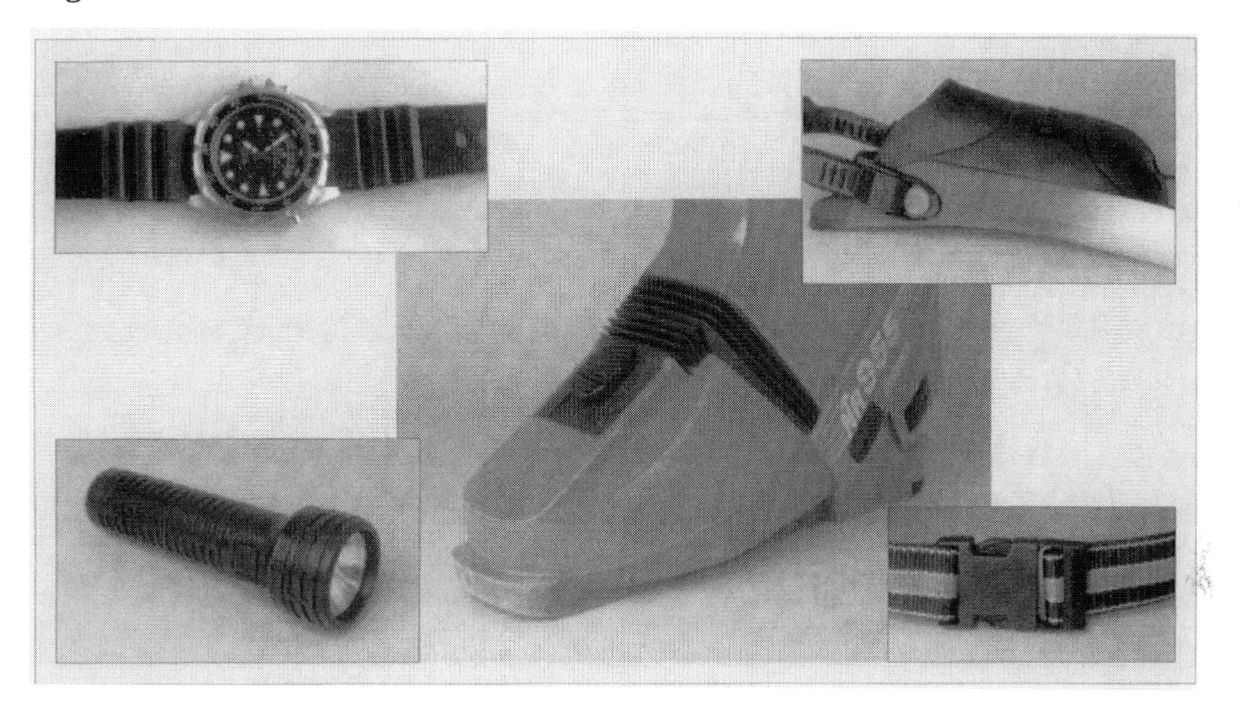

Figure 7.23 Visual theme board for 'fun' mood

Figure 7.24 Styling concepts for the 'adventure' mobile phone

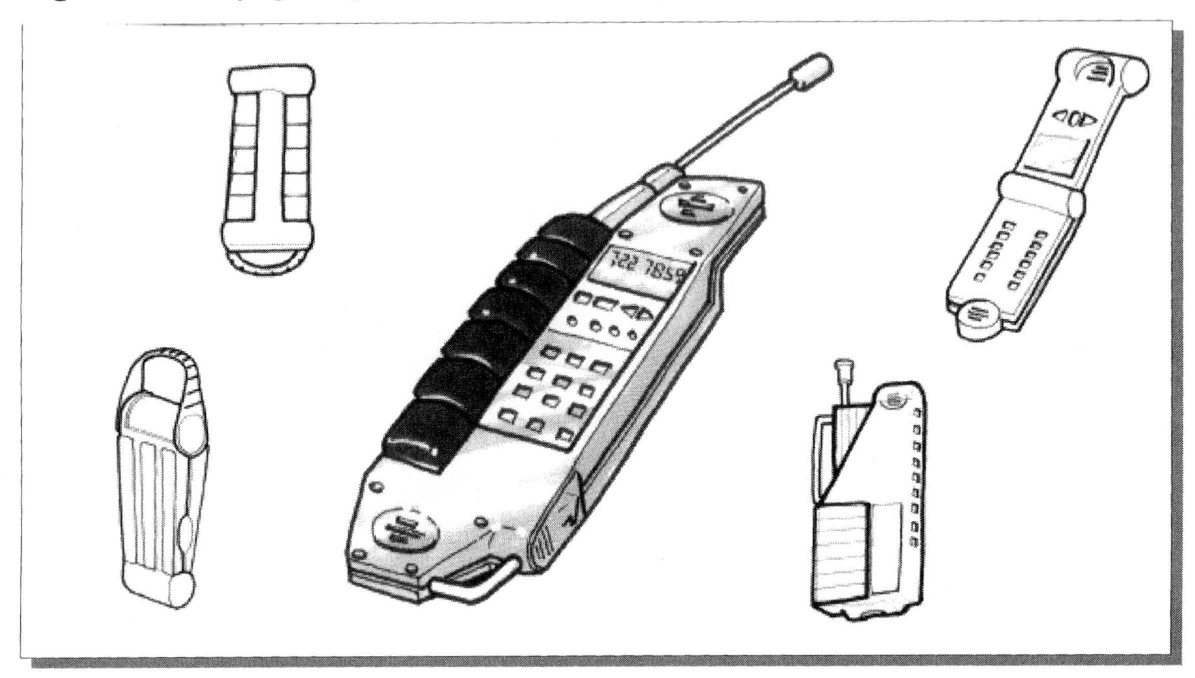

Figure 7.25 Styling concepts for the 'fun' mobile phone

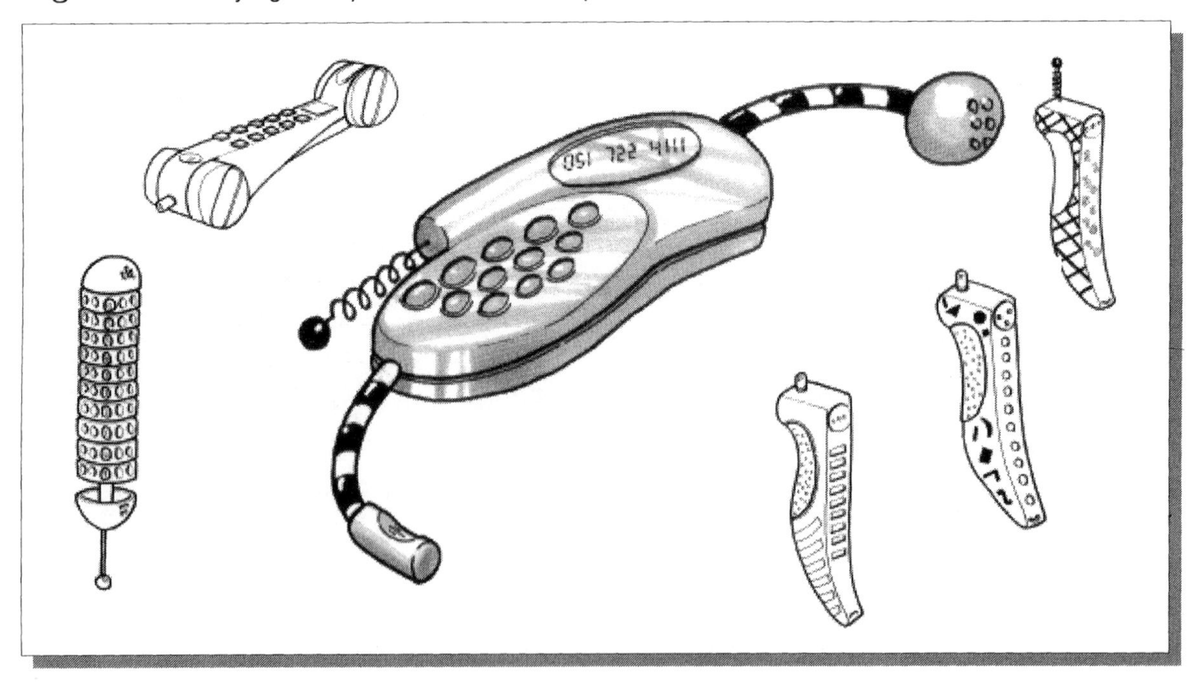

Concept selection

Following concept generation, the final stage of concept design is concept selection. Modern methods of concept selection are due to a large extent to the pioneering work of Stuart Pugh at Strathclyde University in Scotland [8]. Pugh developed the notion of controlled convergence, by which a range of generated concepts are systematically made to converge on a single selected concept. The important feature of controlled convergence is that concept selection is not simply a matter of picking the best of the generated concepts. A great deal of creativity can be incorporated, to combine concepts, amalgamate the best aspects of several concepts and even generate further new concepts. Much value can, thereby, be added to the 'raw' concepts generated earlier. The principles of controlled convergence are shown in Figure 7.26. The first round of concept selection ranks the concepts in relation to a series of selection criteria from the opportunity specification. This is done by means of a concept selection matrix (introduced on p 169 for opportunity selection) in which the concepts are arranged along one axis of the matrix and selection criteria along the other. To make the ranking procedure simple, each concept is judged 'better than' (scored as +1), 'worse than' (scored as -1) or 'the same as' (scored as 0) a reference concept. This reference concept should be the best current competitor to the proposed new product. The outcome of the ranking process will be a single number expressing the relative merit of each concept (a positive number indicates that the concept is better overall than the reference concept, a negative number indicates worse overall than the reference). From these ranks, attention focuses on the better concepts. Now comes the concept hybridisation and generation phase. Essentially this sets out to take all the good features from the different concepts and combine them into a single product. At the same time the weak features should be eliminated. So, look closely at the concepts which were strong

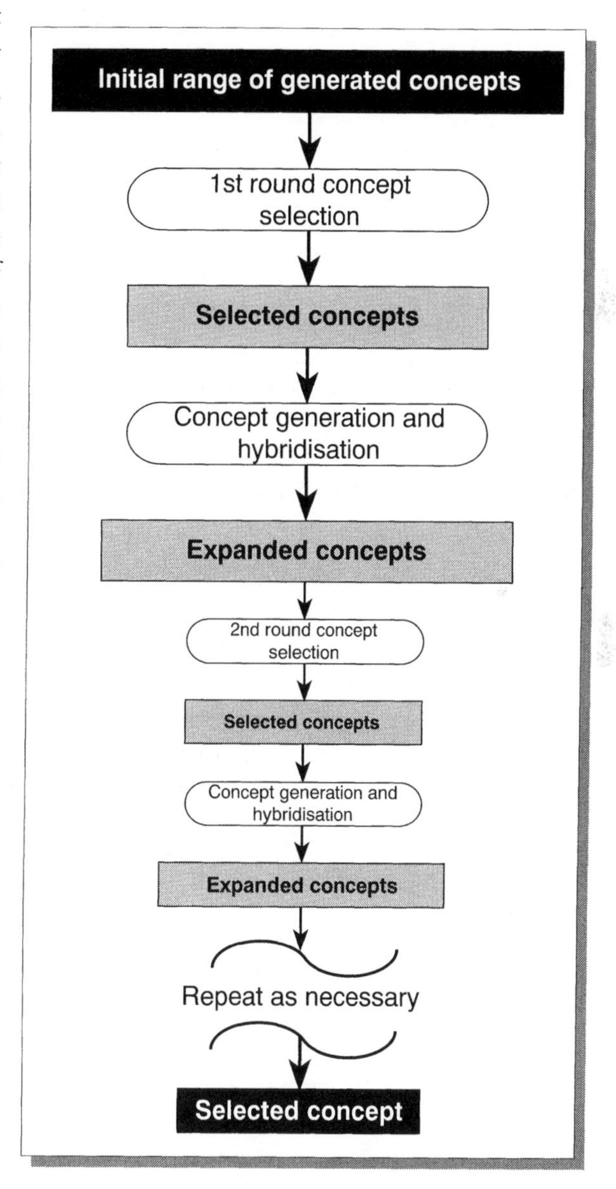

Figures 7.26 The concept selection process [8]

overall but which scored -1 on any of the criteria. Can features of any of the other concepts be 'hybridised' to turn that -1 into a +1, thereby making the strong concepts even stronger? Next examine the strong concept's 0 scores. Can they be enhanced by taking features from any of the other concepts? Once you have done as much concept hybridisation as possible, take some time to think more laterally. Have any new concepts come to mind during the concept selection process? Try running through the creative thinking toolkits in Chapter 4 to generate new concepts. Any that emerge should be added to the concept selection matrix and ranked. Having done your best to improve the existing concepts and devise any new concepts, a judgement needs to be made. If none of the concepts ended up achieving an overall score greater than 0 then none are better than the reference competing product. The concepts need to be improved before you have a competitively viable new product. Design efforts can return to developing a new range of concepts or, alternatively, it may be that the product development should be killed off as non-viable.

Provided at least one of the concepts has a better than 0 overall score, the concept selection procedure should be repeated. Only take the strongest concepts this time (perhaps half of those you started with). Use the highest scoring concept from the last matrix as the reference. This will reveal slightly different strengths and weaknesses in the concepts since the characteristics of the new reference concept are different. Again, try to hybridise the concepts to make the strong concepts even stronger. Again take time to try to think of further new concepts. At the end of this round, you will have one of two results. Firstly, the reference concept may no longer be the best. If any concept achieved a positive overall score then the concept selection procedure should be repeated with a new reference concept. This process should be repeated until no new ideas emerge and no concept reaches a positive score. With a creative and imaginative team involved, this may take many iterations. The benefit will, however, always outweigh the cost of time and effort. Concept selection is a quick procedure. In the space of a few hours, the 'raw' concepts initially generated can be refined, elaborated and developed, often beyond recognition. Given robust selection criteria, concept selection will usually add more value to an emerging product than any other design activity of a few hours duration.

Eventually, however, the second result will emerge in which the reference concept retains its leadership, with all other concepts managing only 0 or negative scores. At this point you have found the best concept out of all those under consideration. Concept selection is complete and you can move on to embodiment design.

Psion case study [9]

We left Psion in Chapter 5 having the decision virtually forced upon them to develop the Psion 3 at breakneck speed. If the new Psion was to be launched in a period of 11 months some key decisions had to be made quickly. The most important yet most difficult decisions in the entire process of product development are, according to the Psion development team, the ones at the 'fuzzy front end'. These are the decisions which make the entire product take shape and determine the magnitude of the problems that you create for yourself in product development. They are also the decisions with the least knowledge upon which to base them. Half way through the development process they will be all too apparent. But at the fuzzy front end, you can only make the best guess with the information available.

When Psion was planning the Series 3, some decisions came easily. It was to be a product targeted at the consumer and business market and sold through high street retailers. It had to be pocket sized and it had to be strongly differentiated from the competing products on offer from Sharp and Hewlett Packard. But there the easy decisions ran out. The first tough decision was just how, exactly, did Psion want this new product to be perceived by customers? Was it an electronic organiser or was it more than that - a pocket sized computer? What basic shape was the product to have? Was it to be 'portrait' shaped like the Organiser 2 and if so was it to be a fold-open format. Or were they going to try to break new ground and develop a 'landscape' shaped product? Were they going to stick with the ABC format keyboard which had brought them success with the Organiser or conform to computer standards with a QWERTY keyboard?

Like all good jigsaws, one key piece fell into place (in June 1990) and the picture of the Series 3 started to become clear. That piece was the conclusion that certain consumer attitudes had changed since the launch of the Organiser in 1984. At that time, computers were still seen as something used by businesses or played with by the kids. The majority of consumers did not feel comfortable with the idea of using a computer. Consequently, to try to get round this techno-phobia, Psion intentionally gave the Organiser a keyboard which looked different from a computer keyboard. They opted for the simpler and more logical ABC layout of keys shown earlier on p 126. By 1990, few consumers had not, at some time, used a computer

Opportunity specification: Psion Series 3

Core benefit proposition:
A pocket sized computer with a full range of basic business software.

Business prospects:
Development costs:
Software – 25 man years @ £25k/man year
Electronics – 2 man years @ £25k/man year
Design and Engineering – 2 man years @ £25k/man year
Tooling and Capital Costs – £250 000
TOTAL – approximately £1 000 000
Projected return:
Sales volume – 5000/ month
Retail price – £250
Psion margin per product – £12.50
Development costs payback – 16 months

keyboard. Most, Psion believed, would no longer be put off by a standard QWERTY layout. The decision was made. The Psion Series 3 was to have a QWERTY keyboard and it was to be a pocket computer. Now the planning for the Series 3 could really begin in earnest. If it was to be seen as a 'real' computer then customers would expect a certain range of functions: a word processor, a spreadsheet, a database, a programming language and communications capabilities. Could they manage to pack all of these functions into so small a package? The smallest competing product to share these features, at the time was at least twice the length, twice the width and twice the depth; in other words it was 8 times the volume of the proposed Series 3. From their background and expertise in developing the Organiser, and, more recently, the MC400 laptop computer, they decided it was 'do-able'.

The business decision to proceed with the Series 3 was based on the information distilled into the Opportunity Specification above. The product had a clear and important advantage over products currently on the market, as described in the core benefit proposition. The business prospects were good. If the product could be made to deliver the anticipated benefits, it should reach its 5000 per month sales target. It should also have a sales life considerably in excess of 16 months. The Organiser had sold up to 20 000 per month and had a sales life of six years.

The design specification (see opposite) for the product, as always, was difficult to define at this early stage. It helped considerably having just developed the MC400 using the same technology platform. Many similar problems had already been tackled and some had been solved in a way that could be transferred directly to the Series 3. The central processor and the multitasking windowing software was the common technology platform upon which both products were based.

The team moved on to concept design. It was decided that the Psion Seies 3 was to have a QWERTY keyboard , which meant that the keyboard layout had to have a landscape format (sketch A – see page 234). If the keyboard was to have a landscape format then it was almost inevitable that it would also have a landscape format screen (sketch B). Thinking about how these would go together revealed that there was really only one solution – a hinge

Design specification — Psion Series 3

Performance specification	Design specification	Status
1. Software		
Full 'business' applications software expected in a PC.	Multi-tasking,windowing operating system.	Exists*
	Word processor	Need
	Database	Need
	Agenda	Need
	Spreadsheet	Want
2. Hardware		
Sufficient to support software.	16 bit (80C86 compatible) (3.84 Mhz)	Exists*
	348 kbyte ROM	Exists*
	256 kbyte RAM	Need
	512 kbyte RAM	Want
	Slots for 2 Solid State Disks (Disks purchased separately)	Need
3. Keyboard		
Qwerty type	Qwerty keyboard	Need
	Size - 62 key	Want
	Additional function keys for applications selection.	Need
4. Display		
Adequate for comfortable word processing	40 character width	Need
5. Size		
Comfortable pocket size	165 x 90 x 25mm	Need

*These features had already been developed for the MC400

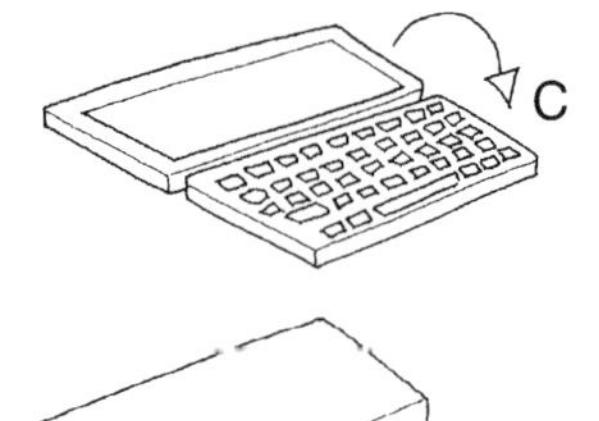

along the long axis between the two (sketch C). With the screen occupying most of one part of thc Psion casing and the keyboard and main circuit board occupying the other half, an obvious proble which remained was where to put the batteries (sketch D). Significant breakthroughs had been made by Psion's electrical engineers to reduce the power requirements to two 'AA' (or penlight) batteries. This reduced the battery space needed but, when space within the Psion was at such a premium, even the size of two small batteries loomed large in their minds. Ideally, they would have liked to include the batteries inside one of the clam shell casings (sketch E). This, however was not possible. Another possibility was to have a cavity moulded into one of the casings, external to the main profile of the Psion (sketch F). This was considered unacceptable because of the uncomfortable bump it would produce in a product intended to be carried in the user's pocket. Having the batteries at the side of the keyboard (sketch G) was not acceptable because this was where the solid disk drives were intended to go. Having them at the side of the screen would have reduced screen size (sketch H).

An unorthodox but highly original solution to this problem was to have the batteries between the keyboard and the screen (sketch I). The concept that turned out to be one of the most distinguishing design features of the entire product started life as a highly practical but prosaic problem. Where do you put the batteries in a product too small and too packed with other features to leave a battery compartment!

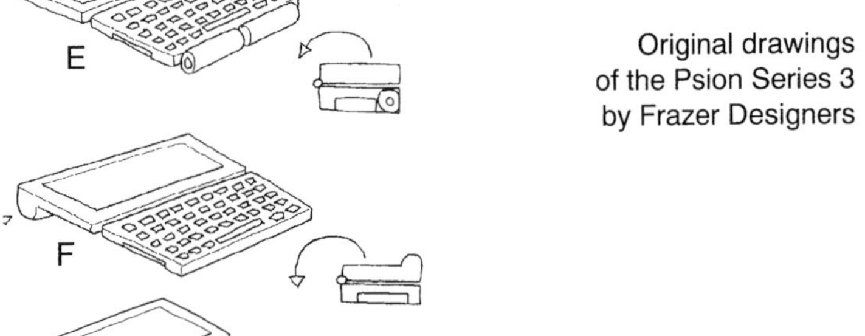

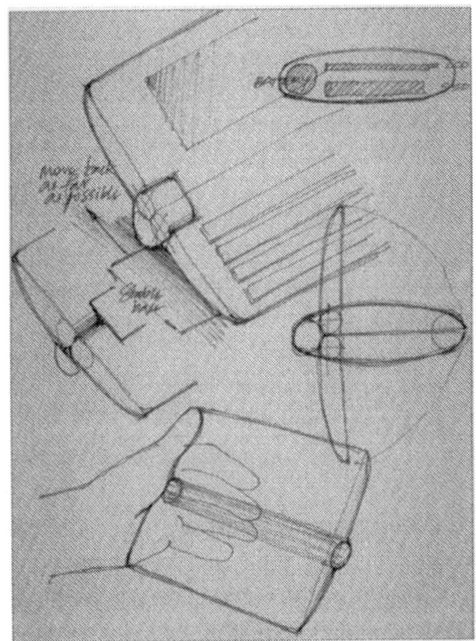

Original drawings of the Psion Series 3 by Frazer Designers

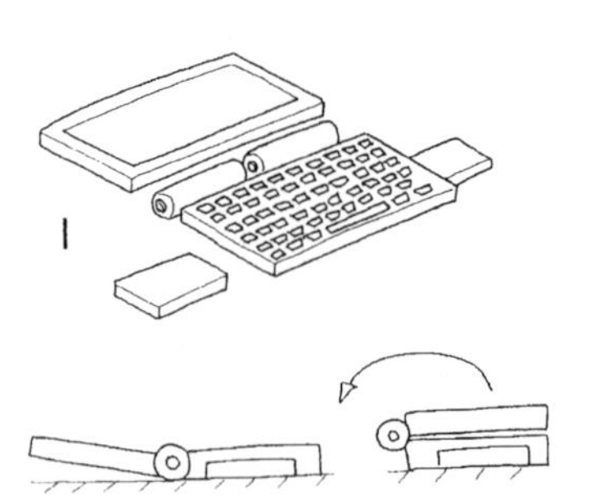

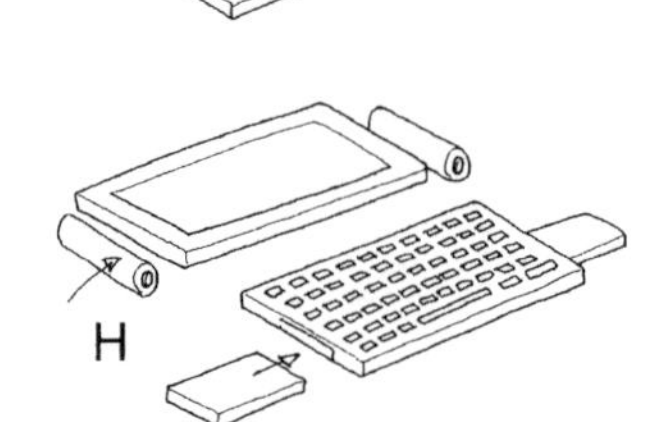

Key Concepts

1. Establish aims and scope for concept design

Different design projects will have very different objectives and constraints determining how radical or incremental the concepts need to be. These must be clearly established, with guidance from the opportunity specification.

2. Generate lots of concepts

Concept design is usually considered to be the creative heart of the design process. As a result, creative idea generation techniques are used most often at this stage. Several techniques exist for the force-generation of new product concepts. These are likely to increase the number of ideas generated from a few to many tens or even hundreds of concepts.

3. Select the best

Concept selection techniques select the best concept against criteria derived from the opportunity specification. Probably more importantly, they provide a framework for hybridising and expanding the range of concepts generated initially. Concept selection can, therefore, comprise a highly creative and invaluable conclusion to the concept design process.

Design Toolkit

Product function analysis

Product function analysis is a method of systematically analysing the functions performed by a product (as perceived by the user). Also known as FAST analysis (Function Analysis Systematic Technique), it is the most basic and probably the most important analytical technique in new product development. All you need for product function analysis is to know how the product will operate in use. You must know, or be able to predict, the functions of the product as perceived by the customer and how the customer rates the relative importance of these functions. It can be applied both to existing products and to those still being designed. Product function analysis provides a detailed understanding of the product from a functional and customer-oriented point of view and presents this understanding in a logical and systematic framework. Its results can be used to stimulate concept generation or can provide the input into further systematic techniques, including value analysis (see page 214) and failure mode and effects analysis (see page 290).

Product function analysis procedure The first step in product function analysis is to brainstorm all the functions the product will serve in the eyes of the customer. The best approach is to write down on individual scraps of paper (Post-it notes are ideal) every single function that you believe your product will perform. This means asking what the product 'does' rather than what the product 'is' - which is how engineers often think. Do this from a customer point of view but make sure that you write down all the functions that the customer values. Do not take any functions for granted (e.g. it is obviously important that a vacuum cleaner has wheels and is able to move across the carpet but customers might take this for granted). Try to keep the descriptions of function to two or three word 'verb-noun' combinations (e.g. contain fluid, break circuit, expel moisture, provide visual indication). Keep brainstorming until you feel you have exhausted all the product functions: most products will have at least 40-60 functions and only the simplest products will have fewer than 20. Next arrange these functions into a 'function tree'. To start the function tree, select the prime function of the product. This is the main reason that the product exists in the eyes of the customer. The prime function of a vacuum cleaner, for example, is to 'remove debris', not to 'suck air'! Once the prime function has been selected, the other functions are grouped logically and hierarchically under it. The next layer under the prime function should describe 'basic functions'. Basic

Product function analysis (Cont'd)

functions relate to prime functions in two ways: i) they are essential to the performance of the prime function and ii) they are direct causes of that prime function occurring. Thus 'suck air' is a basic function of a vacuum cleaner in order to 'remove debris' - its prime function. 'Suck air' is both essential to the performance of 'remove debris' and also is a direct cause. 'Supply electricity', by contrast, is not a direct cause of 'remove debris' and is, therefore described as a supporting function subsidiary to 'suck air'. Keep going through the function tree, asking the question 'How is this function achieved?' At every level the functions should be essential to, and direct causes of, the function above. At the bottom of the tree you should end up with a list of functions which cannot easily or logically be subdivided into subsidiary functions. In most function analyses, these lowest order functions relate directly to single features or components of the product. In other words you have reached the level of description of functions which relate to specific features or components. The 'supply electricity' function, for example, is a lowest order function and relates to the electrical supply cable. Where product function analysis is being performed on an existing product, this provides a useful, though not infallible, check on your function tree. Compare your lowest order functions against the individual features or components of the product. If there are features or components not listed in the function tree then either they should be irrelevant to the customer's perception of the functioning of the product (e.g. a feature to facilitate assembly during manufacture) or you have missed out an entire branch of functions in your function tree!

The more rigorous validation of a function tree is done in two ways: The first is the 'How-Why' validation. Start at the top of the function tree - the prime function. Move down each branch of the tree asking the question 'How?' at each step. How each higher order function works should be explained by functions on the next level down. How does the vacuum cleaner remove debris? It does so by sucking air? How does it suck air? And so on. The functions on the level below should be ***both sufficient and necessary*** to explain 'how' the function on the level above is performed. If the lower functions are not sufficient to explain the higher functions then you have missed some functions out. If some are not necessary then you have either placed the functions in the wrong branch of the tree, or more likely, have placed subsidiary functions too high up the tree. Once you have gone all the way down the tree asking 'how?' and are happy with it, check again by

Design Toolkit

Product function analysis (Cont'd)

moving *up* the tree asking 'why?' Why do you suck air in a vacuum cleaner? In order to remove debris. A function tree validated in both the 'how?' downward direction and the 'why?' upward direction is likely to be a robust representation of the functions of a product. The second validation is, simply, to give the function tree to colleagues who are familiar with the product but have not been involved in the product function analysis. They may question the clustering of functions and may also identify functions which have been missed out. It is important to remember that there may not be only one correct function tree for any product. Different clusterings of functions may be equally logical, equally correct by the 'how?' and 'why?' verifications and equally comprehensive. They are also likely to be equally useful. The important thing is to have the functions fully described and arranged in a systematic order.

Figure 7.27
Product funtion tree for a corkscrew

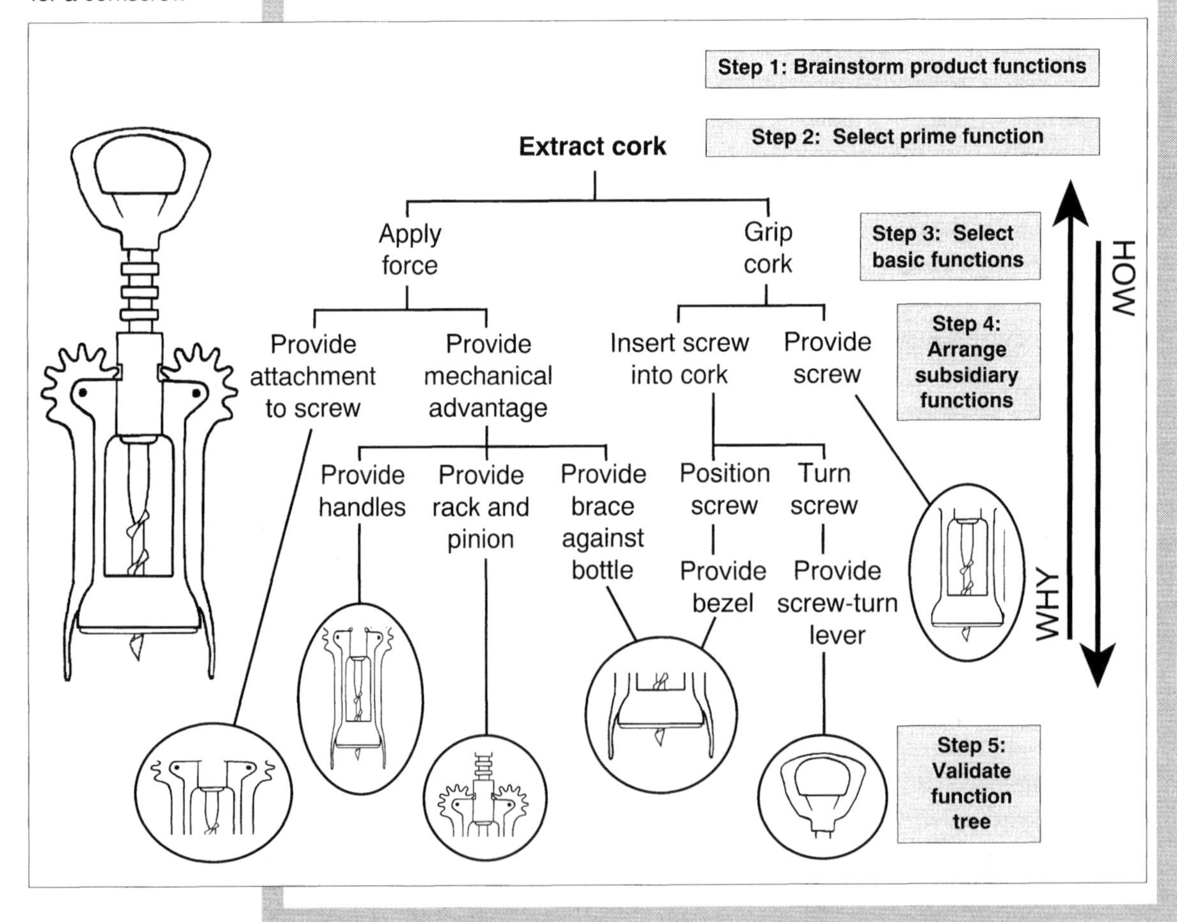

Life cycle analysis

Life cycle analysis is a technique which has largely been 'kidnapped' by designers interested in the environmental impact ('green-ness') of products. Also known as 'cradle to grave' analysis, it looks at the environmental cost of each stage of the life cycle of the product and gives a relative weighting for the manufacture, transport, use and disposal of products. Product development effort is then focused on the stage of the life cycle with the greatest environmental cost. In this type of application, life cycle analysis has proved its value in many applications. There has, however, been a tendency to underestimate the difficulties in weighting different types of environmental impacts and also the difficulties in obtaining sufficiently precise values for environmental costs. These difficulties are a significant challenge to the widespread use of the technique and will be discussed later.

Life cycle analysis can be seen as a much broader analytical technique for exploring opportunities for refining and improving the design of products. In this broader application, the aim is still to explore the product from cradle to grave but attention is not limited to environmental impact. All opportunities for product improvement, including costs, customer value, manufacturing efficiency and ease of transport are considered.

General principles of life cycle analysis Life cycle analysis proceeds in three main steps. Firstly, the product life cycle is described. This should identify the inputs, transformations and outputs for every stage in the product's life cycle. Describing the manufacturing facility reveals what energy and raw materials constitute the inputs, what stages they pass through during the product's manufacture and what products, by-products and waste materials are produced as outputs, once manufacture is complete. Then the storage, distribution and sales procedure should be similarly described. Many products will have several distribution and sales routes and the main features of each should be described. Then typical patterns of customer use of the product should identify any additional resources required for using the product and any pollution resulting as a consequence. Finally methods of disposing of the product, once its useful life is finished, are described. Next, the analytical step in the process commences. This attempts to establish the basic purpose of each step in the life cycle and attributes measures of costs and value to it Then, opportunities are identified for improvement, either environmental improvement or general

Design Toolkit

Life cycle analysis (Cont'd)

improvements in the product's design.

The example of life cycle analysis given on page 213 shows how to set about identifying opportunities for improving the design of a garden plant pot.

Difficulties with life cycle analysis for studying the environmental impact of products

There are two main difficulties in using life cycle analysis to determine the environmental impact of products: one is fundamental, the other practical.

The fundamental difficulty is how to weigh-up different types of environmental impact relative to one another. Imagine, for example, two distinct but incompatible opportunities for product improvement are identified. One has the benefit of reducing the amount of energy required to make the product. The other reduces the amount of pollution into the local water courses. As an analytical technique helping managers to make decisions about product improvement, life cycle analysis should be able to weigh up the benefits of the two opportunities. Unfortunately it can not. The comparison is fundamentally dissimilar (global versus local, energy use versus pollution) and highly dynamic. If the local watercourse was already highly polluted by other factories upstream the reduction in local pollution would be the more critical opportunity. If, however, the upstream factories closed down, the pollution load would drop dramatically, making the energy saving opportunity more critical.

The practical difficulty with life cycle analysis is obtaining adequate data on environmental impact. Consider the difficulties in analysing the environmental impact of different supplies of plastic raw materials. To be thorough, this should consider the cost (energy and pollution) of extracting the base hydrocarbons (oil or coal), the cost of refining the hydrocarbons and the cost of compounding them into polymers. To try to measure these is a monumental task, far beyond the scope of any individual life cycle analysis study. Fortunately, general data are available for different types of plastics from life cycle analysis databases. But unfortunately, these give average figures for a process containing a great deal of variation. If, for example, the hydrocarbons are extracted by offshore oil rigs drilling into deep sea reserves, the energy costs will be huge, compared to the energy costs of open cast coal mining. If the plastics are compounded in Norway, the energy supply is likely to be from hydro-electric power stations, a renewable energy source causing little pollution. If, on the other hand, they are compounded in the heartland of industrial Germany, the energy will come from coal-fired power stations, depleting non-renewable energy sources and causing a higher level of pollution. To say that a raw material is a particular type of polymer and that this polymer has a typical environmental cost of production may be such an inaccurate generalisation that it invalidates the decision reached at the end of the life cycle analysis.

These difficulties do not add up to an argument against the use of life cycle analysis to improve the environmental friendliness of products. Comparing like-with-like environmental costs, using well researched, accurate and reliable data can be a powerful tool for improving product design. Like many fashionable issues, life cycle analysis has tended to be over-sold and the difficulties have been glossed over. Before commencing life cycle analysis (and it can be a lengthy and involved process for even the simplest of products) it is, therefore, important to be aware of the difficulties.

Notes on Chapter 7

1. For general information on ergonomics, Bridger R.S. 1995 *Introduction to Ergonomics.* McGraw Hill Inc. New York offers good up to date coverage of the general subject areainformation on ergonomics. The classic texts are probably Pheasant S. 1987 *Ergonomics: Standards and Guidelines for Designers.* British Standards Institute, Milton Keynes, UK and Galer I. 1987, *Applied Ergonomics Handbook* (2nd Edition). Butterworths, London. More specific sources of information on Anthropometrics include Marras W.S. and Kim J.Y. 1992 Anthropometry of Industrial Populations. *Ergonomics* 36: 371-378 and Abeysekera J.D.A. and Shahnavaz H. 1989, Body Size Variability Between People in Developed and Developing Countries and its Impact on the Use of Imported Goods. *International Journal of Industrial Ergonomics* 4:139-149.
2. Ashby P. 1979, *Ergonomics Handbook 1: Body Size and Strength.* SA Design Institute, Pretoria.
3. This life cycle analysis study was conducted by Tom Inns at the Design Research Centre, as part of an EPSRC LINK research project on use of waste straw as a matrix composite in plastics.
4. This value analysis study was conducted by Chris McCleave and Andy Wilson from the Design Research Centre for Multifabs (Survival) Ltd. of Aberdeen, Scotland (Part of a Teaching Company Scheme, supported by the Department of Trade and Industry). Cited by permission of Mr Niel Allanach, of Multifabs (Survival) Ltd.
5. Photographs provided by Rover, Aston Martin, Honda and Yamaha. Reproduced with permission.
6. Trade marks reproduced with permission of Audi, BMW, Mercedes, and Volkswagen.
7. Photograph of Monsoon bottle provided by Beauty International and reproduced with permission.
8. Pugh S. 1991 *Total Design: Integrated Methods for Successful Product Engineering.* Addison-Wesley Publishing Co. Wokingham, UK pp 74-88.
9. The Psion Series 3 was designed by Frazer Designers. My thanks to both Frazer Designers and Psion for their help in discussing the development of the Psion 3A. The photographs of the Psion 3A and the Psion Organiser were provided by Frazer Designers and are reproduced with permission.

8 Product planning - creating quality, adding value

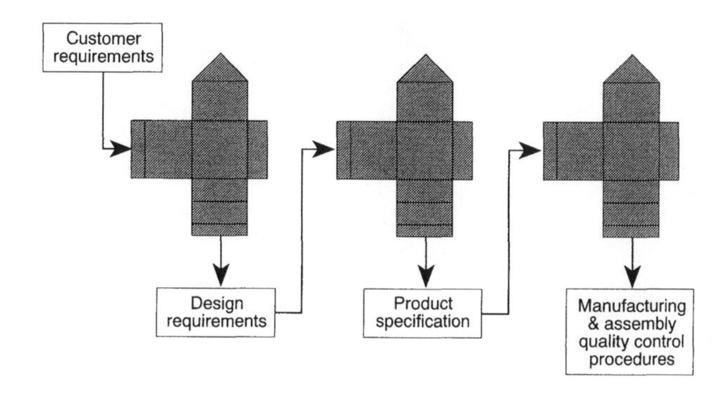

The stage we are at in the design process – preparing the design specification – establishes specific targets for the new product. Factors which are included in the design specification will, hopefully, be delivered to the customer, whereas those which are omitted or overlooked probably will not. It is, therefore, vitally important to get the design specification right in order to make sure that the new product is right. But, what does it mean to say that the product is right? We have seen what it means to get the product right for the company. The company's aims have been explored from its mission statement through to its product development objectives. But we have also seen in Chapter 2 that getting the product right for the company only increases the product's chances of commercial success by a factor of 2.5. Getting the product's market orientation right increases its chances of commercial success by a factor of 5. Market orientation has been considered so far in terms of competing product analysis and preliminary market research to identify the best product opportunity. Now we must focus more closely on market orientation to determine the specific qualities of the product we need to deliver.

Product quality

Product quality means many different things to different people. An engineer thinks of quality in terms of fitness for purpose and builds quality

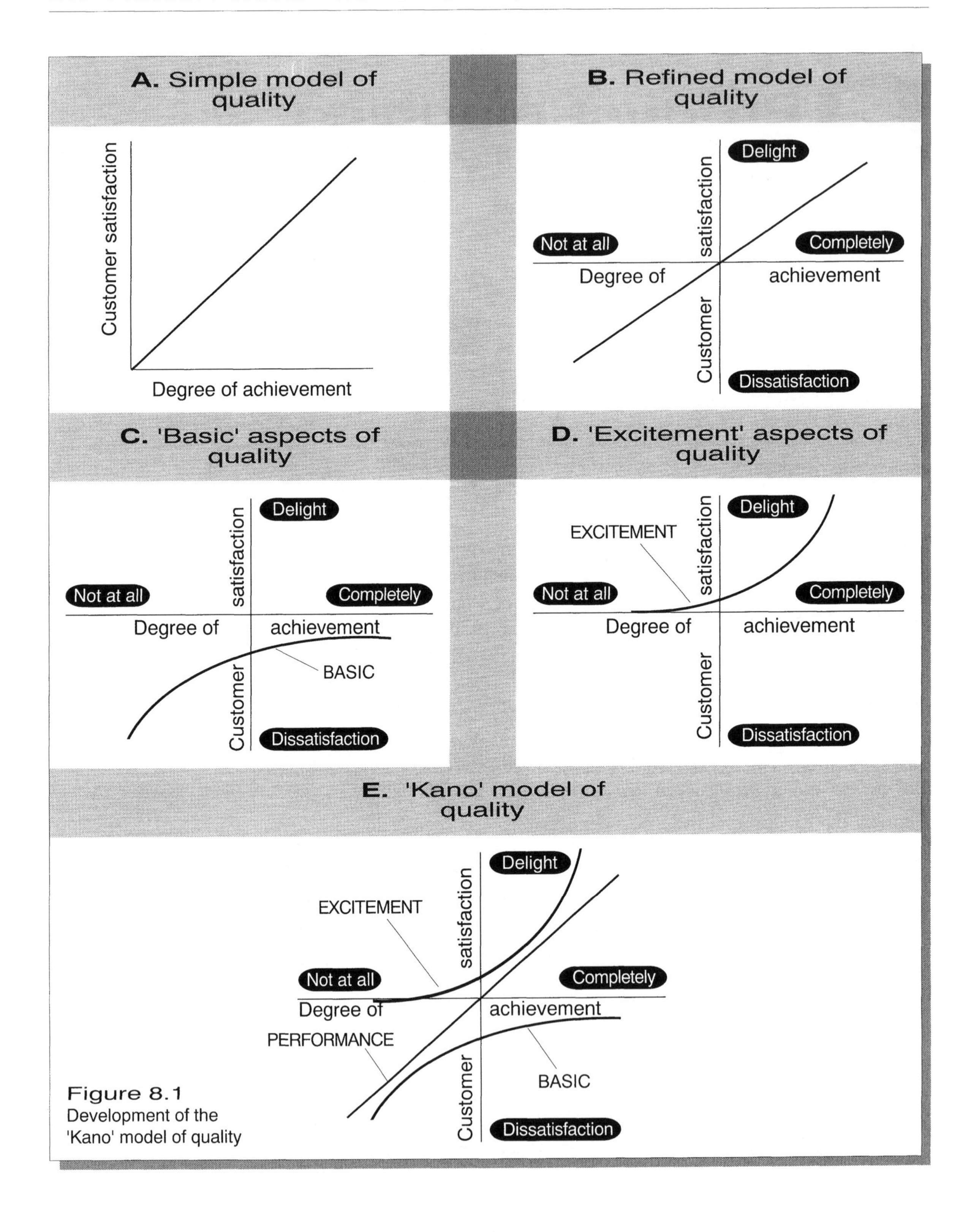

Figure 8.1
Development of the 'Kano' model of quality

into a product by ensuring that the conditions under which that product fails exceed its specified range of operating conditions. A production manager thinks of quality in terms of how easily the product can be manufactured to specification and measures quality by the proportion of products meeting quality standards through routine procedures. A service engineer thinks of quality in terms of the length of the product's maintenance free operating life and the cost or difficulty of maintaining the product when it does break down. These are all vital ingredients of a successful product and, as we shall see, have to be considered when specifying the quality standards which the new product must meet.

The starting point for considering product quality, however, must be more general than this. The most important determinant of quality for a new product is the customer's perception of its quality. The simplest way to think of this is shown in graph A of Figure 8.1. The more a product achieves its intended qualities, the more we would expect customers to be satisfied. Unfortunately, customer satisfaction is not as simple as that. Failure to achieve certain qualities in a new product (such as the products primary function) will not just lead to low levels of satisfaction, as graph A predicts. It is likely that customers will have noticeable feelings of dissatisfaction. This is represented in graph B, where the origin represents no quality 'problems' in the product giving rise to neutral feelings of satisfaction in customers. Even this does not give a detailed enough picture of product quality because not all customer expectations carry the same weight. Customers have certain basic expectations of a product, many of which they may not even be aware of. Failure to achieve these basic qualities will give rise to great dissatisfaction but, because they are unspoken, achieving them does not give rise to any positive feelings of satisfaction. When buying a new car, for example, customers have the basic, unspoken expectation that the car will have wheels. Finding that the car does not have wheels will cause great dissatisfaction but no satisfaction will be achieved by the presence of wheels. These basic expectations of quality are expressed in graph C. At the other extreme there are product qualities, called excitement factors, which give rise to a great deal of satisfaction when they are achieved but their absence causes no dissatisfaction (graph D). The first Sony Walkman contained a number of these excitement factors. Achieving such excellent sound quality from a product which slipped inside your pocket gave rise to customer 'delight'. Before the introduction of the Walkman, however, customers would not have expressed dissatisfaction about the fact that music players did not fit into their pockets. This is because the desire for excitement factors is, like the desire for basic factors, unspoken. The achievement of these excitement factors is said to satisfy 'latent' customer needs. Graph E shows basic and excitement factors together, in the 'Kano' model of quality, named after its inventor, Dr. Noriaki Kano [1]. Between the basic and excitement factors, Kano proposes that there is another determinant of customer satisfaction, which he calls 'performance'. Performance factors cover the range of qualities that customers have come to expect in a particular type of product. The customer's perception of quality varies in direct proportion to the degree to which the maximum or ideal performance of the product is achieved. The ideal car, for example, might be one with a stylish appearance, fast acceleration, good handling, low fuel consumption, minimum maintenance, dual air-bags, central locking, electric windows, CD player, radio and mobile phone. A car with all of these features, as standard, would cause great customer satisfaction. A car with none of them would cause dissatisfaction.

Figure 8.2
Basic, performance and excitement needs

Excitement

- 'Unspoken', latent customer needs and wishes unprecedented in competing products.
- Satisfying genuine needs, they are not just gimmicks.
- Extrapolated from market research aimed at understanding of customer's underlying wishes and their unresolved frustrations with existing products.
- Failure to achieve excitement factors does not lead to customer dissatisfaction.

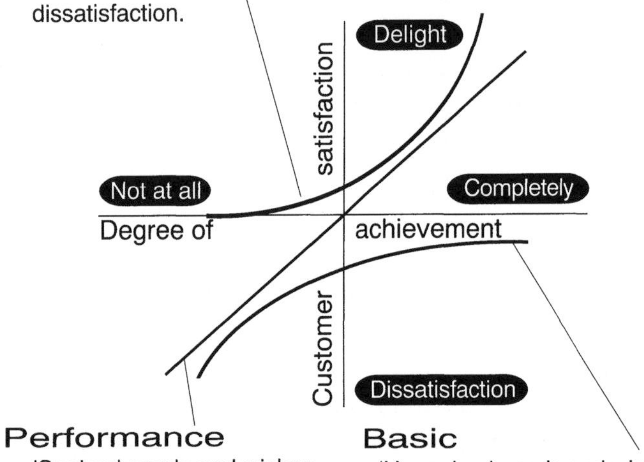

Performance

- 'Spoken' needs and wishes for features familiar from competing products.
- Readily accessible to market research.
- Generally additive in producing customer satisfaction.
- Low achievement of performance factors can give rise to customer dissatisfaction.

Basic

- 'Unspoken' needs and wishes for features typical of, and expected from, competing products.
- Difficult to discover through market research.
- Discovered by competing products analysis
- Failure to achieve any basic needs will lead to customer dissatisfaction.

There are four key features of Kano's model of product quality which we must try to incorporate into the remaining stages of the product planning process (Figure 8.2).

- The unspoken determinants of product quality (basic and excitement factors) are very difficult to identify by means of market research. Customers will not ask for wheels on a car any more than they will ask for a Sony Walkman before it is invented. The best way to identify 'basic' determinants of quality is by means of competing product analysis. 'Excitement' factors must be extrapolated from market research aimed at understanding of customer's underlying wishes and their unresolved frustrations with existing products.

- The achievement of 'basic determinants of customer satisfaction is a pre-requisite for a successful new product. Once these basic needs are satisfied, however, there is no point in trying to develop them any further. The basic needs curve offers a diminishing return of customer satisfaction due to further levels of achievement.
- The achievement of 'excitement' determinants of customer satisfaction, on the other hand shows no such signs of tailing off. The more excitement factors can be achieved the more your customer will be elevated into the heady realms of delight.
- The achievement of performance factors adds customer satisfaction, but not so much as the excitement factors. The Kano model predicts that, provided you have reached a certain minimum level of performance factors (i.e. to the extent that you have avoided customer dissatisfaction) any extra effort or resources would be better spent achieving excitement factors in preference to more performance factors.
- Over time, factors which were initially in the excitement category will drift down into the performance category and eventually become basic needs. In the 1950's a television with colour pictures was definitely 'exciting'. By the mid 1960's it had become a performance factor – it was one of several qualities of a television which contributed to customer's purchasing decision. Now, in the 1990's if you try to find a new black and white television in the shops, you will have a long hard search. Colour has become a basic requirement for customers buying televisions. One consequence of this is that excitement features are only ever exciting once. Once the first manufacturer has introduced an exciting feature in a new product, it soon becomes one of the many performance factors by which customers judge products and are no longer excited by.

Creating quality in a product is, therefore, a matter of achieving the right balance between delivering customer expectations and exceeding them. The value that a customer attributes to a new product happens in two stages. Firstly, products have a baseline value, and achieving product quality beneath that baseline will give rise to customer dissatisfaction. This baseline is determined by the basic, unspoken expectations of customers (cars must have wheels) and a certain expectation of product performance. Adding value above that baseline partly involves achieving higher levels of performance than are found in competing products. The other part is to 'excite' customers with product qualities beyond their expectations. Good product planning incorporates basic, performance and excitement factors into the design specification and aims for customer delight by achieving them all.

Specifying product quality

A product opportunity is a mis-match between the needs of customers and the offerings of competing products. The description of that product opportunity is, and always should be, phrased in terms that customers would understand and that they themselves might use. The new product should be cheaper, have more features, look more attractive or do things that no other product currently does. Describing the opportunity in this way has several advantages. It keeps the opportunity simple, easy to understand and strongly market oriented. It specifies the aims of product development without dictating the means. As we saw in Chapter 6, this offers a basis for commercial commitment without limiting the creative freedom of the product development team to introduce technical innovations. It also provides a common aim for everyone involved in the development of the product. Describing the product opportunity in technical terms, rather than customer-oriented terms, inevitably fragments its description. Some aspects will be of more relevance to production engineers, other aspects will be of relevance to transport and distribution and yet others will be of relevance to marketing.

There comes a time, however, when a technical description of the product is necessary. The eventual outcome of the design process is a specification for the product's manufacture, and this has to be described in language which is fundamentally different from the language of the customer. The product has to be made in materials that would mean nothing to the average customer. Manufacturing processes will be unheard of and engineering tolerances absolutely baffling. The important question is when to make the break from customer-oriented descriptions of the product opportunity to the technicalities of its design.

It is a fundamental principle of effective product development that this translation form customer needs into technical objectives should happen before the start of the design process. Developing a technical design specification from the opportunity specification is a pre-requisite for quality controlling the design and development process. It is only by referring to the design specification that emerging new product can be judged to be progressing satisfactorily, and given direction for its further development. Alternatively, the product is judged to be unsatisfactory and killed off, before further resources are wasted on its development.

Quality control of new product development has, therefore, two functions (Figure 8.3):

- A quality 'guidance' function which targets the development process progressively more closely to the achievement of customer satisfaction.
- A quality 'gating' function which reviews the progress of the new product against targets and only allows through the products meeting the necessary targets.

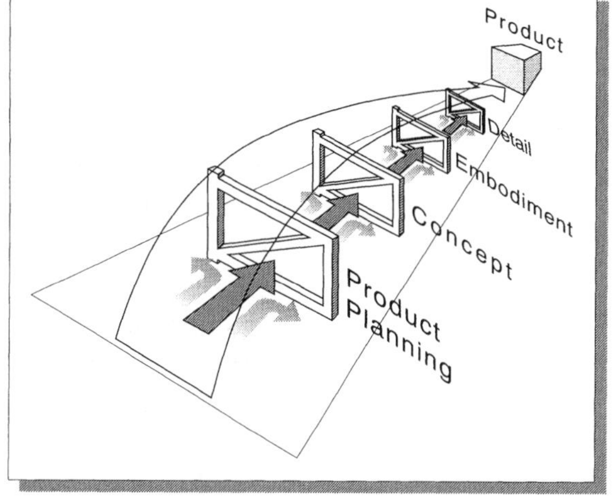

Figure 8.3
The 'guiding' and 'gating' functions of quality control of product development

Translating customer needs into technical objectives

In any work of translation, the main difficulty is achieving the correct balance between utility, accuracy and fidelity. In preparing a design specification, achieving utility means making the specification useful for the purposes of quality controlling product development. The design specification must, therefore, be described with sufficient precision to allow technical decisions to be made. There is no point in having precision, however, without accuracy. These precise descriptions in the design specification must be an accurate reflection of the customer's needs and wishes. And the design specification in its entirety must be a full and comprehensive description of the customer's perception of product value, as described in the opportunity specification. The design specification must have fidelity to customer needs.

To say the least, this is a demanding task. Product designers are required to possess a great many skills, including creativity, visual literacy, technical competence and attention to detail. Humility is not an attribute many designers would choose to write in their curriculum vitae. But at this point, before we go on to what I consider to be the most important few pages of the entire book, I want to pause to consider the case for humility. We have already seen, in Chapter 2 that early planning and specification of a new product is important. This, on its own makes the chances of product success 3 times more likely. I have also emphasised, in several chapters, the importance of a design specification in quality controlling product development. To be blunt, without a thorough design specification, you may as well forget the entire notion of quality control within the product development process. The preparation of this design specification is as difficult as it is important. Get it wrong and you may end up with a product that has been perfectly quality controlled in entirely the wrong direction.

Getting this design specification to translate customer requirements accurately and with utility and fidelity is a fundamentally difficult task. In the language of problem solving, this is a complex (problem solving involves several stages), fuzzy (the problem boundaries are ill-defined), multi-factorial (there are many different variables to consider) problem requiring simultaneous (as opposed to sequential) resolution. It is an unfortunate fact of life that the human brain is particularly poor at this sort of problem solving. Medical diagnosis is a similar sort of problem-solving and it has been shown that for a defined range of diseases, computers make more accurate diagnoses than doctors. They are simply better than people at complex, fuzzy, multi-factorial problems requiring simultaneous resolution. So, coming back to the need for humility in the design profession, the development of a design specification is a task for which you need help. The difficulty people have with complex information processing must be acknowledged and assistance sought for a design task, for which we have been less than perfectly designed ourselves. Designers need help to effectively translate customer needs into a design specification. That help comes in the form of quality function deployment.

Quality function deployment [2]

The mere mention of quality function deployment is enough to send shivers up the spine of many product designers. It is one of the most technical and involved techniques used in product development and it looks horribly complicated. But do not be put off because the complexity is manageable and parts of quality function deployment can be used on their own. An overview of quality function deployment is given in Figure 8.4.

As far as product planning is concerned, there are four main stages to quality function deployment:

- Firstly, a matrix is developed to explore the technical attributes of a product which contribute to customer needs.
- Secondly, competing product analysis ranks the performance of existing products in terms of both customer satisfaction and technical performance.
- Thirdly, quantitative targets are set for each of the technical attributes of the product and
- Fourthly, these targets are prioritised.

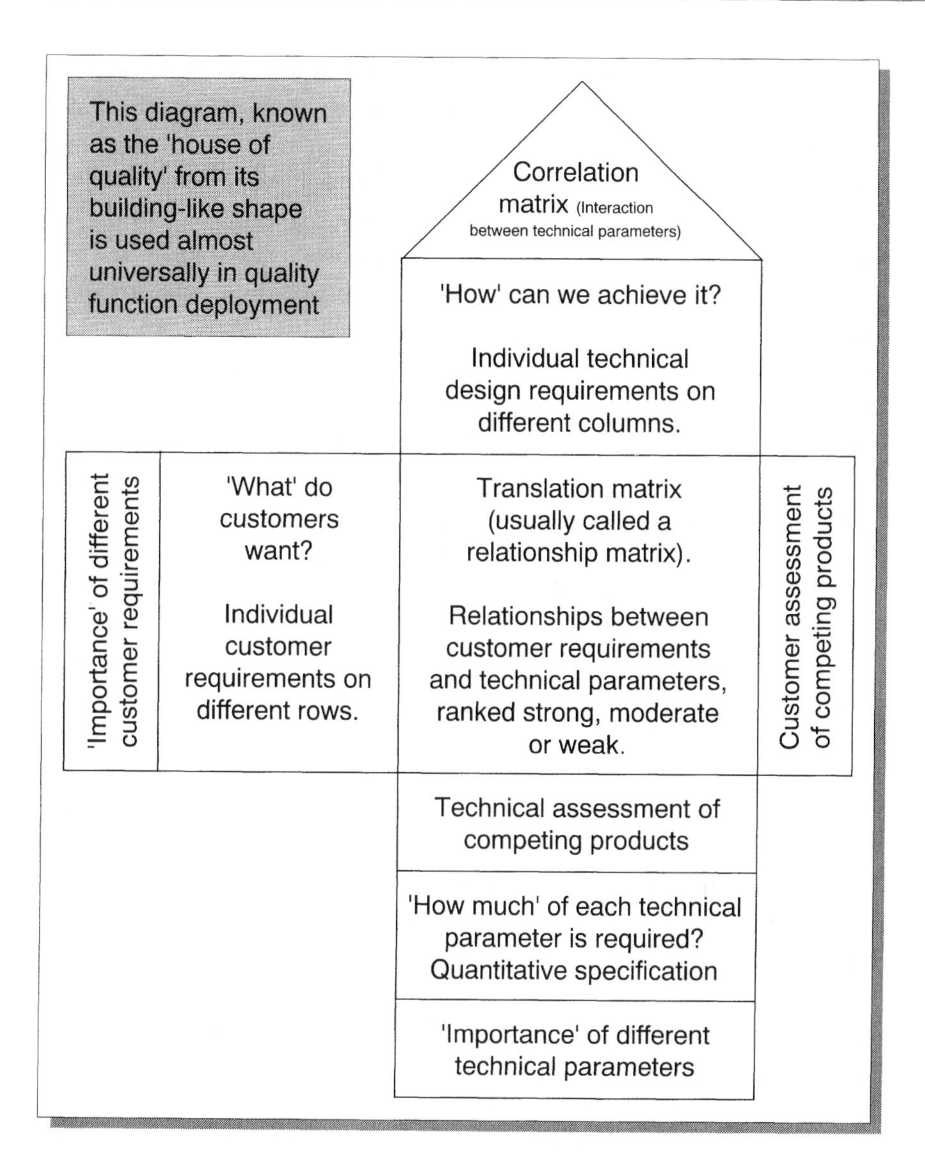
This diagram, known as the 'house of quality' from its building-like shape is used almost universally in quality function deployment
Correlation matrix (Interaction between technical parameters)
'How' can we achieve it?
Individual technical design requirements on different columns.
'Importance' of different customer requirements
'What' do customers want?
Individual customer requirements on different rows.
Translation matrix (usually called a relationship matrix).
Relationships between customer requirements and technical parameters, ranked strong, moderate or weak.
Customer assessment of competing products
Technical assessment of competing products
'How much' of each technical parameter is required? Quantitative specification
'Importance' of different technical parameters

Figure 8.4
The 'house of quality' for quality function deployment

Stage 1: The heart of quality function deployment

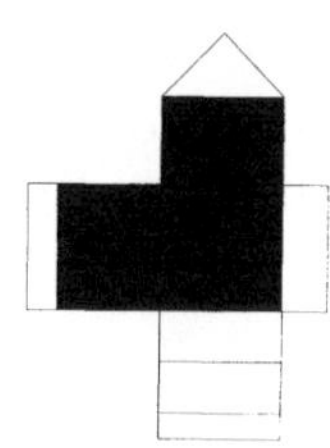

At the heart of both the procedure for quality function deployment and the house of quality diagram is the translation matrix or relationship matrix. This matrix translates individual customer needs into specific technical design requirements by a systematic step-by step process. The process starts by listing all the identified customer needs and

Key:
- Large grey circle: Strong positive
- Small grey circle: Weak positive
- Small black dot: Weak negative
- Large black dot: Strong negative

Customer requirements	Design requirements: + Diameter of head	+ Diameter of pin	+ Strength of join	+ Sharpness of pin
Easy to push into board	Weak positive	Weak negative		Strong positive
Pin does not bend		Weak positive	Strong positive	
Low price	Strong negative	Weak negative		Weak negative

Figure 8.5
The relationship matrix

arranging them in rows down one side of the matrix. Then design requirements for satisfying these customer needs are arranged in columns along the top of the relationship matrix. The extent to which each of these technical parameters relates to each customer need is identified in the matrix. The coding system for the matrix is arbitrary but it should indicate the strength of the relationship and whether it is positive (i.e. makes a positive contribution towards satisfying customer needs) or negative (i.e. detracts from the satisfaction of customer needs). Let us begin with the simplest of examples. A company wants to develop a new thumb tack. After 'extensive' market research they have identified 3 customer requirements: the tack must be easy to push into board, the pin must not bend and the tack must be inexpensive. These are arranged in rows down one side of a relationship matrix (Figure 8.5). Next, think of which design features contribute to satisfying these customer requirements and arrange them in columns along the top of the relationship matrix. Four such design requirements are identified: diameter of tack head, diameter of pin, strength of join and sharpness of pin point. Finally the relationships between the two is coded as strong or weak, within the matrix. Diameter of the tack head has a moderate effect upon ease of pushing the tack into a board and a strong effect on price (because it makes up most of the material in a whole tack). Increasing pin diameter will reduce the risk of bending (only slightly because tacks usually bend at the join between the pin and the head). It will, however, slightly reduce the ease of pushing the tack into the board and slightly increase price. The strength of join will have a strong effect on the tack's resistance to bending. The sharpness of the point will make the tack easier to push into the board but there are likely to be extra manufacturing costs for getting a sharper pin.

Stage 2: Competing product analysis

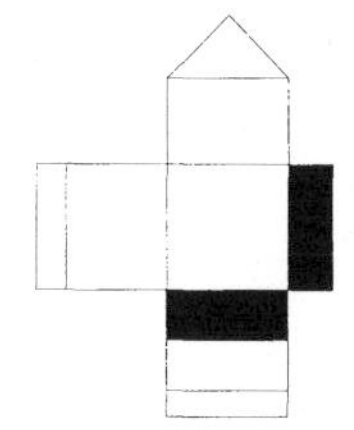

Competitor analysis takes two forms in quality function deployment. Firstly, customers give a rating of competing products on each of their stated customer requirements. Secondly, the design team examines each competing product on each of their design requirements. Two competing thumb tacks were analysed, as well as the company's own existing product (Figure 8.6). The ratings for both customer and design analysis were made on a 1 (worst) to 5 (best) scale.

For a modest amount of effort, we are already beginning to develop some very useful information about thumb tack design. In the relationship matrix it was expected that increasing pin diameter would slightly improve its customer rating on 'pin does not bend'. Customer rating of competing pins, however, revealed that tacks 1 and 2 were judged to be better in pin bending strength than the company' own tack, yet had smaller diameter pins. From

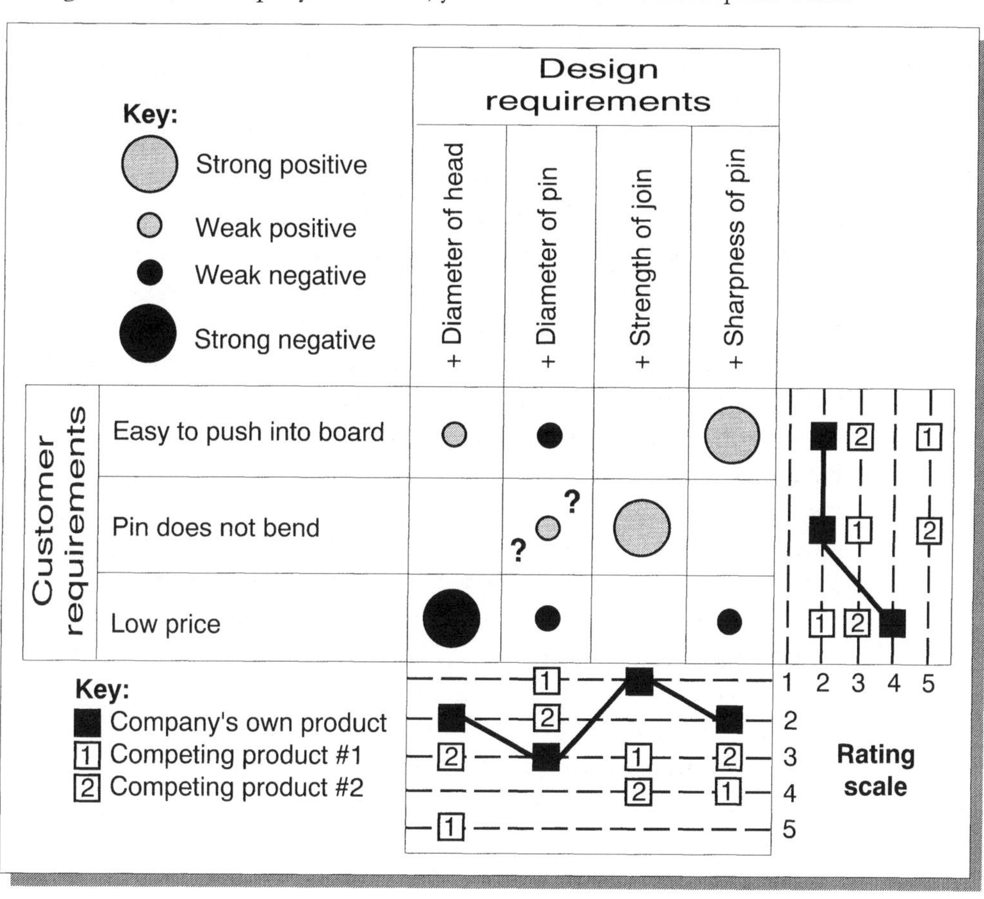

Figure 8.6 Competing products rating

this it would appear that even a small diameter pin is of sufficient strength to resist bending forces and that increasing pin diameter further has no beneficial effect. Tack bending strength appears to be solely determined by 'strength of join'.

Stage 3: Setting quantitative targets

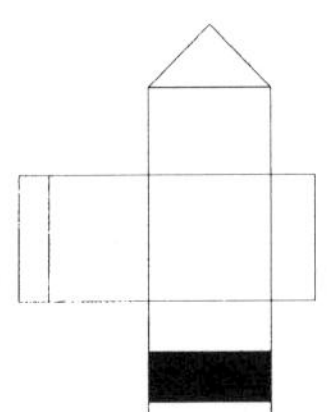

Having seen how competing products compare both technically and in the eyes of customers, we are now in a position to set design targets for the proposed new product. The table (Figure 8.7) shows how the three current products compare on the four design features. The larger the diameter of the tack head the easier customer's find it to push the tack into board. A tack head greater than 10mm is, therefore chosen as the target value for the design of the new tack. Having a large pin diameter is concluded to have little benefit although it adds to cost and makes the tack slightly more difficult to push into board. It is decided to reduce pin diameter from the existing product's 1.1mm to 0.8 mm. The strength of the join between the pin and tack head is tested using a purpose-built test rig. It is estimated that a tack becomes more awkward to use once the pin tip is bent more than 1mm off centre. The test criterion for strength of join is, therefore, the force required to cause a permanent (i.e. plastic) 1mm deformation. A target value of greater than 75N is selected to exceed the best of the two competing products. Similarly, the sharpness of the pin is given a target of less than a 0.1mm radius of curvature to exceed the best competitor. The addition of the target design values to the house of quality diagram is shown in Figure 8.8.

Figure 8.7
Target values from competing products

	Company's own product	Competing product #1	Competing product #2	**Target value**
Diameter of head	7mm	10.5mm	8.5mm	>10mm
Diameter of pin	1.1mm	0.8mm	0.9mm	0.8mm
Strength of join	55N	70N	75N	>75N
Sharpness of pin	0.2mm	0.1mm	0.15mm	<0.1mm

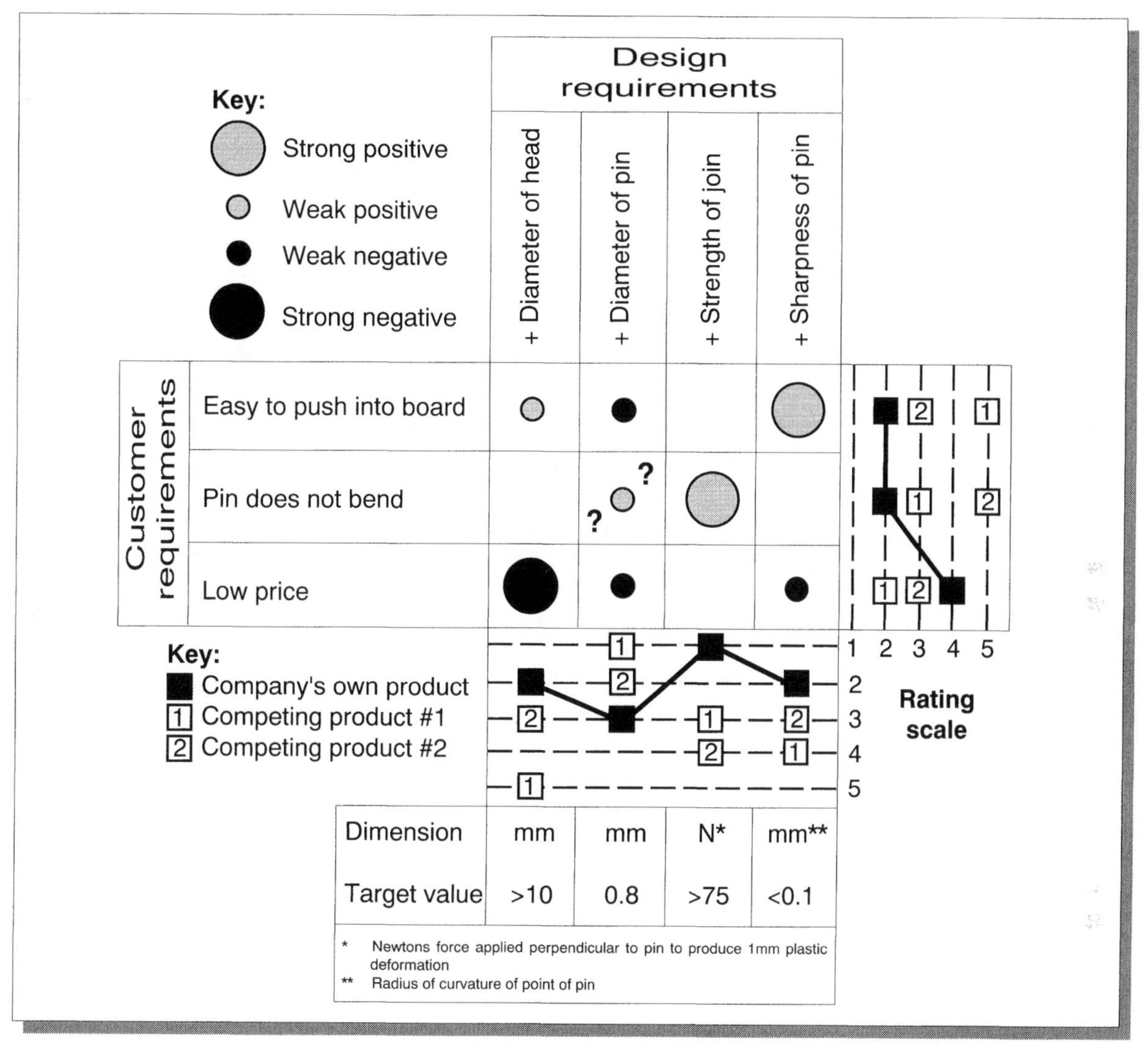

Figure 8.8
Setting target values

Stage 4: Prioritising the design targets

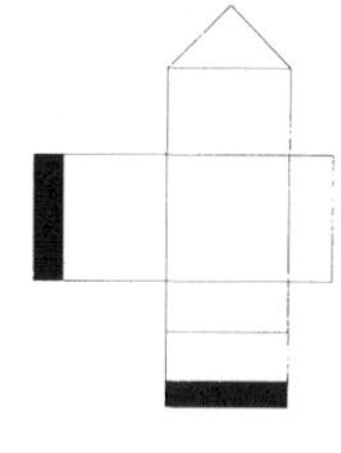

Once the design targets are identified, it is important to know which are the most important so that maximum effort is invested in meeting, or exceeding, the most important ones. It is also possible that certain design targets will have to be compromised in order to resolve conflict between them (e.g. conflict between improved features and reduced cost). In order to do so systematically, each design target is given

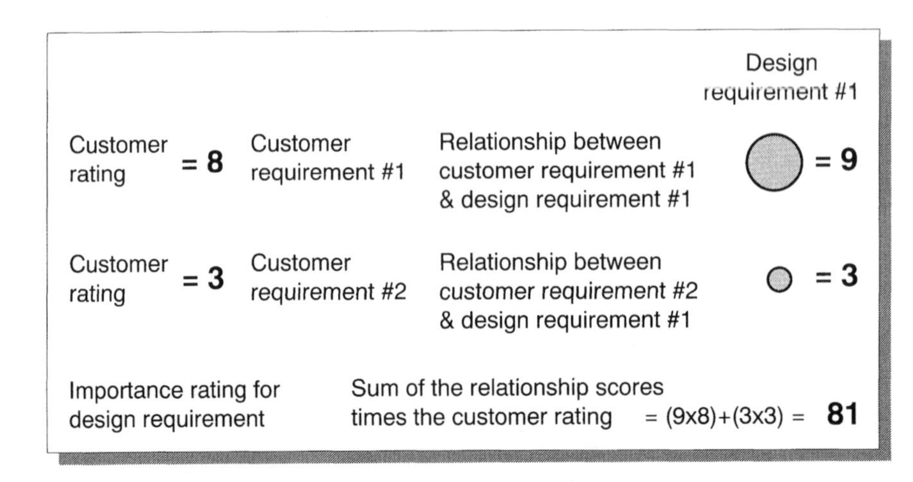

Figure 8.9
Calculating the importance ratings

an importance rating. This rating must be a reflection of their importance for achieving customer satisfaction. Consequently, they are based on the customers' ratings of the importance of the different customer requirements. During the research into customer ratings of competing products, customers would have been asked to give a score from 1 to 10, indicating the importance of each customer requirement. This is entered into the house of quality diagram alongside each of the customer requirements. Then, using a scoring system for each type of relationship within the relationship matrix, an importance rating for each design requirement is calculated (Figure 8.9).

The scoring system used to calculate importance ratings is arbitrary. Scales of either 1 to 5 or 1 to 10 are most common, since this allows a significant differentiation between important and unimportant factors (Figure 8.10). What is important about the scoring system is that it gives importance ratings which intuitively feel right. If they do not feel right, adjust the numbers slightly on the relationship scores to see if a more intuitively reasonable importance rating is produced. The customer ratings should not be adjusted so arbitrarily since they are derived from market research. If, for any reason they do not 'feel' right, the market research should be repeated.

Quality function deployment — beyond product planning

Quality function deployment has been described, so far, only in relation to product planning. In fact, it is a much more powerful technique which can be used to quality control the entire product development process. In effect, the house of quality diagram can be extended into an entire street of quality.

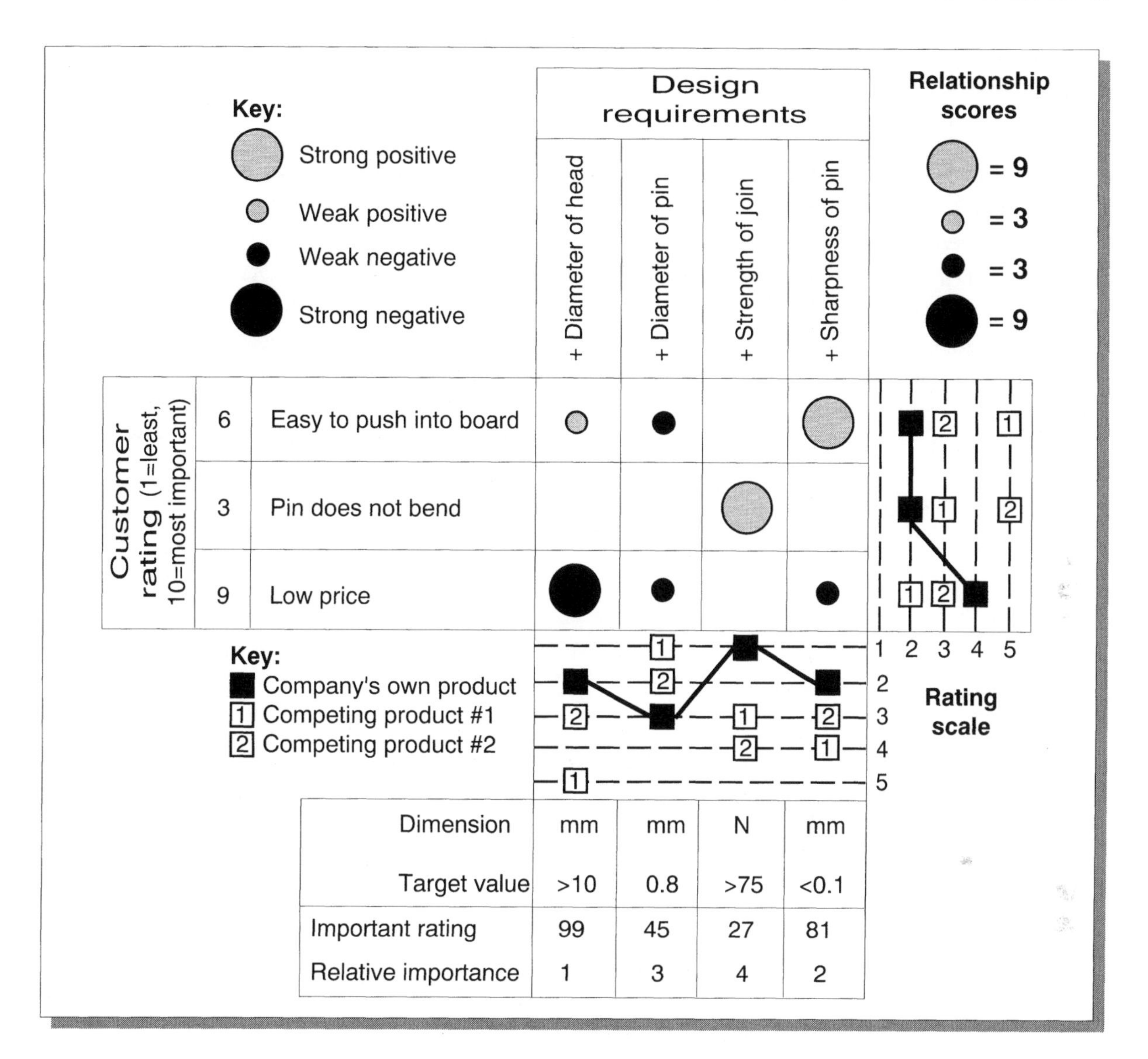

Figure 8.10
Importance ratings for different design requirements

The outputs from one house of quality become the inputs into the next house of quality, as design decisions become progressively more focused on the details of the product's manufacture. This extension of quality function deployment is summarised in Figure 8.11 and can be explored in more detail in the further reading cited in the notes [2].

Figure 8.11 Quality function deployment beyond product planning

Quality function deployment can be used throughout the design process, not just in the product planning stages. The house of quality is turned into a street of quality in which the outputs from one quality function deployment exercise become the input into the next quality function deployment analysis. In this way, quality can be systematically and rigorously steered from product planning to manufacturing and assembly. The quality function deployment procedure we have just been through takes customer requirements as inputs and translates them into design specifications as outputs. These outputs can be taken as inputs into the second house of quality and turned, just as systematically into product specification outputs. So, we concluded in the analysis of the thumb tack that the design specification should include diameter of head, diameter of pin, strength of join and sharpness of pin. Imagine taking these requirements and placing them in the house of quality where we previously put customer requirements. This would give us a relationship matrix from design specification to product specification: how do we engineer the product to meet these design specifications. So, for example, the strength of join might be influenced by the materials used and the detail design of the

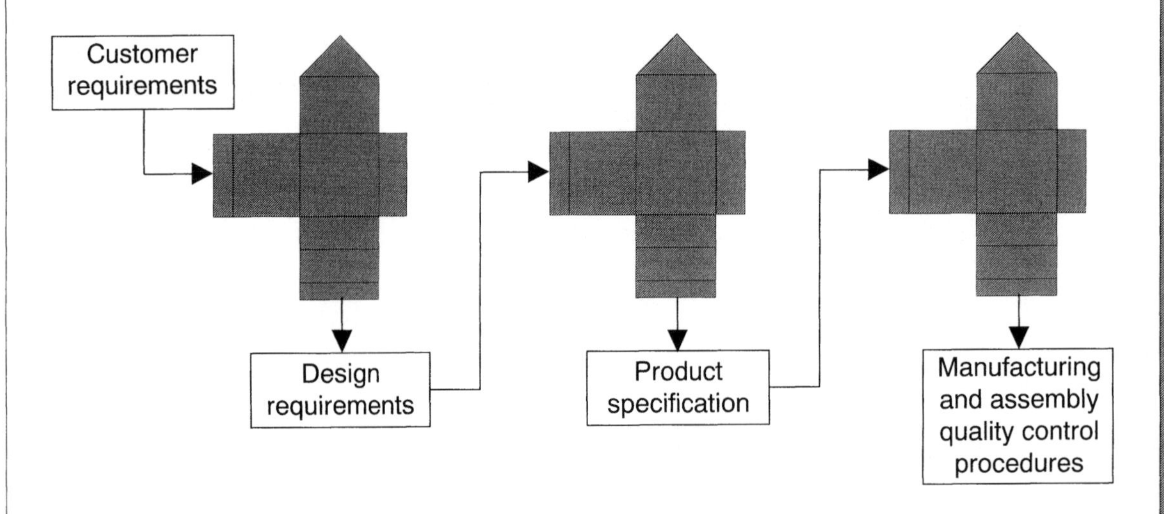

embossed socket into which the pin fits. These would, therefore be two of the columns entered at the top of the relationship matrix. Once a product specification had been established by following the stages of quality function deployment described above, we move to the third house of quality. This translates the product specification into a set of quality control procedures for manufacture and assembly of the product. So the product specification criteria which ensure the correct strength of join would be worked through into manufacturing procedures, such as the weld temperature and contact time and any tests required to ensure that the product has, indeed been designed to specification.

The design specification [3]

Having completed the quality function deployment procedure, customer statements defining the product opportunity should be translated into specific design targets. This is an essential part of the design specification but it would be a mistake to think that it is the entire design specification. There are many important aspects of a product's design which are transparent to customers. How the product is made, how it is distributed, how it complies with the requirements of sales outlets, how it will be serviced: none of these are things that customers can be expected to identify as product requirements. However, overlooking any of them during product development is likely to make the product just as commercially unsuccessful as failing to achieve customer satisfaction.

The aim of a design specification is to try to anticipate everything that could cause a new product to be a commercial failure and specify design targets for avoiding that failure.

We must now look at how to broaden the 'core' specification of customer needs to encompass all the design requirements of the new product. The aim of a design specification is to try to anticipate everything that could cause a new product to be a commercial failure and specify design targets for avoiding that failure. A list of headings for a typical design specification is shown in Figure 8.12. The headings cover the four most important determinants of product success; i) will it sell ii) will it work iii) can it be made and iv) does it comply with legal and commercial obligations? Clearly, the causes of product failure will be different for different products in different markets. It is, therefore important for all companies to develop their own customised design specification format. This is generally worth the initial investment because the format is likely to be applicable to most new products that the company will subsequently develop.

The procedure for preparing and writing a design specification is described in a design toolkit on p 259. This describes the following 4 stage process. Firstly relevant information is gathered from people, both inside and outside the company. This establishes what specific targets relate to each of the design specification headings and determines whether these targets are essential (demands) or desirable (wishes). It should also try to determine which of the 'core' customer requirements are basic, performance and excitement factors, from the Kano model of customer satisfaction (see Figure 8.2). Secondly, this information is written up as a first draft design specification, and a sample format for writing a design specification is also given in the design toolkit (p 259). Thirdly, the draft design specification is then reviewed by the key people who provided the information to begin with. Does it fulfil their requirements? Would a product designed to the bare minimum of this specification be a commercial success? If not, the

Market requirements

Check: Are these sufficient to ensure that the product will sell?

- Target price
- Product performance
- Appearance/image/style

(Target price, Product performance, Appearance/image/style: Opportunity specification)

- Point of sale requirements
  - Labelling
  - Packaging
  - Other marketing material (e.g. leaflets, product stand)
  - Sales information (e.g. barcodes)
- Specific sales outlet requirements
- Transport and storage requirements

Production requirements

Check: Are these sufficient to ensure that the product can be made?

- Target costs of manufacture
- Quantities to be manufactured
- Product size and weight targets
- Bought-in versus manufactured in-house
- Manufacturing constraints/preferences
  - Materials
  - Manufacturing processes
  - Assembly

Life in service requirements

Check: Are these sufficient to ensure that the product will work?

- Target life in service
- Operating environment specification
- Installation/commissioning-for-use requirements
- Product information (e.g. instructions for use, handbook)
- Reliability/durability targets
- Maintenance requirements
  - Access for maintenance
  - Replaceable components
- Disposal/recycling requirements

Conformance requirements

Check: Are these sufficient to ensure that the product will comply with all legal and commercial obligations?

- Statutory requirements
- Industry standards
- Company standards
- Product compatibility requirements
- Safety/product liability requirements
- Testing requirements
- Intellectual property rights
  - Patents
  - Design registration
  - Trademarks

Figure 8.12
Typical headings for the design specification

design specification is inadequate and must be improved. Fourthly, the design specification is written up in its final format, approved by management and then circulated to everyone who will be involved in the product's development.

Product development – project planning

The final stage of product planning is to plan the product development project itself. The first, and often most difficult, aspect of project planning is to establish how the product development process is to be divided up into stages, for the purpose of quality control. Dividing the project into stages is important because checks must be made that the new product remains in conformance with the design specification. As soon as it appears impossible for the product to meet its objectives, the development project must be killed off, to avoid wasting resources on its further development. This can only be done by periodically reviewing progress against the design specification targets. Designers disagree on how the design process is best divided into manageable stages. Some even doubt that it can be effectively and practically divided at all! In practice, pinpointing where one stage of design ends and the other begins is often an arbitrary and subjective judgement.

Let us, therefore, review the main stages of product development to see where useful boundaries can be drawn. Imagine that a company wishes to develop a new adhesive tape dispenser. In concept design, they consider three options: i) a simple, economy, hand held dispenser, ii) a more robust desk-top dispenser and iii) a more elaborate hand held device which assists in the application of the tape (Figure 8.13). By systematically applying concept selection techniques (described in the next chapter) they conclude that the desk top model is best suited to the previously identified opportunity. The principles of the product have, therefore, been established. The product has a base, a holder for the roll of tape and a combined cutter and tape-holding platform. Concept design is complete and the design efforts move on to embodiment design. This begins by examining the holder for the roll of tape. In principle this could be designed any number of ways as shown in Figure 8.14. But,

Figure 8.13
Concepts for an adhesive tape dispenser

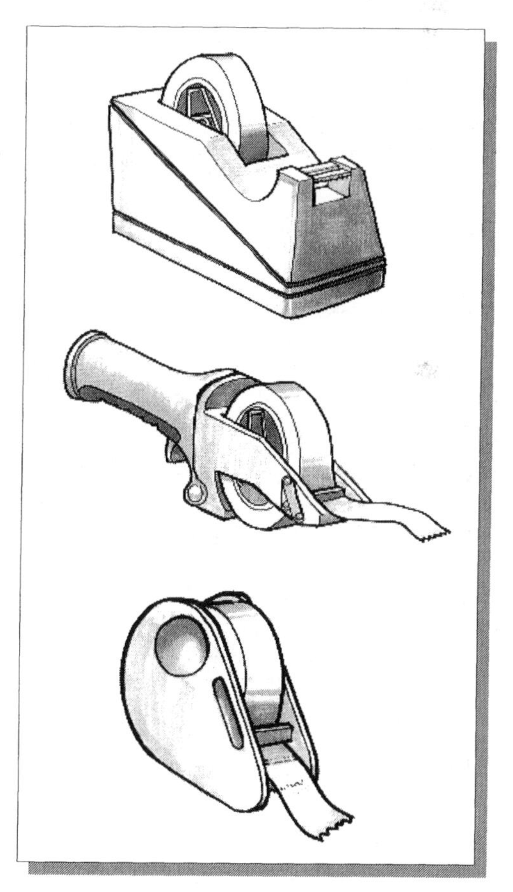

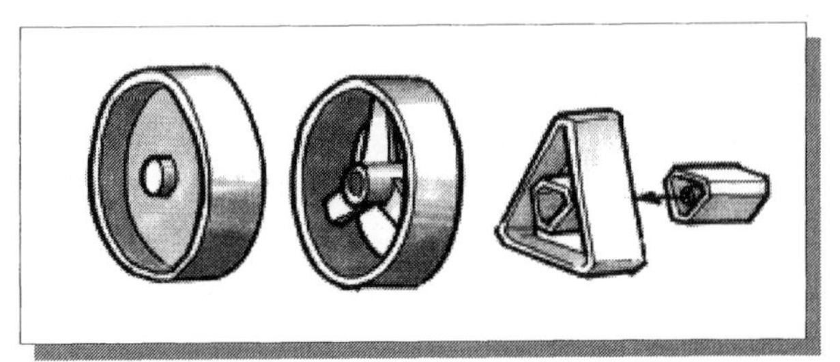

Figure 8.14
Concepts for a component of an adhesive tape dispenser

wait a minute! It is, surely, a conceptual issue to decide the basic form of the tape holder. It must be a matter of design principle whether we choose a tape holder of round or triangular shape. Doesn't embodiment design only decide how that basic shape is made, once its form has been decided? Yet, concept design is not meant to consider product components. So, where exactly are we in this simple concept-embodiment-detail design process? In truth, we are in the conceptual part of embodiment design. I told you it could get complicated!

Sacrificing, for the moment, the simplicity of clear distinctions between concept, embodiment and detail design, we can think of the design process as incorporating different types of design thinking which occur repeatedly at different stages in the development process (Figure 8.15). So, concept design thinks of the design principles but then goes on to include some preliminary thoughts on the embodiment of key components. This is usually essential to prove that the selected concept is viable. Embodiment design then begins by considering how the product can be divided into components for manufacture. This is normally known as the product's architecture and it is first explored in principle. In other words, product architecture is studied at a conceptual level, using the same techniques as concept design; exploring a variety of forms and functions for each component and systematically selecting the best. Then it is worked out how each component will be made. In doing so embodiment design inevitably considers preliminary ideas for detail design. This includes the materials and the manufacturing process by which the components will be made. Finally detail design examines the principles of the detailing of each component. This considers, at a conceptual level, whether the strength of a particular element of the product, for example, will best be achieved by increasing its wall thickness or adding ribs and bosses. It then goes on to prepare engineering drawings and a full specification describing the materials and processes to be used for its manufacture. Each stage of the design process, therefore, includes a mini-cycle of design-in-principle then design-in-detail as you progress to more specific aspects of product design.

Although the distinctions between concept, embodiment and detail design have been blurred, they have not disappeared altogether. We can still think of concept design as producing a set of functional and styling principles for the entire product. Concept design can be considered formally complete when these principles show how the proposed new product will meet the opportunity specification. In other words, the concepts should show how the product meets customer needs and achieves differentiation from competitors. This can be checked in a quality control review and a decision made on that basis as to whether or not the product development should continue. Concept design does not concern itself with product architecture - how many parts and how do they go together, nor with the design of any of the component parts. That is the role of embodiment design. Embodiment design takes the preferred concept and determines how it can be made. This decides not only product architecture and component design but also the generic materials and manufacturing processes by which the product will be made. Embodiment design only goes as far as making a full working prototype. The quality control check at the end of embodiment design checks the product's fitness for purpose and confirms that the product is 'manufacturable', although it does not yet specify precisely how it will be made. Thus, embodiment design will say that a part will be injection moulded and detail design will determine the draw angles needed to remove the part from the tool. Embodiment design will say that a part will be stamped from steel sheet and detail design will determine what finishing is required to remove any sharp edges. Embodiment design will say that a part will be made from cast aluminium and detail design will determine what grade of aluminium to specify. Detail design, therefore, translates the working prototype into a full set of engineering drawings and the final

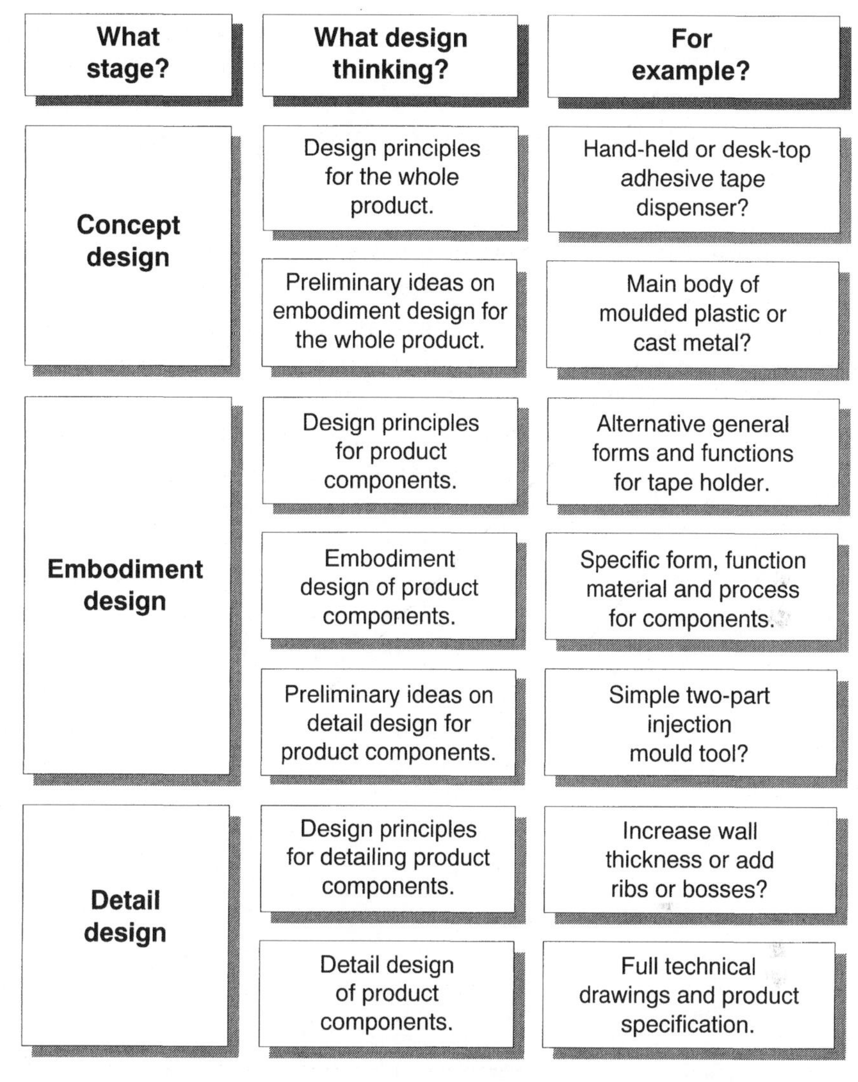

What stage?
What design thinking?
For example?
Concept design
Design principles for the whole product.
Hand-held or desk-top adhesive tape dispenser?
Preliminary ideas on embodiment design for the whole product.
Main body of moulded plastic or cast metal?
Embodiment design
Design principles for product components.
Alternative general forms and functions for tape holder.
Embodiment design of product components.
Specific form, function material and process for components.
Preliminary ideas on detail design for product components.
Simple two-part injection mould tool?
Detail design
Design principles for detailing product components.
Increase wall thickness or add ribs or bosses?
Detail design of product components.
Full technical drawings and product specification.

Figure 8.15 Stages of the design process

product specification for manufacture. This should be a sufficient set of instructions to allow the manufacture of the product to commence. The quality control check at this point constitutes the 'sign-off', approving the product for manufacture. The main distinctions between concept, embodiment and detail design are given in Figure 8.16.

The important requirement for product development is that quality control reviews periodically occur. It is not of critical importance when they occur. In particular development projects, it may be that the end of concept, embodiment and detail design , as we have just defined them, is not the most useful times for quality control reviews. The working principles of a new product may, for example critically depend upon a single component. It may, therefore, be best to delay the review following concept design until the design of this component has been examined more closely in embodiment design. As a general rule, then, the first step in project planning is to decide what achievements in the design process will trigger a quality control review. With the design process thus divided into stages, the next task is to timetable each stage. Obviously, the later stages of the design process will be more difficult to timetable since it is not yet known what design activities will be involved. Previous experience of product development projects is usually the best way of estimating this.

Product development projects can be thought of as either being time constrained or resource constrained. A time constraint is often imposed by deciding upon a product launch date. This is usually required so that sales and marketing efforts are synchronised with product development. In a time-constrained project, resource flexibility is important to ensure that difficulties late in the development project can be tackled by expanding the product development effort. In a resource-constrained project, timing should be flexible to allow difficulties to be tackled with the same develop-ment effort over a longer time period. Often, managers insist that development projects are both time and resource constrained. This is unrealistic since it assumes that the problems to be tackled are estimated with

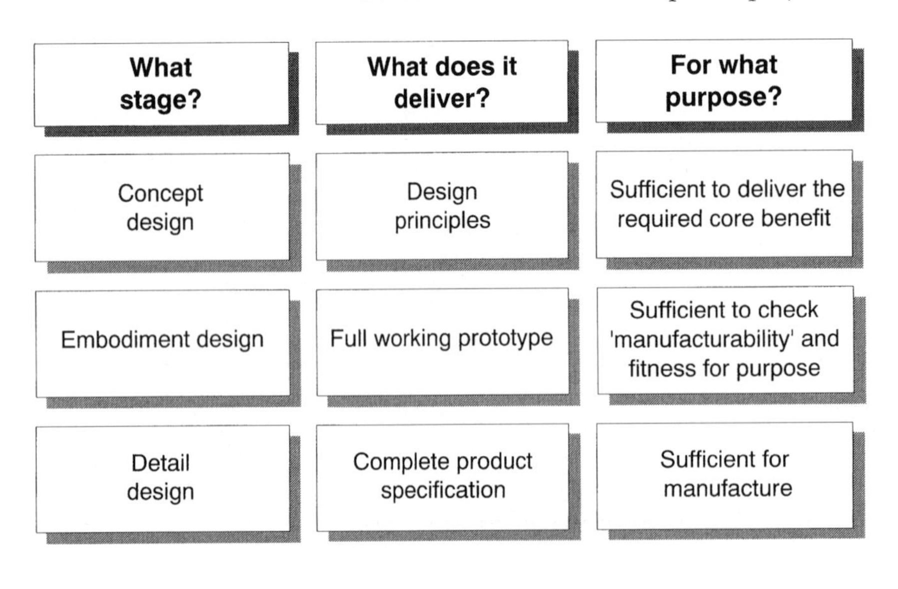

What stage?
What does it deliver?
For what purpose?
Concept design
Design principles
Sufficient to deliver the required core benefit
Embodiment design
Full working prototype
Sufficient to check 'manufacturability' and fitness for purpose
Detail design
Complete product specification
Sufficient for manufacture

Figure 8.16 The stages of design and what they aim to achieve

complete accuracy before the development work has begun. It is probably for this reason that product development performance in getting products designed to specification and on time is so poor in most companies (see Figure 5.3). Product designers themselves must also shoulder a significant responsibility for this. It is a common mistake in product planning to be over-optimistic about both the time and cost involved in getting a product to market. Often this is done in all innocence. The difficulties that will inevitably emerge later are simply not apparent at the start. It can, however, also be tempting to be over-confident about the development of a particularly cherished new product, in order to secure management approval for its development. It is, therefore good practice to give a best and worst estimate of both the time and cost of development.

Time planning of a development project is usually described adequately by a Gantt chart (Figure 8.17). This shows simply and effectively the duration of each stage of product development and the quality control review points. For more complex development projects, more complex time planning may be necessary. The programme evaluation and review technique (PERT) identifies the linkages between more specific tasks and shows which tasks must be completed before others can begin. As its name suggests, PERT offers more than simply a time plan of activities. It shows the sequence of events sufficiently clearly that the development process can be evaluated, reviewed and refined as part of project planning. Figure 8.18 shows a sequence of events which might initially be thought of as a linear sequence in time. These same events when planned to run concurrently using PERT analysis are shown in Figure 8.19.

Figure 8.17
Gantt chart for product planning

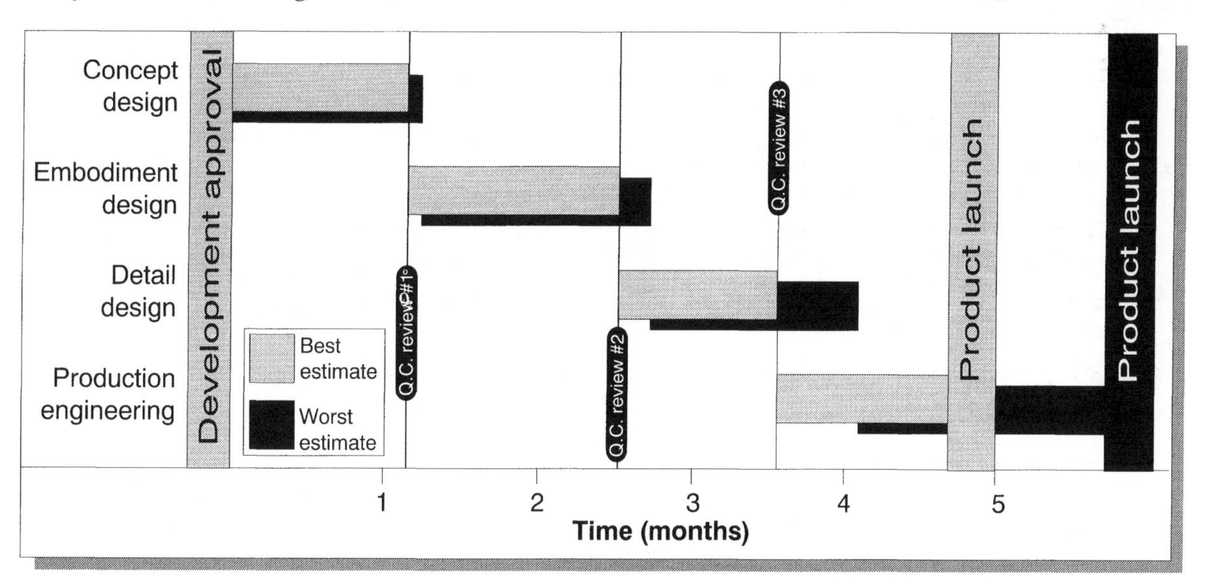

Figure 8.18
Linear sequence of events in product development

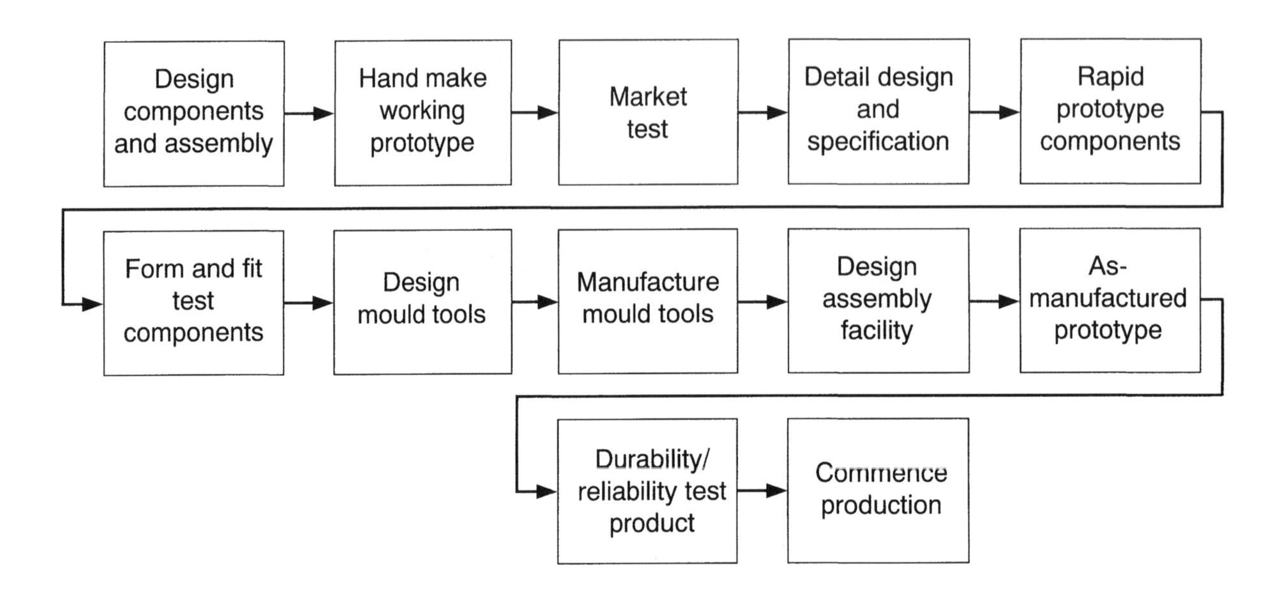

Figure 8.19
The same events run concurrently and planned by PERT analysis
(Note that the thick line indicates the critical path for these events)

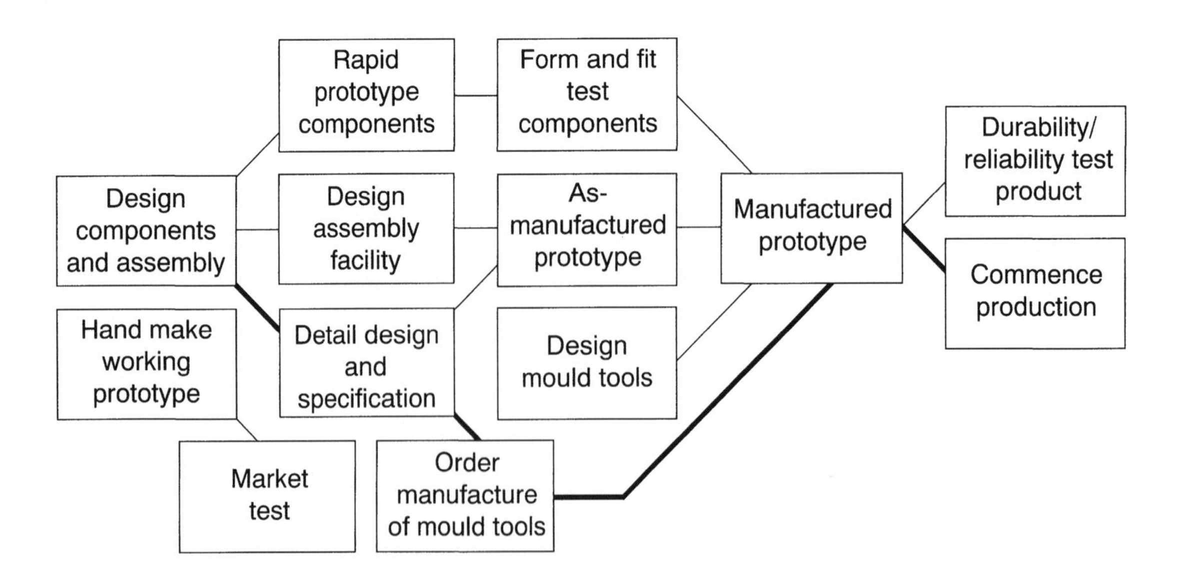

Key Concepts

1. Deliver customer value

The key to successful product design is delivering customer value and the key to delivering customer value is to specify what the product must do to offer value to the customer.

2. Kano model of quality

Customer satisfaction with a new product can be broken down into three component parts, according to the Kano model of quality: basic, performance and excitement factors.

3. Quality function deployment

Quality function deployment translates customer needs into design requirements using a relationship matrix. It then goes on to quantify and prioritise these design requirements by means of competing product analysis and importance weighting.

4. Design specification

Producing a design specification requires a thorough working understanding of these customer requirements, but, in addition, it also covers aspects of the product which may be transparent to the customer. A full design specification describes market requirements, production requirements, life in service requirements and conformance requirements.

5. Project planning

The final stage of product planning is to prepare a time plan for the development project. This breaks the product development process into stages for the purposes of quality control and estimates the time required for each stage.

Design Toolkit

Design specification

The design specification is the key quality control document for product development. It determines what the product must and should do and it sets the criteria by which the product is either allowed to continue on its development or be killed off during quality control reviews. It should also be a useful and helpful guide to the design team to ensure that nothing is 'left out' of the product's design during its development.

There are 4 main stages in the preparation of a design specification:

1. Review and finalise the commercial objectives for the product from the opportunity specification.
2. Conduct design specification briefings with the key staff involved in product development and subsequent product management after the product is launched. Establish from them what is required of the product.
3. Draft the design specification
4. Review the draft design specification with the key staff briefed earlier and finalise it.

The briefing session to collect information on the design specification is intended to make sure that all aspects of the product's performance will fit with the company and be conducive to its sales success. Thus staff should be included from marketing, production, storage and distribution and sales. With each of them, go through the main headings of the design specification asking for any relevant requirements they might have for the new product. Every time a new requirement is identified, it should be clear whether this is a demand or a wish. Is it, in other words, a requirement essential to the sales success of the product. If so it is a demand and the product development should be killed off as soon as its is clear that such a requirement cannot be met. Otherwise it is a wish. For any requirements relating to customer needs, try to establish whether they are basic, performance or excitement factors. In addition, every new requirement should be associated with the individual who proposed it. This will allow you to go back to that person to ask them for further details or clarification of its importance. It will also tell you, long after the design specification has been completed, why that requirement was proposed in the first place. A format for a briefing sheet is given on the opposite page to make all this information gathering easier. Once the briefing sheets are completed by all relevant staff, the next stage

Design specification

Design Toolkit

is to establish which level of the design specification the different requirements relate to. A design specification can be thought of as a hierarchy of needs ranging from performance requirements – what is required of the product) to design requirements (how the product should do it) to design specification (what, quantitative criteria the product should meet). For market requirements, this hierarchy is worked through using product function deployment. The example below shows how a draft design specification is prepared, using this hierarchy of requirements as a logical and systematic procedure.

Figure 8.20
Briefing form for preparing the design specification

Proposed product: Potato peeler
Information from: Ian (Marketing)
Designer: John
Date: 18th January '95

Product requirement	Demand or wish	Type of requirement	Basic/perf./ excitement
Must look hygienic	Demand	Market	Basic
Must feel comfortable to hold	Demand	Market	Performance
Should be super-sharp	Wish	Market	Excitement

Figure 8.21
Format for a draft design specification

Proposed product: Potato peeler
Date: 19th January '95
Designer: John

Performance requirement	Design requirement	Design specification
Must remove the eyes from potatoes	Must have a gouge Must be designed so that the gouge can be used without having to grip the handle	Gouge must be positioned close to the handle Gouge should be used without having to change the grip on the peeler

Notes on Chapter 8

1. Kano's model of quality is concisely summarised in the American Supplier Institute Inc. 1992, *Quality Function Deployment: Practitioner Manual.* American Supplier Institute, Dearbourn, Michigan pp 5-6.
2. Quality function deployment has been written about extensively in the engineering and quality management literature. Two of the most concise reviews are Sullivan L.P. 1986 Quality Function Deployment. *Quality Progress* (June) 39-50 and Hauser J.R. and Clausing D. 1988, The House of Quality. *Harvard Business Review*, 66: 63-73. A more in-depth treatment which takes the reader step by step through the QFD process is the American Supplier Institute Inc. 1992 (see 1 above).
3. The pioneering account of the design specification is Pahl G.and Beitz W. 1988, *Engineering Design: A Systematic Approach*. Design Council, London. An interesting survey of the extent to which design specifications are used by manufacturing companies is provided by Walsh V., Roy R., Bruce M. and Potter S. 1992, *Winning by Design: Technology, Product Design and International Competitiveness.* Blackwell Business, Oxford, UK, pp198-206.

9 Embodiment and detail design

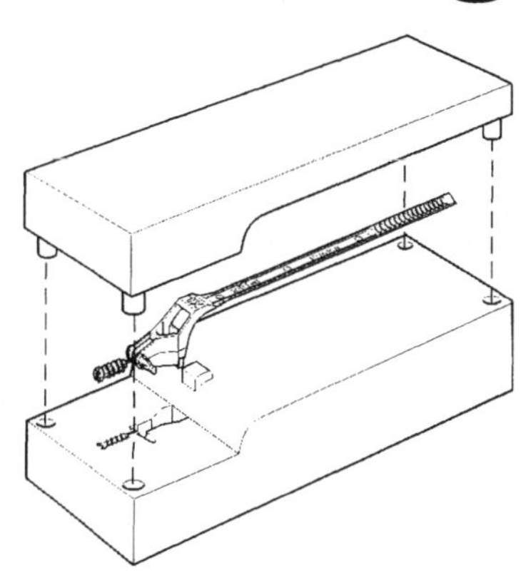

Embodiment design begins with a preferred concept and ends with a fully developed and tested working prototype. It comprises four main stages:

- Idea generation, exploring all the possible ways to make the product.
- Idea selection, picking the best of these ideas against the design specification
- Failure modes and effects analysis, to assess the ways in which the product might fail
- Prototyping and testing, to develop, refine and eventually either accept or reject the preferred design.

As with all design activities, these four stages do not always occur in a neat orderly fashion. FMEA may be a useful tool to help with the selection of potential embodiment designs. Some degree of prototyping may be needed during the generation of ideas. And there will often be a need for iteration through several stages until the design becomes satisfactory.

The main difference between embodiment design and concept design, in terms of design activities, is the introduction of a significant measure of product testing and evaluation. The reason is that this is the first time we have anything testable to work on. There is only a limited amount of testing you can do with a set of working principles before they have been turned into a prototype of some sort. Figure 9.1 shows the inputs and outputs involved in the embodiment design process. The inputs are simply the outputs from

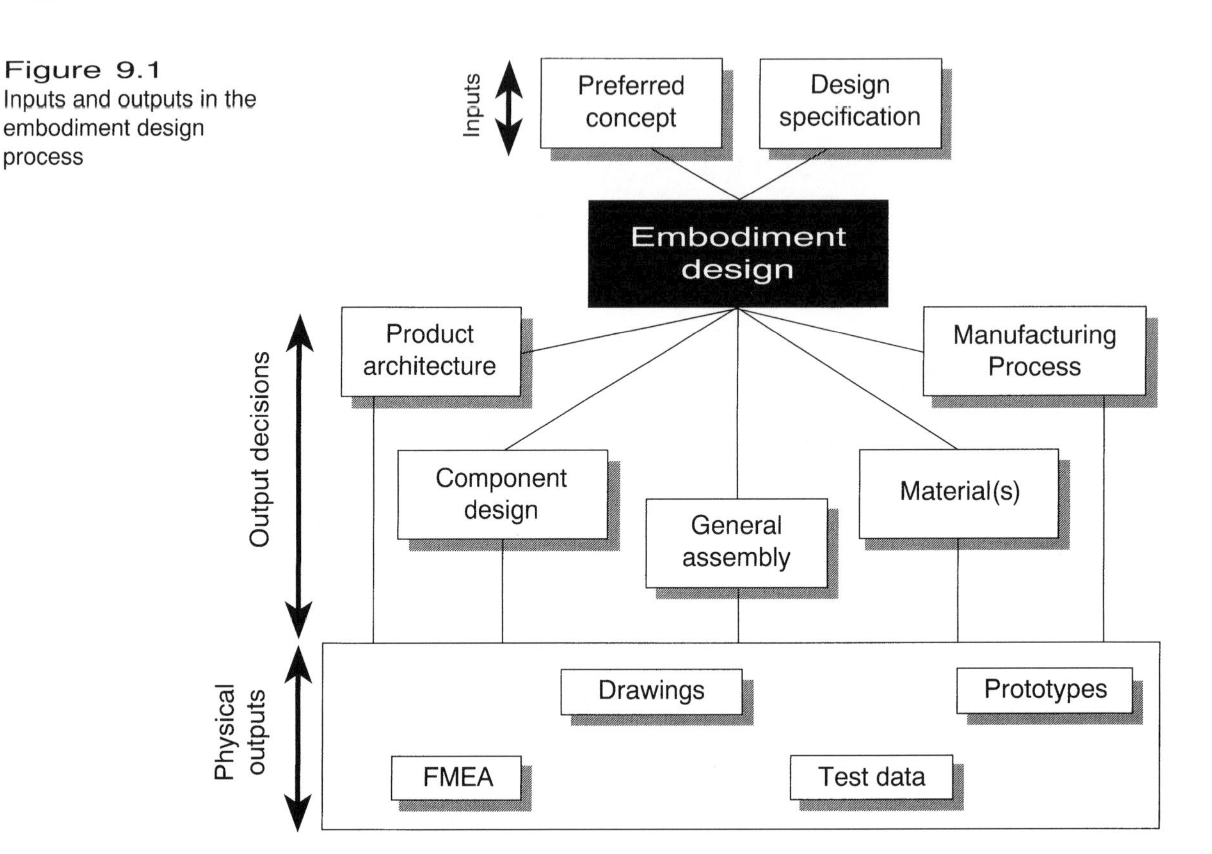

Figure 9.1
Inputs and outputs in the embodiment design process

the preceding concept design activities and the all-embracing design specification. The decisions which need to be made by the end of embodiment design include the architecture of the product (how is the product broken down into discrete components for manufacture) the form and function of each of these components, their assembly and the type of material(s) and manufacturing process(es) by which each will be made. The physical outputs recording these decisions include drawings and models, as well as a failure mode and effects analysis and data from any prototype tests.

Detail design takes these outputs from embodiment design and determines, in detail, how the product will be made. This involves deciding which components will be bought in (i.e. as standard catalogue items), which will be manufactured, either in-house or by sub-contractors. A specification must be prepared for each component, describing the design of that component in sufficient detail for the manufacturing tools to be specified and grade of materials to be specified. The output from detail design is a full 'product specification'. This is a set of instructions for manufacture of the product, derived from the design specification. Where the design specification set targets for the product's performance and appearance, the product

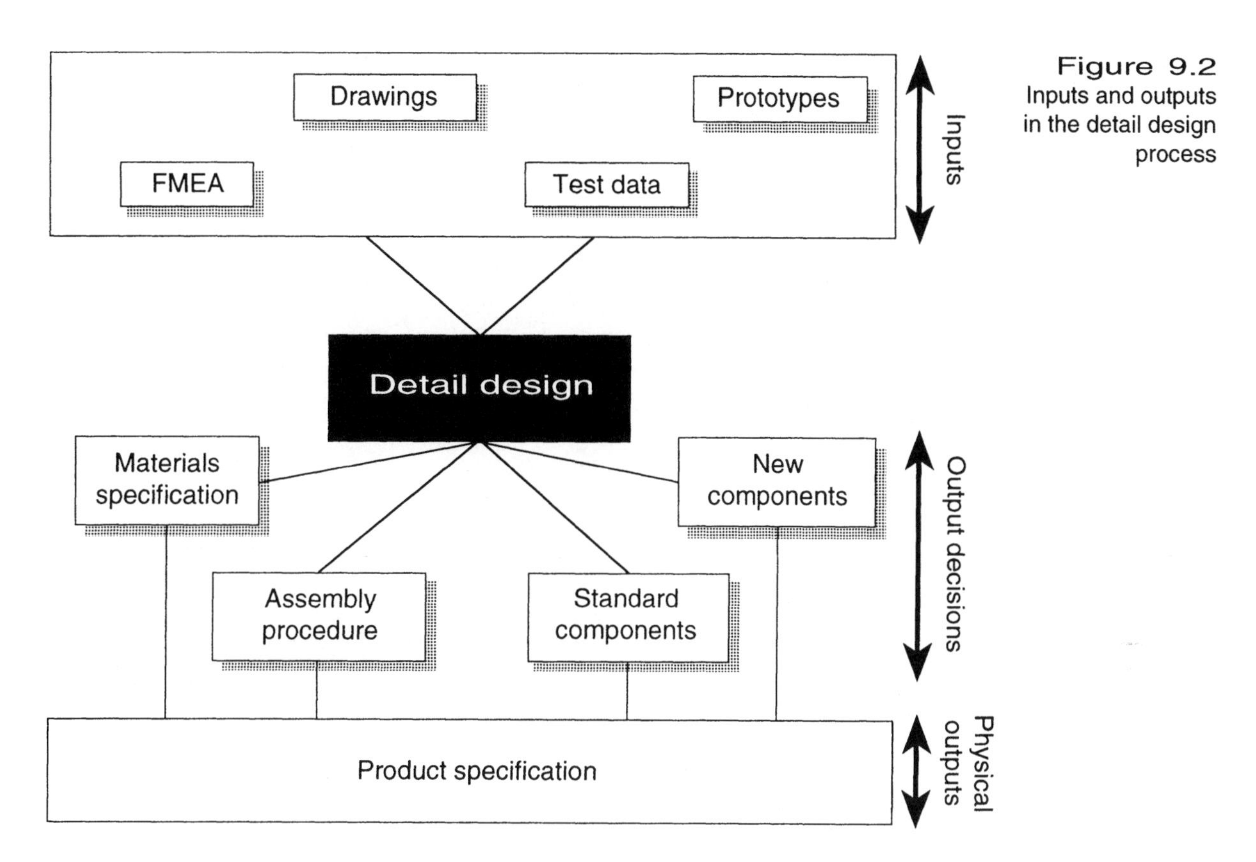

Figure 9.2 Inputs and outputs in the detail design process

specification has drawings, specifications and quality control procedures to show how this is achieved during production.

As we move from concept design, through embodiment design and into detail design, the design principles become progressively more specific to the materials and manufacturing processes involved. This presents a significant problem for a general book on product design, such as this. The only way to get round the problem is to present design techniques and procedures which are general to most types of product and give specific examples.

Product architecture

Concept design has left us with a set of functional and styling principles for the whole product. Embodiment design must now explore the arrangement of the elements of that product as a first step towards breaking the product down into components which can be individually manufactured. As always, in design, there is a temptation to rush in too quickly to specific details. Again, as always, this can wreck the gradual and progressive risk management process which takes us one step at a time from a general business opportunity to a specific manufactured product.

Feature abstraction

Figure 9.3 shows the procedure for avoiding this temptation. The styling and functional principles from concept design are firstly broken down into their key elements. This is a process of abstraction, similar to problem abstraction described in Chapter 4 (page 84). For the purposes of embodiment design, we no longer need to abstract the functions of the product: the basic function of the product has been decided in concept design. Now we need to consider product features which deliver that function. To continue with the example of the potato peeler, concept design selected a new peeler which has a fixed-type blade, with a swivel handle. The peeler gloves, potato tenderiser and potato grater, for example, have all been considered as functional options and rejected. Now we need to abstract the main features of the swivel handle peeler and explore how these features could be arranged into a working embodiment.

Five main functional features are abstracted from the swivel handle peeler (Figure 9.4). These are a blade, a depth limiter, a handle, a gouge and a swivel interface. Thinking through these features, it is decided that the blade and depth limiter are inextricably linked and together can be thought of as a single feature. Furthermore, it is decided that the swivel interface must be positioned between the handle and the blade and thus may be considered simply as part of the handle feature. This reduces the number of peeler features to three: a handle, a blade and a gouge.

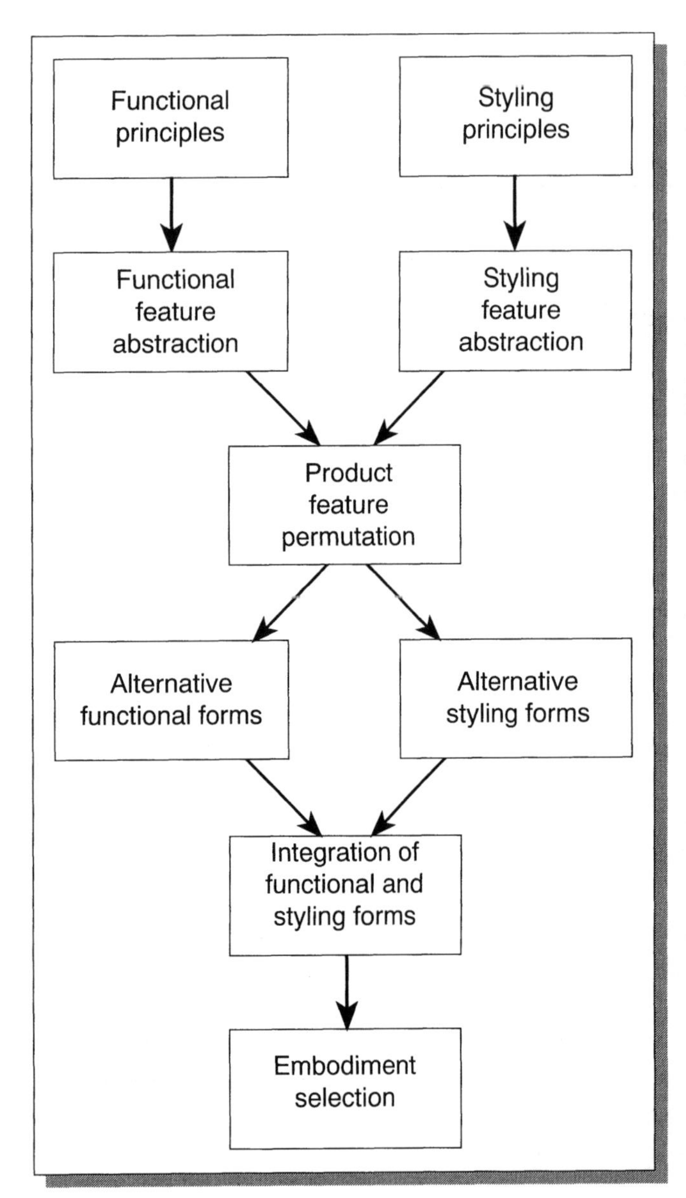

Figure 9.3
The process of embodiment design

Product feature permutation

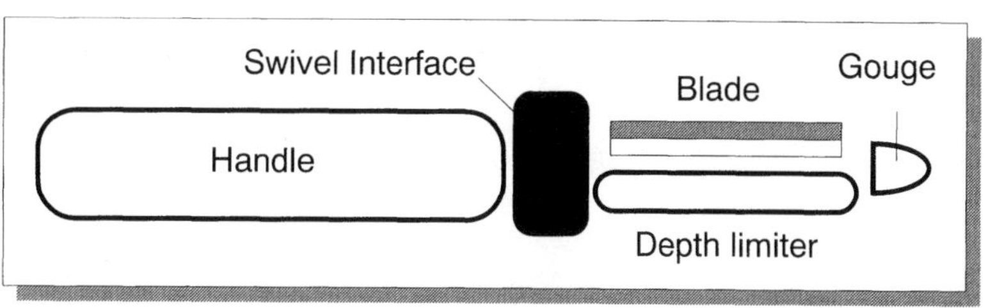

Figure 9.4
Feature abstraction for potato peeler

Product feature permutation is a technique which systematically rearranges the features of a product according to all their possible permutations. Pioneered by the Danish designer Eskild Tjalve [1] this simple, albeit laborious technique is a powerful idea generator. Figure 9.5 shows how the three peeler features, handle (H), blade (B) and gouge (G) can be arranged in all their various permutations. Not all are workable as a peeler; how, for example, could you embed a handle inside the blade and still be able to hold it? There are, however, 6 handle-blade-gouge permutations which look like they would work. These are expanded in Figure 9.6 to

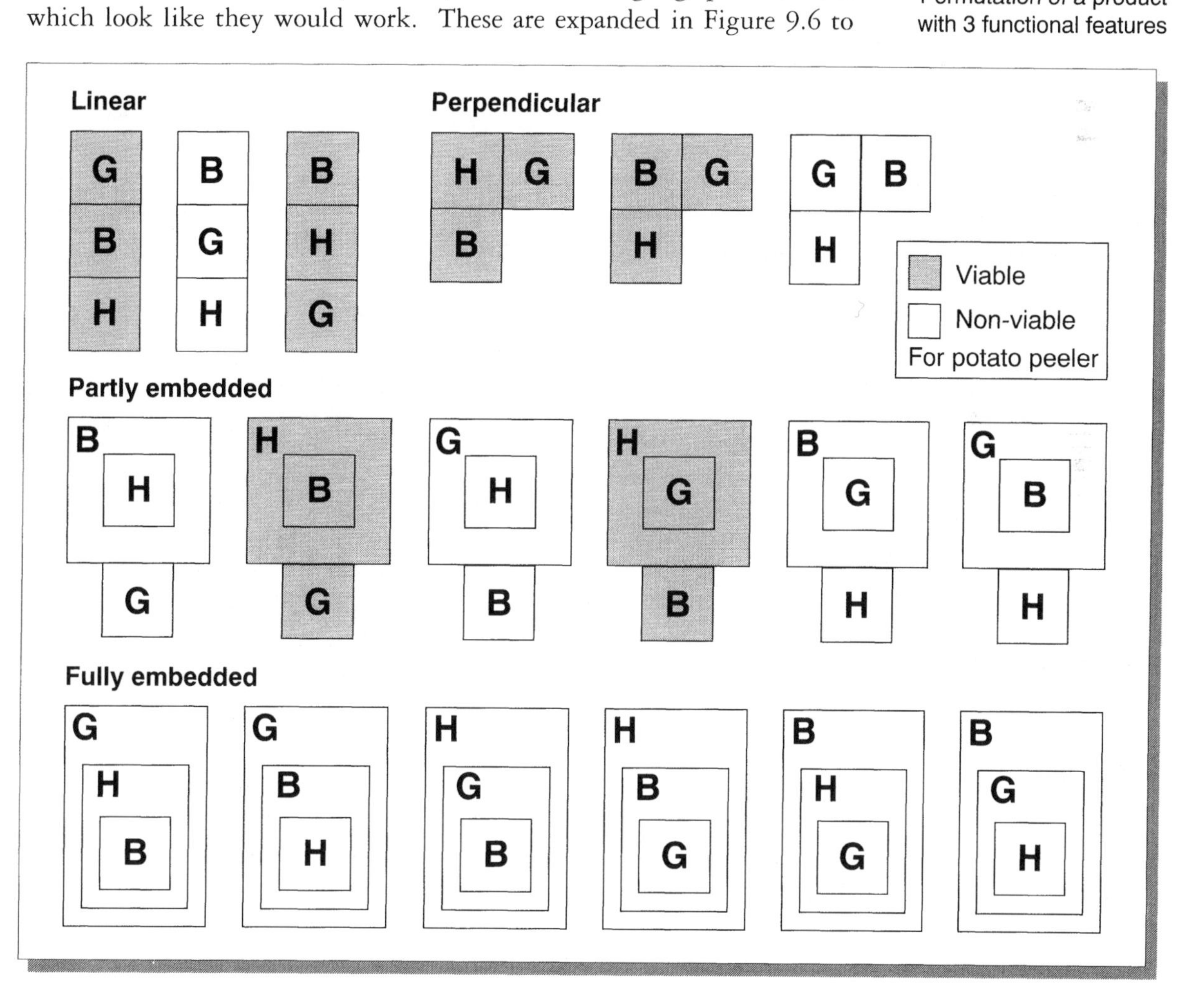

Figure 9.5
Permutation of a product with 3 functional features

consider the arrangement of parts as well as their configuration.

If we consider that existing peelers also can be designed with swivel blades, with either one ended or two ended supports, product feature permutation has generated a total of 48 different peeler embodiments (the 16 shown in Figure 9.6 with either fixed, one ended swivel or two ended swivel blades). Since Plasteck have constrained the concept design to a fixed blade swivel handle peeler, they are not able to make full use of this range of ideas. They select the 9 embodiments marked with a '*' for later consideration. With a more open-ended problem for embodiment design, however, product feature

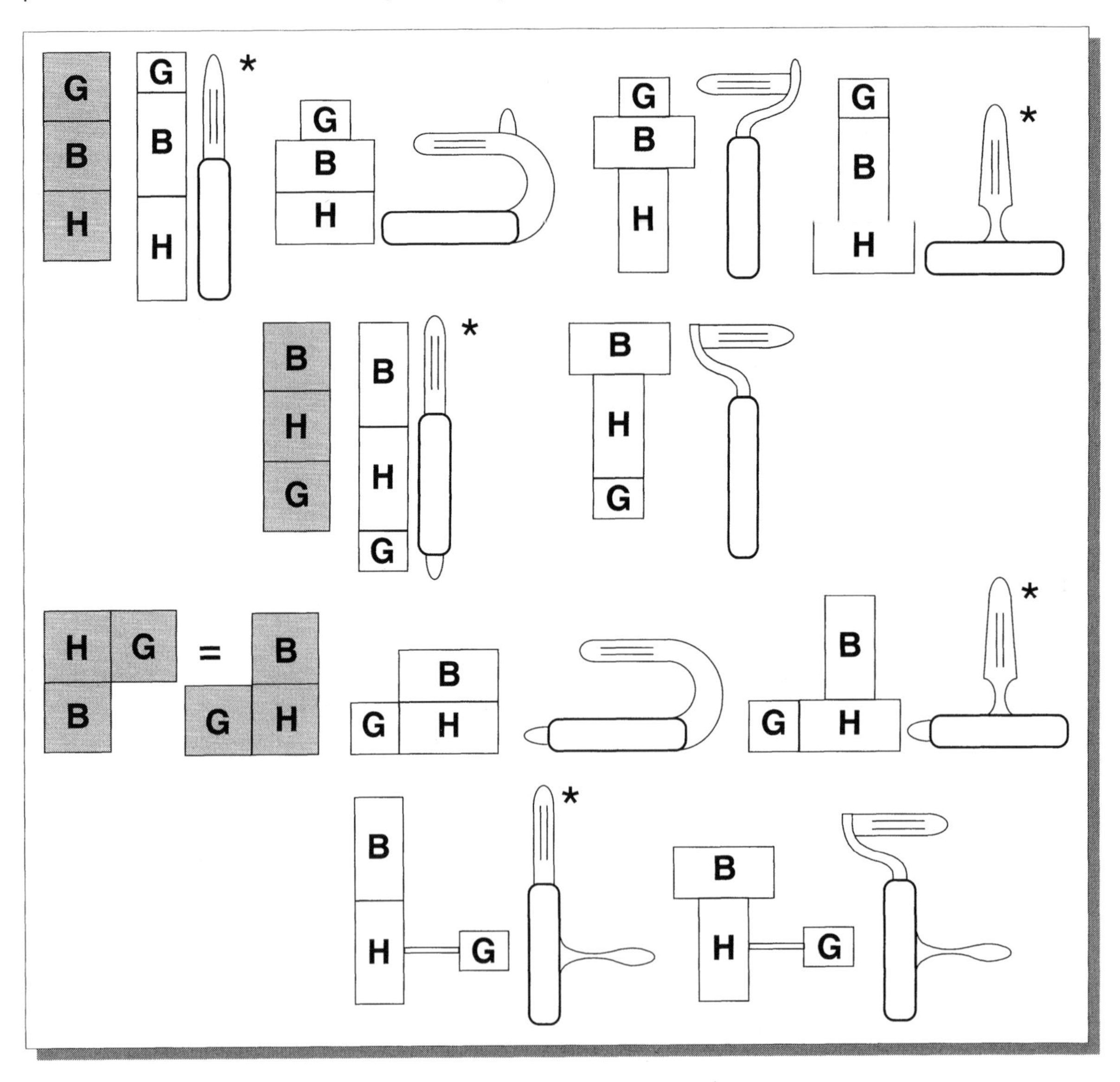

Figure 9.6 Permutation and arrangement of peeler features

permutation can be a powerful problem solving tool.

Another useful way of stimulating ideas about embodiment design is to use a product improvement checklist. This suggests general ways of changing any type of product and a variety of different checklists has been produced. One of the simplest is known by its acronym 'SCAMPER' (design toolkit page 90). This stands for 'substitute, combine, adapt, magnify or minify, eliminate or elaborate and rearrange or reverse'. This technique is most commonly applied when relatively minor changes to an existing product are sought. Imagine that Plasteck already manufactured a potato

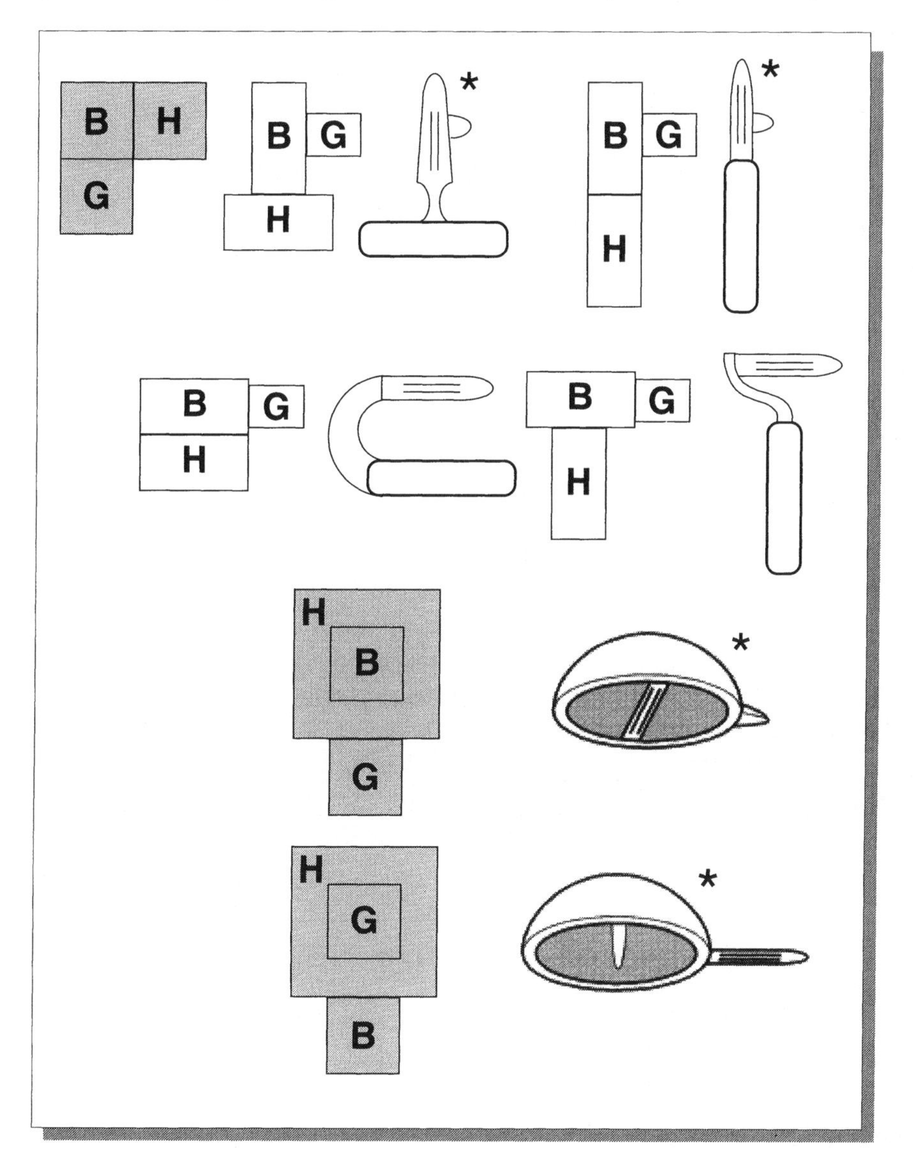

Figure 9.6
(cont'd)

Figure 9.7
The use of SCAMPER to develop embodiment ideas

SCAMPER	Peeler improvement	Benefit
Substitute	Different handle material	Rubber handle, more comfortable to hold
Combine	Combine peeling with other functions	Peeler with potato scrubbing brush
Adapt	Adapt peeler for other uses	Potato, carrot and asparagus (?!) peeler
Magnify	Longer cutting edge on blade	Better for peeling large potatoes
Minify	Fold away blade	Safer when stored in drawer
Eliminate	Eliminate plastic – have one piece metal peeler rolled to form handle	Striking all-metal appearance
Elaborate	Curved blade	Fits round curved surface of potato
Rearrange	Have blade at 120º angle to handle	Better ergonomics for peeling(?)
Reverse	Gouge at base of handle instead of at tip of blade	Easier to use

peeler but they had decided they needed to improve it. They might have particular customer needs or even customer complaints in mind which would direct their attention to particular types of improvement. Let us take the example of a simple product differentiation exercise. Customers are faced with as many as 12 different peelers to choose from in any particular shop. How could Plasteck differentiate their peeler so that it stands out more from the 'crowd'? Scamper takes you through the 9 different types of product improvement with the aim of stimulating ideas in 9 different directions (Figure 9.7).

Design integration – the heart of embodiment design

Now that we have ideas for embodiment design generated from product feature permutation and SCAMPER, we come to the serious stuff! Real design is all about getting to the point that something can be made, and that time has now arrived. All the ideas about customer needs, functional and styling concepts and possible embodiments have got to be brought together and integrated into a single embodiment. Something you can touch feel, test and probably throw away and start again. But getting to the stage of making is the bit that most designers feel is the real achievement in product development: especially if you have exercised the self-control required to get through the previous chapters without saying 'risk management my ass, I'm just going to make it!'

The concept of design integration can be seen at two different levels. At its broadest, design integration refers to the design of a single embodiment which contains both the functional and styling principles derived for the new product. At its narrowest, design integration means building all the required features into the simplest and most elegant embodiment possible. Let us begin by exploring the broad interpretation of design integration , by examining the Plasteck peeler.

We have three functional principles we wish to embody in the new Plasteck peeler: we want it to have a fixed type blade; we want it to have a swivel handle; we want it to have a gouge that does not require the user to grip the blade to use it. We also have three styling principles: we want it to look clean and hygienic; we want it to look sharp; we want it to look functional and utilitarian. Embodiment mapping is a technique which takes all of the required principles and encircles an otherwise blank page with them. The space they leave in the middle of the page is where you start sketching embodiments which satisfy all of the surrounding requirements. It often helps if you take this in stages. Try firstly sketching or describing the elements of the product which meet the individual requirements. Then take these elements and integrate them into a single whole product. Figure 9.8 shows these two stages for the potato peeler. Of course, this is not expected to be a 'right first time' procedure. Once you have the functional and styling principles (the outermost boxes in Figure 9.8) laid out on the page, photocopy it 10 or 20 times and try to generate 10 or 20 different integrated embodiments which satisfy all the functional and styling principles. Use product feature permutation and SCAMPER to help you come up with different embodiments. Then use a selection matrix (see p 169) to fit the one(s) which match the design specification best.

Design integration in its narrower sense turns an effective and workable embodiment into an elegant and streamlined design. The potato peeler is not a good example here, because, with an injection moulded handle and a stamped steel blade, it is already well integrated. A simple and highly elegant example is, however, provided in the design of an accessory for a mountain bike. Mountain bikes are currently hardly ever sold with a bicycle

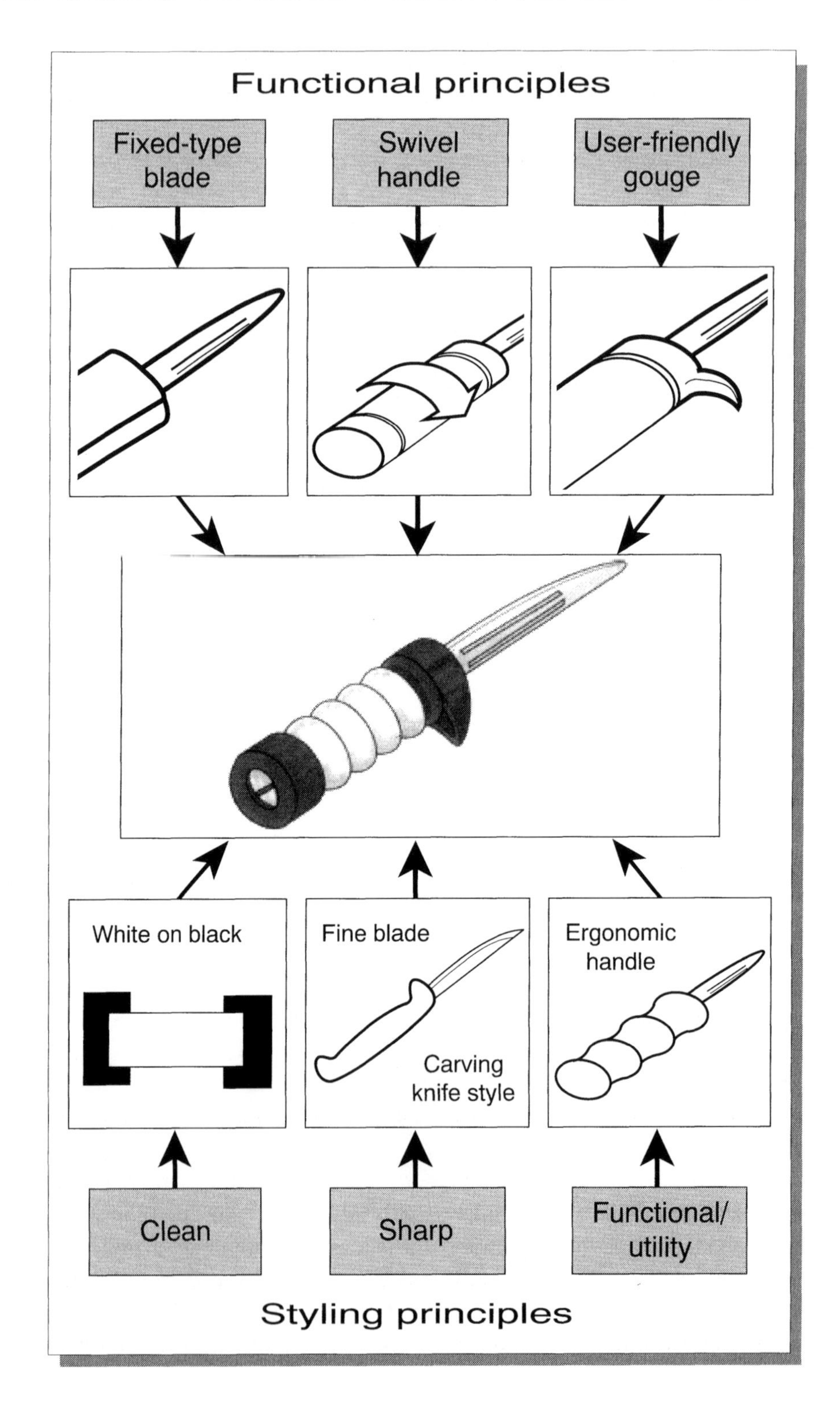

Figure 9.8
Embodiment mapping to integrate functional and styling principles

pump. The pump is an added accessory by which the cycle industry extracts even more money from its customers! As a result, mountain bikes are not designed to have any means of attaching the pump to the bike. For manufacturers of pumps, some means of attaching the pump to the bike is, therefore, necessary.

At a conceptual level this requirement is simple. A mechanism of some sort is required to grip the tubular frame. This must be adjustable, in order to fit the different diameters of tubular frames. Attached to this gripping mechanism there needs to be a spike, to fit into the hole at the end of the pump. And there need to be two of these devices to support either end of the pump. These requirements are shown as a conceptual principle in Figure 9.9.

The first embodiment sketched out to meet these conceptual principles uses a jubilee clip to attach a pair of spikes to the bike frame. This is found to work well, but it is thought that the embodiment design could be better integrated to provide a more elegant solution to the problem. Analysing the jubilee clip with spike attachment, it is found to consist of 5 components (Figure 9.10). As a first step towards design integration, this can be reduced

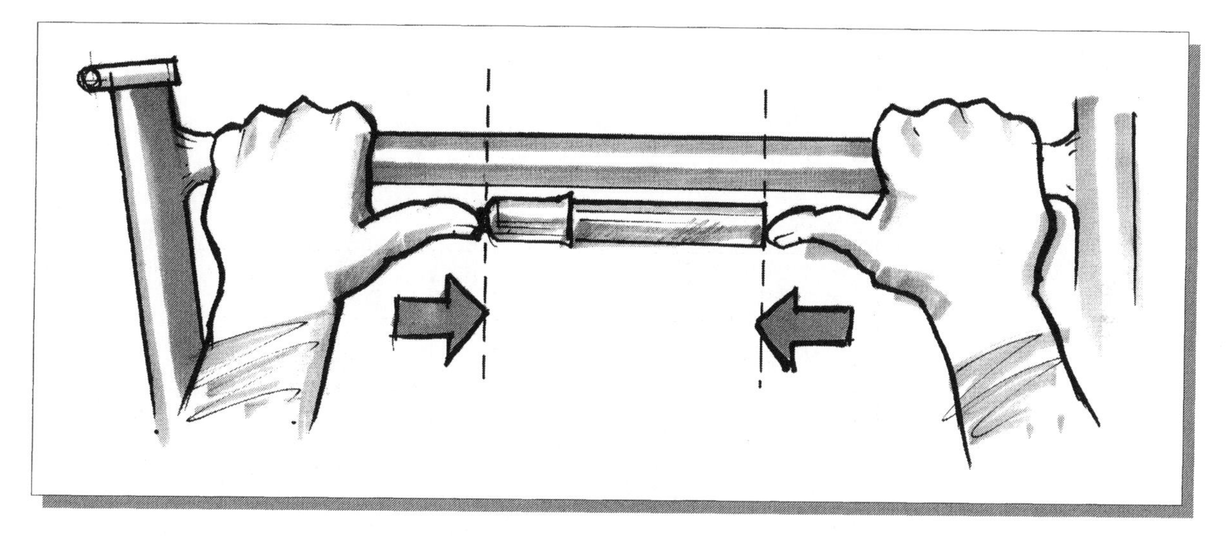

Figure 9.9 The conceptual principle for a pump clip

to three components by having a moulded screw housing with an integral spike. Developing the idea further reveals that this integral screw housing and spike could probably be extended so that the strap is also part of the same moulding. Five components have been reduced to two. The final integrating step would be to mould the screw as well and produce the entire product as a single injection moulding. This is achieved as shown in Figure 9.11. The single moulding produces the strap, screw housing and screw. To attach the pump clip, the tail of the strap is rolled round the bike frame and inserted

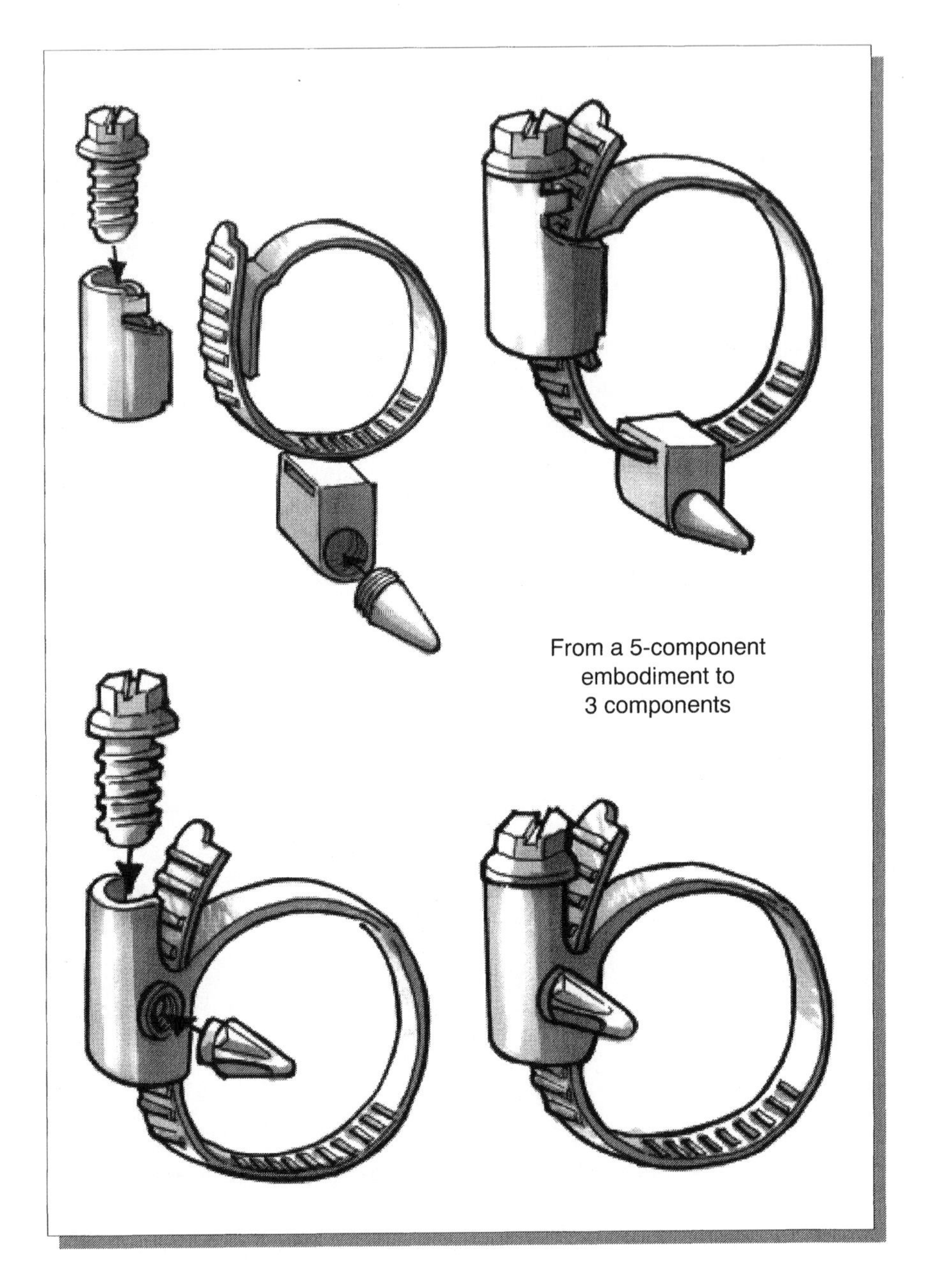

Figure 9.10
The first step in design integration of the bicycle pump clip

into the slot in the screw housing. The screw is then snapped off the moulding, inserted in its locating threads and tightened into position. As a final but very important aspect of design integration, analysis of the detail design of the product revealed that it could be manufactured using a simple two part injection mould tool (Figure 9.12).

So, by design integration, a moderately complex and comparatively expensive

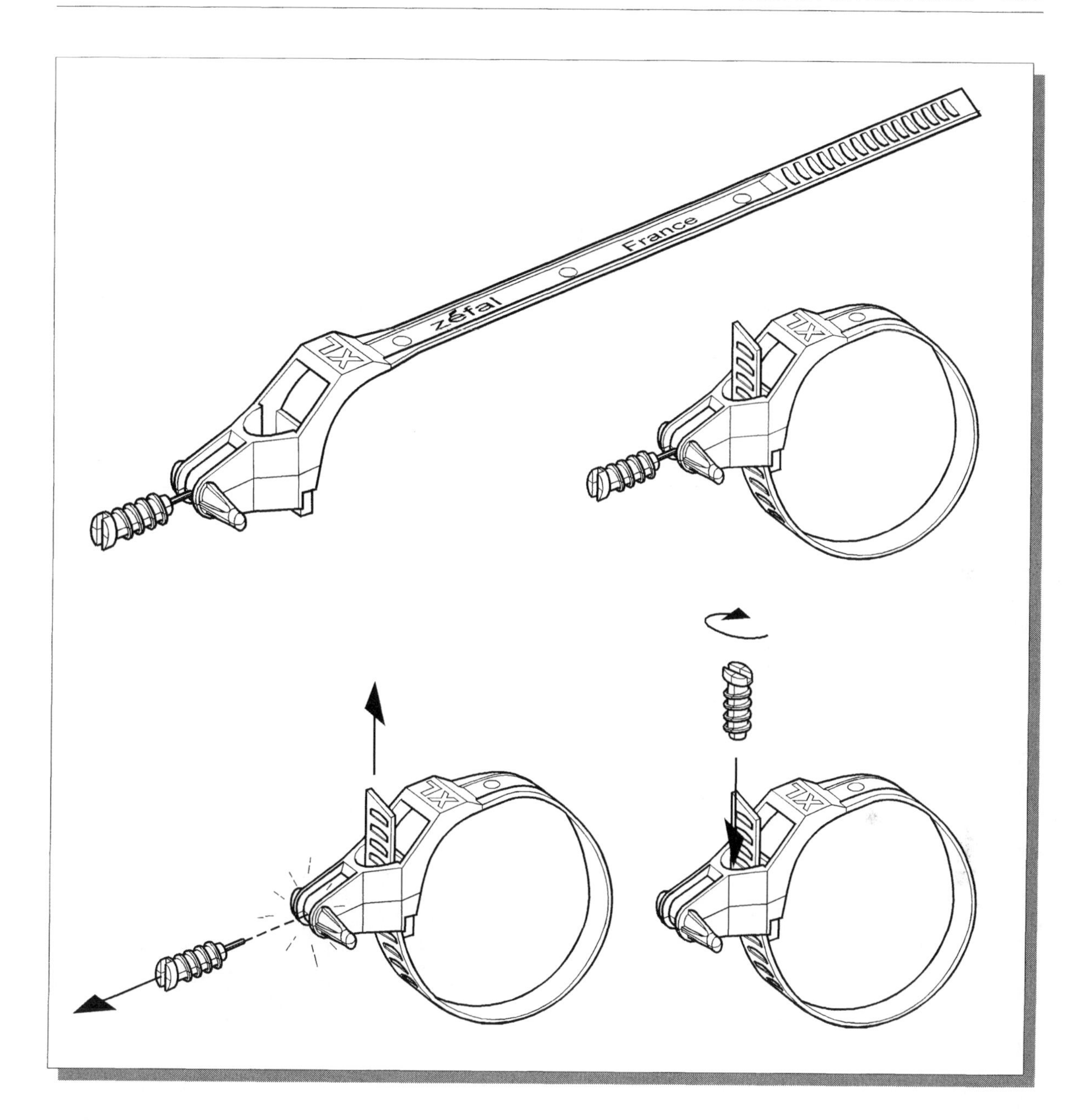

Figure 9.11
Reducing the product to a single component

product has been reduced to an elegantly designed, inexpensive product. Clearly, the process of design integration will become very much more involved, and hence more difficult to do, with more complex products. Often, complex products have to be broken down into sub-systems before design integration can be tackled effectively. The principles are, however, the same, taking us back to the risk management funnel. The opportunity

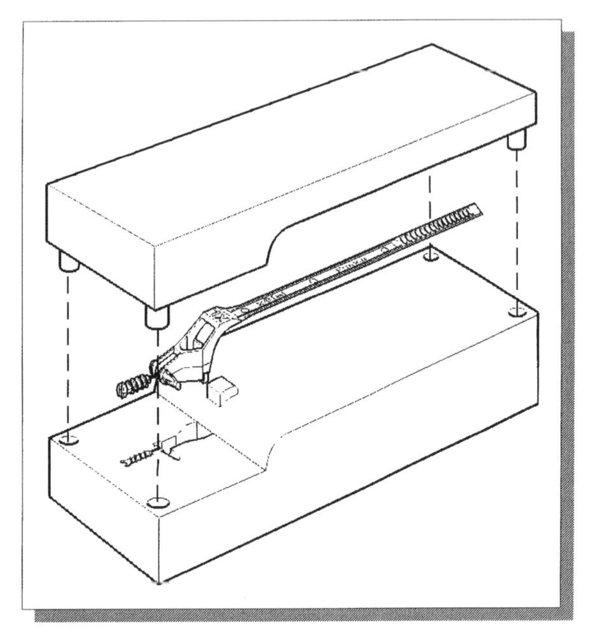

Figure 9.12
Two part mould for pump clip

and design specifications remain the key sources of targets and quality control criteria for embodiment design. These, however, are translated into a set of functional and styling principles in concept design. These conceptual principles provide the input to the process of embodiment design, which goes through three main stages. Firstly, embodiment ideas are generated, using techniques such as product feature permutation and SCAMPER. Secondly, the ideas generated are integrated, broadly to pull together both functional and styling requirements. Then, thirdly, the features of embodiment design are integrated to turn a workable and manufacturable embodiment into the most simple and elegant solution possible.

Figure 9.13
Finished product in use, as manufactured by Zefal, France.

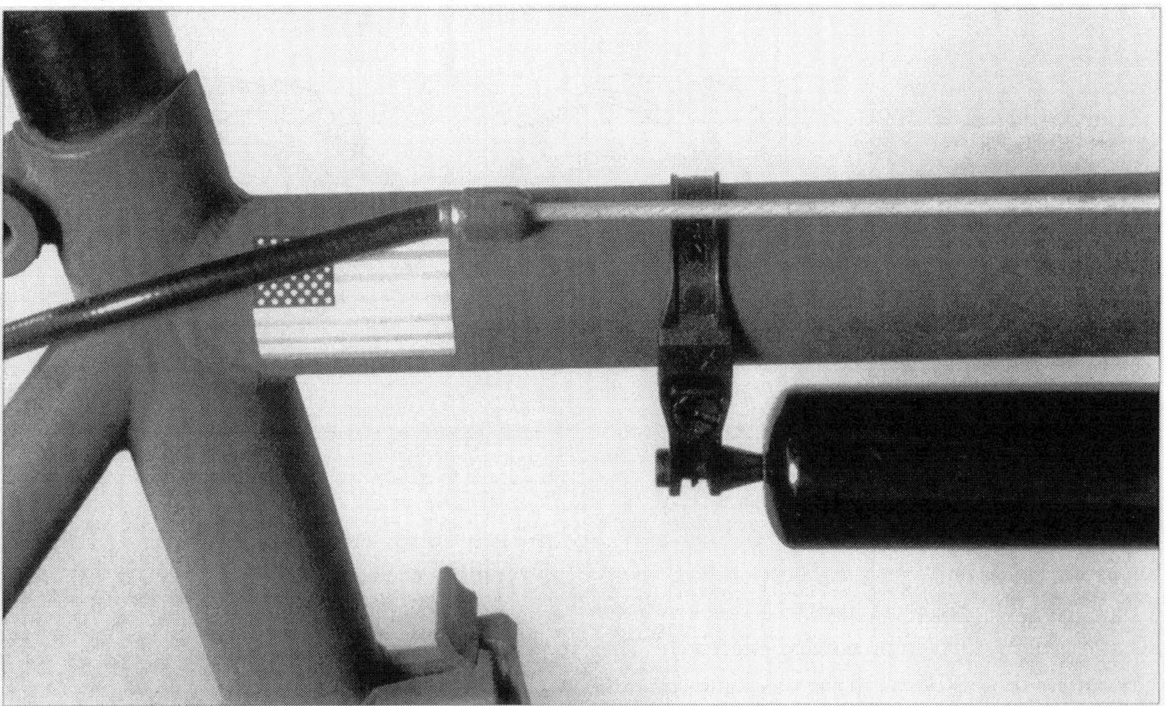

Embodiment prototyping and testing

Having reached the point at which an embodiment solution has been found, it is now time to check that this embodiment is fit for its intended purpose. This involves prototyping and testing the new product. Prototyping is a key aspect of product development but it is an activity that can take up a disproportionate amount of time relative to the value which it adds to the design. Designers often get into the habit of developing prototypes at particular stages of the design process, as a way of marking their arrival at a milestone (e.g. completing concept design). Clearly, if these prototypes are not essential, they could be wasting a lot of valuable time.

Prototypes fulfil several functions in product development (Figure 9.15). They can act as communication media for discussions with clients, potential customers or other company staff. They can help the designer to develop new product ideas, especially when the new product is complex and highly three-dimensional and, therefore, difficult to visualise on paper. And they can be used to test products or components in order to verify designs. Obviously, different types of prototype are required for these different functions. A simple card or foam model may be sufficient for design development purposes and for some types of communication. These models would, however, be completely inappropriate for functional testing. A categorisation of different types of prototypes is given in Figure 9.16.

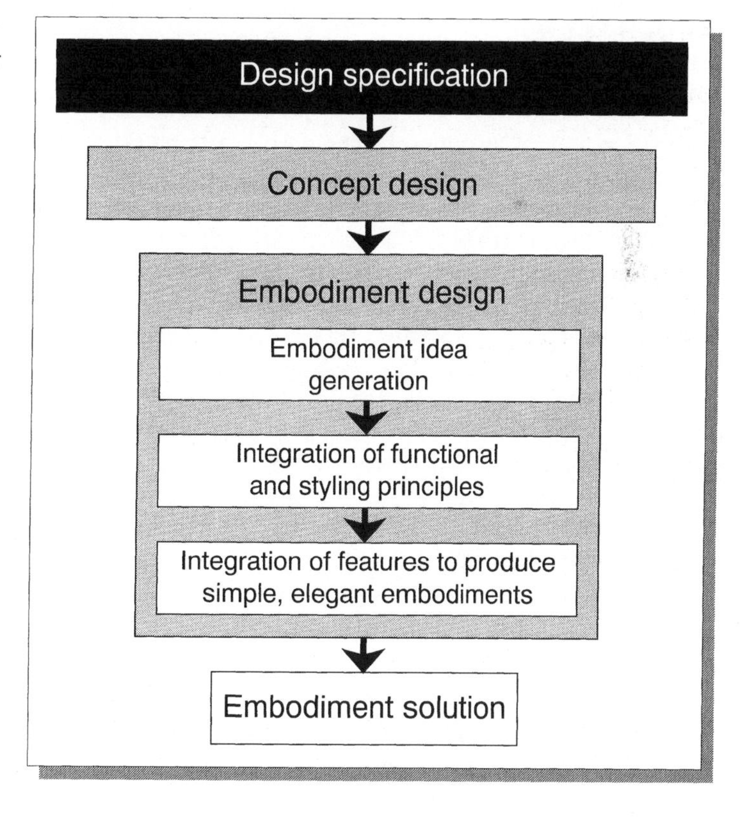

Figure 9.14 Risk management within embodiment design

Principles for prototype development

The general rule in developing prototypes is 'only do so when necessary'. Prototyping, as we have already said, is a time consuming activity and inevitably diverts effort from other activities which add more value to the design of the emerging new product.

Purpose of models	Applications
Communication • with customers • with managers • with other members of the product development team	Novel applications
Design development • developing design ideas by working in three-dimensional forms with models • especially during design for manufacture and assembly	Complex geometric forms or complex working principles
Design testing and verification • short-term - normal operation • short-term - extreme operation • long-term - lifetime operation	Risk reduction based on Failure Modes and Effects Analysis

Problems with words:

The term 'model', in general technical use, is a physical or mathematical representation of an object or of a more abstract system, such as the weather or human consciousness. In product design, a model ***should*** refer to any representation of a product or part of that product. In fact, the word 'model' is usually taken to refer to only computer models (e.g. a 3D rendering of a product in a CAD or graphics programme, a finite element analysis programme which models the strength of a product) or physical representations of the visual appearance of products. These appearance models are also sometimes called maquettes, the french word for the small, preliminary wax or clay model used by sculptors

A prototype is literally the 'first of a type'. In the early days of manufacturing, the prototype was the master copy of a product which was later mass produced. In product design, the word prototype refers to two types of representations of products. The first, and most accurate use of the word refers to physical representations of the product as it will eventually be manufactured. These are pre-production and production prototypes. The second use of the word prototype refers to any physical representation developed for the purpose of physical testing.

Figure 9.15
Why model and prototype?

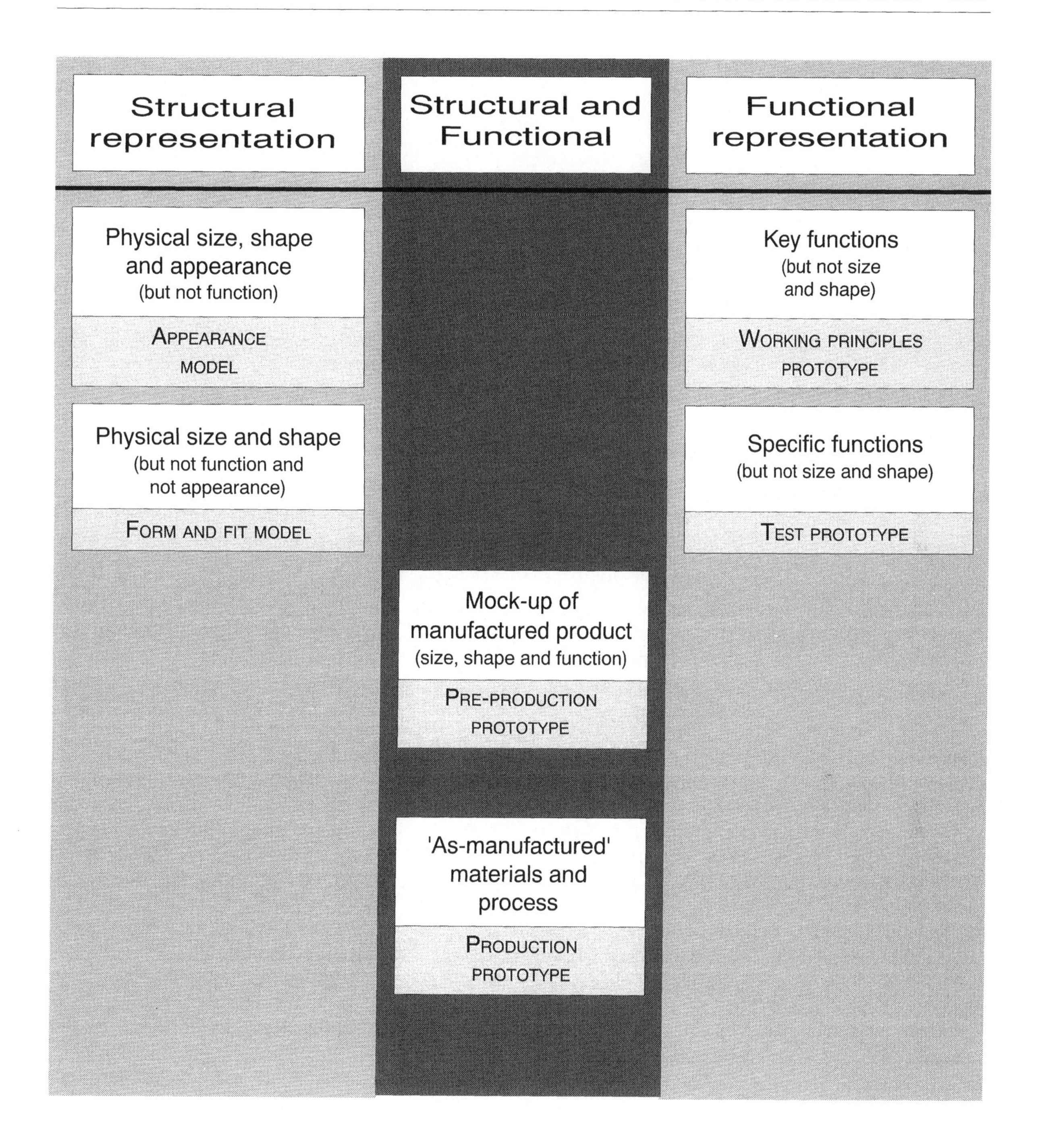

Figure 9.16
Types of models used in the design process

- Only prototype when you cannot get the necessary information from other sources.
- Never prototype when sketching or rendering would do instead.
- Only ever develop prototypes to the minimum degree of complexity and sophistication required to obtain the answers you need.

In Chapter 2 Robert Cooper's 'gambling rule' was introduced (p 10). This suggested that the stakes should be kept low when the risks are high and that the stakes should only be increased when the risks reduce. In relation to product prototyping, this means keeping the prototypes as simple and inexpensive as possible during the early stages of the design process, until you can be confident that you have a commercially viable product. This simplicity has an added benefit early in the design process. During the early assessment of a new product idea, you wish to know in principle whether customers, management or production engineers think the product is a good idea. To do so you need only give them an outline of what the product will do and how this is different from other products on the market. This could be achieved by means of a sketch, rendering or simple block model. If you present a very detailed prototype made to look like a manufactured product, the response is likely to relate to the details of its design. Will it be that colour? Is this part strong enough? Isn't it a bit big? Of course, these are exactly the sorts of response you do ***not*** want when trying to find out if the principle is viable.

For reasons of both risk management and obtaining relevant answers to the questions being asked, the complexity and sophistication of prototypes should increase progressively as the product develops. Figure 9.17 shows this progression and illustrates it with a series of prototypes used in the development of a typist's copy stand.

Testing for product failure

The most difficult decisions about the use of prototypes concern their use for testing product failure. How do you know what to test and what sort of failures to look for? Deciding whether or not to prototype for communication or design development purposes is comparatively straightforward. You know who you need to communicate with and what you need to inform or persuade them of during that communication. You are, therefore, in a good position to determine if the effort required to make a prototype is worthwhile. Similarly, you know where you are trying to get to in developing a new

Stage in the design process	Assessment required	Results sought	Test materials
Concept design	Market testing of functional or styling principles	Is the concept understood and are the underlying principles appreciated?	Sketches or renderings of product concepts
		Is the novelty of the proposed concept valued?	Story-boards of point of sale information or of product use
		How does it compare with the underlying principles of existing products?	Simple block models
Embodiment and detail design	Design development	Form and fit of component parts	Card, foam, wood or plastic models
	Product failure modes	Evaluation of predicted product failure modes and development of design or production changes required to avoid them	Physical and structural prototypes of parts or functions of products, designed for specific tests
	Manufacture and assembly testing	Can the product be made and assembled to specification?	As-manufactured components – often by means of rapid prototyping techniques
	Market testing of product	Does the product offer sufficient value for its price?	Either as-manufactured prototype or production prototype.

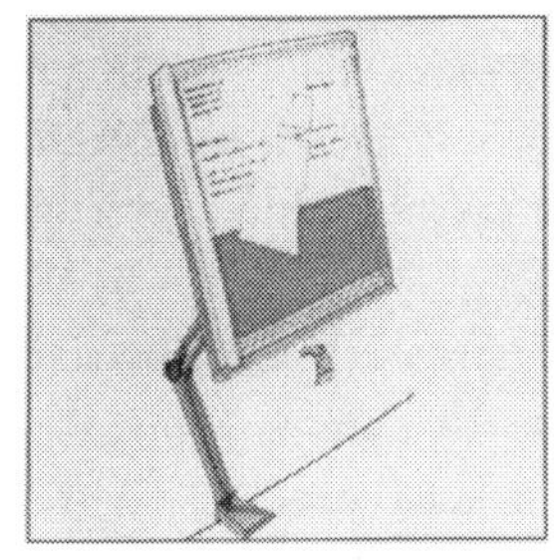

Concept rendering for initial market testing of principles

Foam model for form and fit of components

Card model for early embodiment design

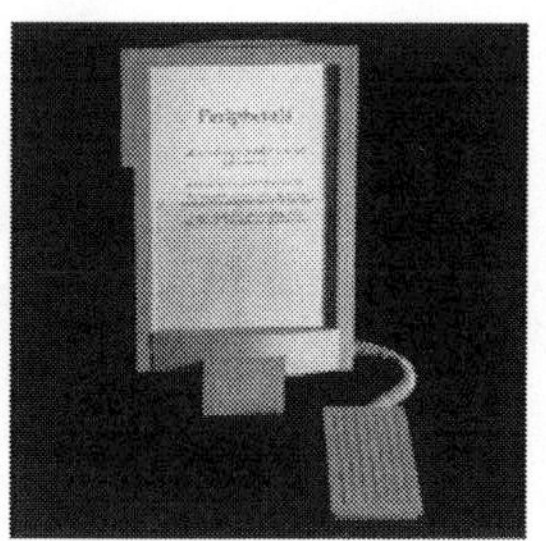

As manufactured prototype for market testing of product

Figure 9.17
The progression from simple to detailed prototypes as products develop [3]

product and can work out whether a three-dimensional prototype is likely to get you there faster or more creatively than working with two dimensional media. Failure testing is much less under your own control. Materials can fail in all sorts of ways, leading to different failures in the individual components. The interaction (e.g. joint) between components can also fail in several different ways and adding up all these various specific failures can result in a multitude of different failures of the whole product. In addition, the product may remain exactly as it was designed to be yet fails in ways not anticipated in the design specification. The approach to failure analysis that I have just described is a classic bottom-up analysis. It starts with the smallest details of the product and tries to work up to predict failures of the whole product as a consequence. This type of analysis can lead to several practical difficulties. Firstly, failures are always dependent upon the use to which the product is being applied. This is difficult to integrate into a failure analysis that begins with the potential failures of a particular type of material or specific component. In other words, looking at product use is a top-down analysis and it is difficult to fit with the sort of bottom-up analysis we are discussing. Secondly, a bottom-up analysis may predict failures which have little or no consequence for the customer. There may, for example be features of a product which are important for its assembly but are unimportant once the product is assembled. Failure of that feature is, therefore, of no relevance to the customer.

Failure modes and effects analysis

It is because of these very difficulties that Failure Modes and Effects Analysis (FMEA) was developed. It is a top-down failure analysis which starts with the functions of the product valued by the customer. It then goes through a number of systematic stages in which the potential failures of these functions are identified as well as their causes, the occurrence and severity of these failures (as they affect the customer) is estimated and then the likelihood or ease of detection of the failure mode is identified. From all of this a single number is calculated to reflect the importance of each individual failure mode as perceived by the customer. This is called a risk priority number and, as the name suggests, it allows the designer to prioritise action in order to detect or prevent the most important failures. An important part of the action resulting from failure modes and effects analysis will be to prototype the product to verify the expected failure modes and discover how to avoid them. Failure modes and effects analysis will therefore, play a key role in determining what prototyping is required for a product during its development.

Let us examine the failure modes and effects analysis for the proposed new potato peeler. Since a full analysis is a lengthy process and ends up with a lengthy document, this example will include only the failures expected from one of the peeler's functions: the new swivel handle.

The first step is to conduct a product function analysis. This has already been done for the peeler (Chapter 7 p 211) and used for both concept generation (p 212). From this, the function 'swivel handle' (described as swivel blade in the original function analysis before the idea of the swivel handle had been generated) is selected.

Figure 9.18 Failure modes and effects analysis for a potato peeler

Function	Potential failure mode	Cause	Occurrence	Effect	Severity	Design verification	Detection	Risk priority number
Swivel handle	Swivel jams	Moulding flash	8	Handle fails to swivel	4	Precise tooling, check mouldings	2	64
		Tolerances too tight	3	Handle fails to swivel	4	Prototype and test	4	48
		Swivel blocked by dirt	6	Handle fails to swivel	4	Prototype and test	9	216
	Handle pinches user	Tolerances too loose	3	Pain to user	8	Prototype and test	4	92
	Handle unhygienic	Tolerances too loose	3	Peeler unusable	6	Prototype and test	4	72
		No dirt flushing route	6	Peeler unusable	6	Prototype and test	9	324
	Swivel casing breaks	Low impact resistance	2	Loss of swivel	4	Calculate impact strength	3	24
		Moulding flaws	2	Loss of swivel	4	Mould flow analysis	3	24

From embodiment to manufactured product

With the embodiment of the product designed, evaluated against the design specification and 'failure-proofed', it only remains to design the product for manufacture. At this point, a book such as this, with its emphasis on design for all types of manufacturing, outlives its usefulness. Design for manufacture, by its very nature, has to be specific to the manufacturing process in mind. Of course, the intended manufacturing process should have been considered in the latter stages of concept design and the early stages of embodiment design (as discussed on p 272). But now, with embodiment design completed, detail design must be almost entirely specific to the manufacturing process.

To help with this aspect of the design process there are two types of information available. Firstly, there are general books on manufacturing systems (e.g. Kalpakjian [5]). Secondly, there are books focusing specifically on detail design and design for manufacture. The 'bible' amongst these latter type is Boothroyd, Dewhurst and Knight's 'Product Design for Manufacturing and Assembly [6].

Key Concepts

From concept to working prototype

Embodiment design takes a set of functional and styling principles (the outputs from concept design) and develops them to the point at which a full working prototype is (or could be) made.

Idea generation then idea selection

The stages of embodiment design follow the key principles of all design innovation: generate lots of ideas for possible solutions and then select the best to meet the design specification.

Design integration

The key to successful embodiment design is design integration. This occurs at two levels. Firstly there is the integration of functional and styling principles into a single product. Secondly there is the integration of all the functional and styling features of the product into a design with the minimum of complexity, minimum of parts, minimum of cost and maximum ease of manufacture

Prototyping guided by FMEA

Once the best embodiment design is selected, a FMEA is conducted, to examine the potential failure modes for the design. Then prototypes are developed to test the identified failure modes and establish how the design can be improved to avoid them.

Design Toolkit

Failure modes and effects analysis

Failure modes and effects analysis is a method of systematically appraising the potential failures of a product and ranking their importance. As the name suggests, this analysis considers separately the modes of product failure and their effects on the customer. It is known as a 'right first time' design method and sets out to anticipate potential failures during the design process before it is too late to do anything about them. The output from FMEA should be seen as a prioritised menu for change to be implemented during the subsequent stages of the product development process. The action suggested directly from the FMEA can include design changes and improved failure detection procedures. Other less direct consequences of FMEA can include revision of the design specification or, more dramatically, the termination of the entire product development project because of previously unforeseen and unresolvable product failures.

In principle failure modes and effects analysis uses functions identified within the product function analysis to work out all the different ways in which a product could fail to perform its intended function. This set of potential failures constitutes the failure modes for the product. The causes of each failure mode are identified and their likelihood of occurrence scored on a 1-10 scale. Then the effects of each of the failure modes are listed and their severity ranked on a 1-10 scale. Then methods of design verification to try to avoid or prevent the failure are examined and the effectiveness of these is scored on a 1-10 scale. From the three scores for occurrence, severity and detection, a 'risk priority number' is calculated by multiplying the scores together. Finally, recommendations for action can be drawn up to overcome the causes of failure, taking into account the risk priority of each.

The elements of FMEA?

1. Product function analysis. The failure of a product, broadly speaking, is the non-performance of a function. The first step in any failure modes and effects analysis is, therefore, to conduct a product function analysis using the method described in the design toolkit on p 236.
2. Failure modes. The functions identified within the product function analysis are then examined one by one and failure modes identified for each. A failure mode is written down as a simple two or three word statement of one way in which the function could fail to be performed. Failure modes are described in physical terms - what does the product 'do' or 'not do' in order to fail. For many product functions there are likely to be several failure modes. The warning buzzer on an electronic instrument,

Failure modes and effects analysis

for example, has the function 'give warning'. It could fail by 'failure to activate', 'always activated', 'insufficiently audible' or 'failure to reset'.

3. Cause of failure There may be several causes of a mode of failure for a product. Returning to the example of the warning buzzer on a piece of electronic equipment, its failure mode of 'always activated' could be caused by the sensor being continuously triggered, by incorrect wiring or by a short circuit. These causes should be described in physical terms and as briefly as possible. In general, the causes of product failure are of two types. Firstly there are failures caused by the product or component being manufactured or assembled incorrectly (i.e. not to specification). Secondly there are failures which occur in a product made entirely to specification.
4. Occurrence of failure The likelihood of occurrence of the different causes of product failure are ranked on a 1 to 10 scale, in a similar way to failure severity (see above). A likelihood of occurrence table is drawn up, appropriate to the product being analysed. A high likelihood of occurrence is given a high ranking, since it constitutes a more serious product

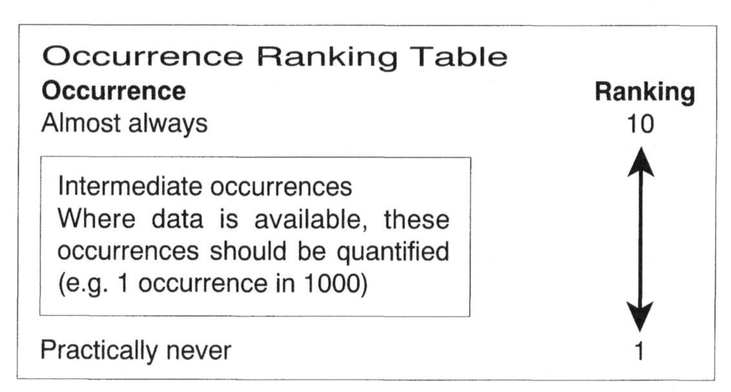

failure. A generalised likelihood of occurrence table is given below.

5. Failure effects The next stage in FMEA is to identify the effects of each failure mode. An effect of failure is its consequence, particularly as perceived by the customer. There are many sorts of failure effects and these can be considered in a roughly hierarchical order; i) its effect on a part or component of the product (e.g. it vibrates), ii) its effect on the entire product (e.g. it ceases to operate) iii) its effect on the customer (e.g. safety risk or customer dissatisfaction) iv) its effects beyond the customer (e.g. fire risk, environmental pollution or non-compliance with standards or

Failure modes and effects analysis

regulations). The description of the failure effects should be as concise as possible although sufficient information must be given to allow the severity of the failure to be judged. An overheating electrical component, for example, may 'risk burning the user' but does this mean a slightly burnt finger or setting fire to the whole house?

6. Failure severity The severity of each failure effect is ranked on a 1-10 scale with minor effects ranked 1 and serious effects ranked 10. This is usually achieved by preparing a simple and common-sense table of failure effects of different severity. These will inevitably differ enormously for different types of product; consider the most sever consequences of failure for a paper clip on the one hand and a nuclear power station on the other! A generalised severity ranking table is given below.

Severity Ranking Table

Effect	Ranking
Serious personal injury (requires medical assistance)	10
Moderate personal injury (requires first aid)	9
Minor personal injury (no attention required)	8
Irreparable failure of prime product function	7
Serviceable failure of prime product function	6
Intermittent operation	5
Inconvenient/uncomfortable to use (inhibits use)	4
Irreparable failure of secondary product function	3
Serviceable failure of secondary product function	2
Inconvenient/uncomfortable to use (irritation)	1

7. Design verification The significance of product failure can be very different if the problem is likely to be noticed before the product is sold to the consumer. This gives the option of either rectifying the problem or replacing the product. This part of the FMEA outlines the design verification procedures which may detect the different failure modes or their underlying causes. Design verification may be a desk exercise in which the design can be evaluated against known performance or safety criteria. The strength of different materials, for example, is widely available. Whether or not a component is likely to fail because of the forces imposed upon it may, therefore, be verifiable by standard engineering calculations and reference to the technical specification of the chosen material. More commonly, design verification will comprise a testing procedure, conducted before the product leaves the factory. This could be

Failure modes and effects analysis

anything from a simple operation test (e.g. switch it on and see if it works) to a lengthy and complex durability test (how many times can the product be used before it fails). The design verification procedures entered into the FMEA must be realistic rather than idealistic. There is no point in describing every imaginable test that could be performed if they are never going to be.

8. Failure detection This ranks the likelihood of detection of the failure resulting from the stated design verification procedures. As with failure severity and failure occurrence, a ranking from 1 to 10 is used, with 'almost impossible to detect' having a ranking of 10 and 'almost certain to be detected', a ranking of 1.
9. Risk priority number The summation of all the information collected on product failure, during FMEA is expressed as a single number - the risk priority number. This number represent a simple logic; the risk to the manufacturer, of anything from customer dissatisfaction to product liability prosecution, increases with increasing severity of the effects of the failure, increasing likelihood of occurrence of product failure and decreasing likelihood of detection of product failure. The risk priority number is calculated by multiplying the severity ranking, occurrence ranking and detection ranking together. The higher the risk priority number the greater the risk to the manufacturer and the greater the need for action to rectify the problem.

The written format of any FMEA is kept as simple and compact as possible. The normal arrangement is to compile a table with the elements of FMEA along the top of the table and the different failures described in columns down the page. The table below shows an example layout of a FMEA and an example is given on page 290.

Function	Failure Mode	Effect of Failure	Severity	Cause of failure	Occurrence	Design verification	Detection	Risk priority number	Action
Description	*Description*	*Description*	*Rank (1-10)*	*Description*	*Rank (1-10)*	*Description*	*Rank (1-10)*	*Calculation*	*Description*

Notes on Chapter 9

1. Still the most thorough and comprehensive text on embodiment design, even although it is now somewhat dated, is Pahl G. and Beitz W. 1988, *Engineering Design: A Systematic Approach.* Design Council, London.
2. Product Feature Permutation is the creation of Tjalve E. 1979, *A Short Course in Industrial Design.* Newnes-Butterworths, London (p 19 onwards).
3. The sketches and prototypes shown were developed at DRC, based on an idea from an inventor within Brunel's 'Inventor Programme'. They are of a typist's copy stand and range from the initial rendering to a full visual and working prototype.
4. Failure Modes and Effects Analysis is described in O'Connor P.D.T. 1991, *Practical Reliability Engineering* (3rd Edition). Wiley, Chichester, UK. For an account of its application see Dale B.G. and Shaw P. 1989, *Failure Modes and effects Analysis: A Study of its Use in the Motor Industry.* Manchester School of Management, Manchester, UK.
5. Klapakjian S. 1992, *Manufacturing Engineering and Technology*, (2nd Edition) Addison-Wesley Publishing Co. Reading, Massachusetts.
6. Boothroyd G. Dewhurst P. and Knight W. 1994, *Product Design for Manufacturing and Assembly.* Marcel Dekker Inc. New York.

References

Abeysekera J.D.A. and Shahnavaz H. 1989, Body Size Variability Between People in Developed and Developing Countries and its Impact on the Use of Imported Goods. *International Journal of Industrial Ergonomics* 4:139-149.

American Supplier Institute Inc. 1992, *Quality Function Deployment: Practitioner Manual.* American Supplier Institute, Dearbourn, Michigan.

Angier R.P. 1903. The aesthetics of unequal division. *Psychological Monographs*, Supplements 4: 541-561.

Ashall F. 1994. *Remarkable Discoveries.* Cambridge University Press, Cambridge.

Ashby P. 1979, *Ergonomics Handbook 1: Body Size and Strength.* SA Design Institute, Pretoria.

Beeching W.A. 1974. *Century of the Typewriter.* Heinemann Ltd. London.

Belbin M.R. 1994, *Management Teams: Why they Succeed or Fail.* Butterworth Heinemann Ltd. Oxford.

de Bono E. 1991, *Serious Creativity.* Harper Collins Publishers, London.

Boothroyd G. Dewhurst P. and Knight W. 1994, *Product Design for Manufacturing and Assembly*. Marcel Dekker Inc. New York.

Booz-Allen and Hamilton Inc. 1982, *New Product Management for the 1980's*. Booz-Allen & Hamilton Inc, New York.

BSI 1989, Guide to Managing Product Design, BS 7000. British Standards Institution, London.

Bridger R.S. 1995 *Introduction to Ergonomics.* McGraw Hill Inc. New York.

Bruce V. and Green P 1990, *Visual Perception: Physiology, Psychology and Ecology* (2nd Edition). Lawrence Erlbaum Assoc. London.

Churchill G.A.Jr. 1992, *Basic Marketing Research* (2nd Edition). The Dryden Press, USA.

Cook T.A. 1914. *The Curves of Life.* Constable & Co. London.

Cooper R.G. 1993, *Winning at New Products*, Addison Wesley Publishing Co. Boston.

Crozier R. 1994, *Manufactured Pleasures: Psychological Responses to Design*. Manchester University Press, Manchester UK.

Dale B.G. and Shaw P. 1989, *Failure Modes and effects Analysis: A Study of its Use in the Motor Industry*. Manchester School of Management, Manchester.

Davis G.A. and Scott J.A. 1971, *Training Creative Thinking.* Holt, Rinehart and Winston Inc, New York.
Design Council 1994, *UK Product Development, A Benchmarking Survey.* Gower Publishing, Hampshire, UK.
Diamantopoulos A. and Mathews B. 1995, *Making Pricing Decisions: A Study of Managerial Practice.* Chapman & Hall, London.
Fechner G.T. 1876. *Vorschule der Aesthetik*. Breitkopf & Hartel, Leipzig.
Fox M.J. 1993 *Quality Assurance Management,* Chapman & Hall, London.
Freeman C. 1988, The Economics of Industrial Innovation (2nd Edition). Francis Pinter (Publishers), London.
Galer I. 1987, *Applied Ergonomics Handbook* (2nd Edition). Butterworths, London.
Gordon, W.J.J. 1961. Synectics. Harper & Row, New York.
Gould S.J. 1980, The Panda's Thumb: More Reflections in Natural History. Penguin Books, London.
Greenhalgh P. 1993, *Quotations and Sources on Design and the Decorative Arts*. Manchester University Press, Manchester, UK.
Greunwald G. 1988, *New Product Development.* NTC Business Books, Lincolnwood, Illinois.
Griffin A. and Hauser J.R. 1993, The Voice of the Customer. *Marketing Science*, Vol 12.
Hargittai I. and Hargittai M. 1994. *Symmetry: A Unifying Concept*. Shelter Publications Inc, Bolinas, California.
Hauser J.R. and Clausing D. 1988, The House of Quality. *Harvard Business Review*, 66: 63-73.
Henderson S., Illidge R. & McHardy P. 1994, *Management for Engineers.* Butterworth Heinemann Ltd. Oxford, UK.
Henry J. and Walker D. (eds) 1991, Managing Innovation. Sage Publications Ltd. London.
Higbee, K.L. and Millard, R.J. (1983) Visual imagery and familiarity ratings for 203 sayings. American Journal of Psychology 96: 211-222.
Hinde, R. A. 1987, The Evolution of the Teddy Bear. Animal Behaviour.
Hollins B. and Pugh S. 1987, *Successful Product Design.* Butterworth & Co. London.
Huntley H.E. 1970, *The Divine Proportion: A Study in Mathematical Beauty.* Dover Publications Ltd, London.

Ingrassia P. and White J.B. 1994, *Comeback: The Fall and Rise of the American Automobile Industry.* Simon & Schuster, New York.
Inwood D. and Hammond J. 1993, *Product Development, An Integrated Approach*. Kogan Page Ltd. London.
Johnson G. and Scholes K. 1993, *Exploring Corporate Strategy.* Prentice Hall, London.
Kanizsa G. 1979, *Organisation in Vision: Essays on Gestalt Perception*. Preager, New York.
Klapakjian S. 1992, *Manufacturing Engineering and Technology*, (2nd Edition) Addison-Wesley Publishing Co. Reading, Massachusetts.
Kenjo T. 1991. *Electric Motors and their Controls*. Oxford University press, Oxford.
Khatena J. 1972. The use of analogy in the production of original verbal images. *Journal of Creative Behaviour* 6: 209-213.
Koestler A. 1964. *The act of creation.* Hutchinson & Co., London.
Koffka K. 1935. *Principles of Gestalt psychology.* Harcourt Brace, New York.
Kohler W. 1947. *Gestalt psychology: an introduction to new concepts in modern psychology.* Liveright Publishing Company, New York.
Krippendorf K. and Butter R. 1993. Where meanings escape functions. *Design Management Journal*, Spring 1993, 29-37.
Lawson B. 1994 *How Designers Think.* (2nd Edition) Architectural Press, London.
Lewalski Z.M. 1988 *Product Esthetics: An Interpretation for Designers.* Design & Development Engineering Press, Carson City, Nevada.
Lorenz C. 1986, *The Design Dimension.* Basil Blackwell, Oxford.
Lorenz K. 1971, *Studies in Human and Animal Behaviour,* Vol 2. Methuen & Co. London.
Marroquin J.L. 1976. *Human visual perception of structure*. MSc thesis, Massachusetts Institute of Technology, Boston.
Marras W.S. and Kim J.Y. 1992 Anthropometry of Industrial Populations. *Ergonomics* 36: 371-378.
McCulloch W.S. 1960. *Embodiments of Mind.* MIT Press, Cambridge, Massachusetts.
Mingo J. 1994. How the Cadillac Got its Fins. Harper Business, New York.
Mintel Market Intelligence 1992 *Cookware*. July 1992, p 8-9, Mintel International Group, London.

Moore W.L. and Pessemier E.A. 1993. *Product Planning and Management: Designing and Delivering Value.* McGraw Hill Inc. New York.

Morito A. 1991. Selling to the World: the Sony Walkman Story. In Henry J. and Walker D. (eds) *Managing Innovation.* Sage Publications Ltd. London.

Morris, W.C. and Sashkin, M. (1978) Phases of integrated problem solving (PIPS). In Pfeiffer, J.W. and Jones, J.E. (eds) *The 1978 Handbook for Group Facilitators.* University Associates Inc. La Jolla, California. pp 105-116.

Nayak P.R., Ketteringham J.M. and Little A.D. 1986 *Breakthroughs!* Mercury Business Books, Didcot, Oxfordshire, UK pp 29-46.

O'Connor P.D.T. 1991, *Practical Reliabilty Engineering* (3rd Edition). Wiley, Chichester, UK.

Palmer A. and Worthington I. 1992, *The Business and Marketing Environment.* McGraw Hill, Berkshire, UK.

Page A.L. 1991, *New Product Development Practices Survey: Performance and Best practices*. Paper presented at PDMA's 15th Annual Conference, October 1991.

Pahl G. and Beitz W. 1987, *Engineering Design: A Systematic Approach*. Design Council, London.

Pheasant S. 1987 *Ergonomics: Standards and Guidelines for Designers.* British Standards Institute, Milton Keynes, UK.

Pickford R.W. 1972. *Psychology and visual aesthetics*. Hutchinson Educational Ltd, London.

Proctor T. 1993, Product Innovation: the pitfalls of entrapment. *Creativity and Innovation Management* 2: 260 - 265.

Pugh S. 1991 *Total Design: Integrated Methods for Successful Product Engineering.* Addison-Wesley Publishing Co. Wokingham, UK.

Reinertsen D.G. 1983, Whodunnit? The search for new product killers. *Electronic Business*, July 1983.

Schnaars S.P. 1994, *Managing Imitation Strategies.* The Free Press, New York.

Smith P.G. and Rheinertsen D.G. 1991, *Developing Products in Half the Time*. Van Nostrand Reinhold, New York.

Starck P. 1991 (Text by Olivier Boissiere) *Starck.* Benedict Taschen & Co. Koln, Germany.

Sullivan L.P. 1986 Quality Function Deployment. *Quality Progress* (June) 39-50.

Thamia, S. and Woods, M.F. 1984. A systematic small group approach to creativity and innovation. *R & D Management* 14:25-35.
Thurston, J.B. and Carrahar, R.G. 1986. *Optical illusions and the visual arts*. Van Nostrand Reinhold, New York.
Tjalve E. 1979, *A Short Course in Industrial Design.* Newnes-Butterworths, London.
Ulrich, K.T. and Eppinger, S.D. 1995. *Product Design and Development.* McGraw Hill, New York.
Urban G.L. and Hauser J.R. 1993, *Design and Marketing of New Products* (2nd Edition). Prentice Hall Inc, Englewood Cliffs, New Jersey, pp 357-378.
Uttal W.R. 1988 *On Seeing Forms*. Lawrence Erlbaum Associates Inc, London.
Van Gundy A. 1988 *Techniques of structured problem solving.* 2nd edition. Van Nostrand Reinhold, New York.
Von Oech R. 1990, *A Whack on the Side of the Head: How you can be More Creative.* Thorsons Publishing Group, Wellingborough, UK
Wallas G. 1926. *The Art of Thought*. Harcourt, New York.
Walsh V., Roy R., Bruce M. and Potter S. 1992, *Winning by Design: Technology, Product Design and International Competitiveness.* Blackwell Business, Oxford, UK.
Waterman R.H. Jr. 1990. *Adhocracy: The Power to Change.* Whittle Direct Books, Knoxville, Tennessee.
Wheelwright S.C. and Clark K.B. 1992, *Revolutionising Product development: Quantum Leaps in Speed, Efficiency and Quality.* Free Press, New York.
Which? 1993 *Sitting Safely*. August 1993, pp34-38.
Zangwill W.I. 1993, *Lightning Strategies for Innovation.* Lexington Books, New York.

Index

Analogies (toolkit) 91-92
Ansoff matrix 117-118
Anthropometry 208
Archimedes Eureka! 63
Apollo syndrome (team performance) 123
Attractiveness and styling 55-57
    Four faces of attractiveness 55
Augmented product 1435

Bisociation and styling 50
Bisociation and creativity 68-71
Brand identity (styling) 175, 219
Buckyballs (creativity) 65-66

Car styling 45, 49-50, 53, 54, 218-219
Cliches and proverbs (toolkit) 94-95
Collective notebook (toolkit) 87
Company sales due to new products 2
Competing product analysis 152-154, 253-254
    Plasteck's 163-164, 179-181
Competitor analysis (toolkit) 134-135
Concept design 12-13, 15, 19, 201-234, 263-264
    Aims and scope of... 202-205
    Process of... 201-203
    Psion case study 231-234
Concurrent design 17
Core benefit proposition 147-150, 198
Core business function 115
Corporate strategy 11, 17, 101-126, 150, 205
    Plasteck's 161-163
    Psion case study 124-126
    SWOT analysis (toolkit) 128-129
Cost-plus financial planning 172
Costs and benefits of design stages 28
Costs of development problems 105, 136-138

Creativity 62-79
    First insight 62-63
    Guide to design toolkits 70
    Incubation & illumination 65-66
    Preparation 64-65, 71-73, 97
    Problem abstraction 84-86
    Problem digression 74
    Problem expansion 74
    Problem reduction 74
    Psychology of... 62-71
    Reviewing the creative process 79, 96-99
    Stairway to... 62

Dot sticking (toolkit) 93
Delphi technique (toolkit) 189-190
Design integration 278-284
Design specification 21, 259-261
    Toolkit 268-269
Design toolkits
    Analogies 91-92, 212
    Brainwriting 81
    Cliches and proverbs 94-95
    Collective notebook 87
    Competitor analysis 134-135
    Delphi technique 189-190
    Design specification 268-269
    Dot sticking 93
    Failure modes & effects analysis 294-297
    Life cycle analysis 239-240
    Market needs research 191-196
    Opportunity specification 197-199
    Orthographic analysis 88-89
    Parametric analysis 82-83
    PIPS creative process review 96-99
    Problem abstraction 84-86
    Product function analysis 236-238
    Risk audit 136-138
    SCAMPER 90

SWOT analysis 128
Tracking study 131
Detail design 12-13, 15, 19, 263-264

Embodiment design 12-13, 15, 19, 263-264, 271-292
Environmental design 82-83, 239-240
Ergonomics 206-207
Eureka (Archimedes) 63
Equiangular spiral 48

Face perception 44-46
Failure modes and effects analysis 290-291
Toolkit 294-297
Faraday (creativity) 64-65
Fashion (clothes) 51-52
Fibonacci series (mathematical) 46-47
Financial commitment 15
Financial planning 15, 172
Force generation of ideas 205-206, 275

Gantt chart (time plan) 265-266
Gestalt rules of visual perception 36-41
Proximity 38
Similarity 38
Good continuance 38
Figure-ground 39
Golden ratio 47
Golden section 47-50

House of quality 251-258

Idea generation 73-78
Concept design 210-228, 234
Force generation of ideas 205-206, 275
Potato peelers 209-210, 212, 275-278
Procedures for... 75-78
Brainwriting 81
Reviewing procedures... 97-98
Idea selection 78-79
Concept selection 229-230
Reviewing procedures... 98-99
Matrix 169, 229-230
Investment:return ratios 27
Innovation management matrix 102
Iterative design activities 19

Kano model of quality 244-246

Lateral thinking 68-71
Life cycle analysis 213-214
Toolkit 239-240

Market needs research 154-158
Qualitative 195
Quantitative 196
Toolkit 191-196
Market research 17, 150
Competitor analysis (toolkit) 134-135
'Morito' factor 157
Plasteck's 165-166, 180-184
Tracking study (toolkit) 131
Marketing mix 135
Michaelangelo (Adam's creation) 49
Mission statement 112-114, 120, 150, 205
Plasteck's 162
Models 285-290

New product opportunity 12

Opportunity specification 21, 141-187
Justification of... 148-149, 185-187
Opportunity selection 166-171
Plasteck's 177-187
Styling in the... 177
Toolkit 197-199
Orthographic analysis 88-89

Parametric analysis (toolkit) 82-83

Parthenon 49
PERT analysis (timeplan) 265-266
PEST analysis (toolkit) 130
PIPS creative process review (toolkit) 96-99
Plasteck Ltd 161-166
Business strategy 161-163
Competing product analysis 163-164
Concept design 208-212
Embodiment design 275-280, 291
Market research 165-166
Mission 162
Opportunity specification 177-187
Product planning 163-166
Price-minus financial planning 172-174, 186, 199
Price of semiconductors 160
Price positioning new products 171-174
Price;value map for new products 172-173, 186, 198
Primal sketch (visual perception) 34
Primary global precedence (visual perception) 33
Problem abstraction (toolkit) 84-86
Feature abstraction 274
Problem definition 64-65, 71-73, 97
Problem gap 72, 117, 202-205
Product architecture 273-284
Product development costs 143, 186
Product development strategy 106-122, 142, 145
Plasteck's 161-163
Process... 110-122
Product maturity analysis (toolkit) 132-133
Resource requirements and...106-110
Risk audit (toolkit) 136-138
Product development teams 102, 122-123, 139
Product feature permutation 275-277
Product function analysis 210-212, 215
Product function tree 211, 215, 237-238
Toolkit 236-238
Product maturity analysis (toolkit) 132-133
Product planning 141-187, 243-266
Break-even 187
Plasteck's 163-166
Process of... 142-146
Quality targets 248-249
Profit forecast 187
Project planning 261-266
Sales projections 185, 187
Product predecessors (styling) 175
Product re-design 133
Product styling 31-58, 174-177, 217-228, 274
Benchmarking 175-176
Brand identity 175
Contextual factors 175-176
Lifestyle, mood & theme boards 221-226
Planning of... 174-177
Process of... 57,
Product semantics 218
Product symbolism 218-221
Social, cultural & business effects ... 51-54
Product testing 188-291
Product trigger 142, 151-152
Products
Adhesive tape dispenser 261-262
Aircraft carrier (take-off platform) 68-69
Aston Martin DB7 219
Bicycle pump clip 281-284
Ballpoint pen 109
Boeing 777 143
Child's car seat 169-171, 173-174
Chrysler Concorde (car) 143
Computer printer 143
Corkscrew 237-238
Cosmetics packaging 82-83, 220, 223
Ford Scorpio (car) 45
Home security equipment 92

Lawn mower 85-86
Leisure boat 112-113
Locking bolt 90
Mickey Mouse 45
Mobile phone 40, 223-228
Motorbike 219
Nissan QX (car) 49-50
Offshore survival suit 215-217
Pet food 95
Plant pots 213-214
Post-it notes 67
Potato chips 88-89
Potato peelers 165-166, 177-187, 208-212, 275-280, 291
Psion Organiser 124-125
Psion Series 3 124-126, 231-234
Range Rover (car) 219
Rover MG RV8 (car) 222
Screwdriver (Stanley) 143
Sony Walkman 157, 245
Starck's lemon squeezer 51
Teddy bears 45
Thumbtack 252-257
Typewriter 41
Prototyping 285-290

Quality control 20-25
Quality function deployment 250-258
Quality targets 23-25, 83, 248-249, 254-256

Risk audit (toolkit) 136-138
Risk management 2, 14
Risk management funnel 11-21
Analogy 16
Example 13
& Quality control 21
Stages 14

SCAMPER (toolkit) 90, 278-279
Strategy gap analysis 118
Success and failure of new products 2, 9-11
Success of product development 103-105
SWOT analysis (toolkit) 128-129
Synectics 77

Tangible product 135
Task analysis 206-209
Potato peeler 208-209
Technology forecasting 160
Technology monitoring 159
Technological opportunities 158-161, 189-190
Tracking study (toolkit) 131
Two stage visual processing 32-33

Value analysis 214-217
Visual perception 32-39
Visual simplicity 41-44
Berlyne's theory 42-44